*Mass Spectrometry
in Drug
Metabolism and
Pharmacokinetics*

Mass Spectrometry in Drug Metabolism and Pharmacokinetics

Edited by

Ragu Ramanathan

A JOHN WILEY & SONS, INC., PUBLICATION

Library of Congress Cataloging-in-Publication Data:

Mass spectrometry in drug metabolism and pharmacokinetics/edited by
Ragu Ramanathan.
     p.; cm.
  Includes bibliographical references and index.
  ISBN 978-0-471-75158-8 (cloth)
  1. Drugs—Metabolism—Research—Methodology. 2. Drugs—Design. 3. Mass Spectrometry.
  4. Pharmacokinetics—Research—Methodology. I. Ramanathan, Ragu.
    [DNLM: 1. Pharmaceutical Preparations—metabolism. 2. Drug Design. 3. Drug Evaluation,
Preclinical—methods. 4. Mass Spectrometry—methods. 5. Pharmacokinetics. QV 38 M414
2008]
    RM301.55.M37 2008
    615′.7--dc22

             2008021424

Printed in the United States of America

10 9 8 7 6 5 4 3 2 1

Contents

Preface

Within the pharmaceutical industry, the mass spectrometer was long considered a useful and challenging analytical tool largely limited to the specialist user. The steady movement from specialist use to general use gained considerable speed in the 1990s, particularly due to the development of practical, sensitive liquid chromatography–mass spectrometry (LC–MS) interfaces and advances in the microelectronics. The rapid proliferation of quadrupole ion trap, linear ion trap, orbitrap, quadrupole mass filter, time-of-flight, and other types of mass spectrometers has impacted the industry from the earliest stages of disease determination through the final stages of clinical testing. This book, based on an American Society for Mass Spectrometry (ASMS) session, which I was fortunate enough to chair, will examine several of the ways in which mass spectrometry continues to have a profound influence on the direction and speed of drug discovery and development, especially in the area of drug metabolism (DM) and pharmacokinetics (PK).

To facilitate introduction to the topics contained in this book, the first chapter considers briefly the broader processes of drug discovery and development within the pharmaceutical industry. The specific roles of DM and PK, the applications considered throughout this book, are defined as well as major terms and concepts in mass spectrometry. Finally, the role of mass spectrometry in DM and PK is developed and the ensuing chapters introduced. For the experienced professional, this final section of the first chapter may represent the appropriate starting point in reading this book.

Chapter 2 systematically defines some of the important PK parameters and guides the reader through the types of quantitative LC–MS experiments performed to elucidate the PK parameters necessary to move a drug through discovery, preclinical development, and clinical stages. Chapters 3, 4, and 5 respectively introduce the readers to quadrupole mass filters and liner ion traps, time-of-flight mass spectrometers, and Fourier transform (FTICR and Orbitrap) mass spectrometers and their applications in the area of DM and PK. The high-resolution LC–MS mass defect filter (MDF) approach is considered in Chapter 6. Today the MDF approach has been adapted by all the major mass spectrometer vendors to help accelerate drug discovery and development. Chapter 7 elegantly describes the utility of high-sensitivity radioactivity and mass spectrometry techniques for drug metabolism studies. While online electrochemical–LC–MS techniques available for generating metabolites are discussed in Chapter 8, Chapter 9 describes some of the LC–MS tools and techniques available for detecting and characterizing isomeric metabolites. Chapter 10 is dedicated to online sample processing and turbulent-flow LC–MS techniques. Finally, Chapters 11 and 12 present some of the laser desorption–based mass spectrometry applications in the DM and PK arena.

This book would have never been possible without the efforts and dedication of more than 35 co-authors and the editorial staff at Wiley. I am very grateful to Kevin B. Alton, Honggang Bi, Jimmy L. Boyd, Swapan K. Chowdhury, John R. Eyler, Michael L. Gross, W. Griffith Humphreys, Steven Michael, Richard Morrison, Noel Premkumar, Laszlo Prokai, Rasmy Talaat, Poonam Velagaleti, and Ronald E. White for their continued mentorship throughout my professional career. I am also very grateful to my parents, brothers, aunts, uncles, and grandmother for supporting my education and career. Finally, my deepest gratitude goes to my wife, Dil, and Vishan and Eshal for continuously supporting all my endeavors.

RAGU RAMANATHAN, PH.D.

New Jersey, USA
September, 2008

About the Editor

Ragu Ramanathan received a B.Sc. in Chemistry from the University of Southern Mississippi and a Ph.D. in Physical Chemistry/Mass Spectrometry from the University of Florida. His graduate research focused on coupling of electrospray ionization (ESI) to Fourier transform ion cyclotron resonance (FTICR) mass spectrometer. After spending three years as a postdoctoral research fellow with Professor Michael L. Gross at the Washington University, St. Louis, Missouri, Dr. Ramanathan managed the Center for Advanced Mass Spectrometry at the Analytical Bio-Chemistry Laboratories, Columbia, Missouri. In 1998, Dr. Ramanathan joined Schering-Plough Research Institute's (SPRI) Drug Metabolism and Pharmacokinetics (DMPK) Department and completed his tenure as a senior principal scientist in 2008. While at SPRI, Dr. Ramanathan was involved in the application of LC–MS for profiling and characterization of metabolites of drug candidates in the preclinical development and clinical stages. Dr. Ramanathan was with Pfizer Global Research and Development from 1999 to 2002 as a group leader of the Ann Arbor site biotransformation group. Dr. Ramanathan is currently an associate director at the Bristol-Myers Squibb, Co. and is responsible for elucidating biotransformation pathways of development drug candidates. Dr. Ramanathan's accomplishments include 35 peer-reviewed papers, 10 book chapters, and over 60 oral/poster presentations. He also served as a chairperson for the North Jersey ACS Mass Spectrometry Discussion Group and as a chairman for DMPK sessions of the American Society for Mass Spectrometry and Eastern Analytical Symposium meetings.

Contributors

IAN N. ACWORTH, Vice President, ESA Biosciences, Inc., Chelmsford, MA

KEVIN B. ALTON, Senior Director, Schering-Plough Research Institute, Department of Drug Metabolism and Pharmacokinetics, Kenilworth, NJ

JOSE M. CASTRO-PEREZ, Laboratory Manager-Metabolite Profiling, Waters Corporation, Milford, MA

SWAPAN K. CHOWDHURY, Director, Schering-Plough Research Institute, Department of Drug Metabolism and Pharmacokinetics, Kenilworth, NJ

LUCINDA H. COHEN, Director, Merck Research Laboratories, Department of Drug Metabolism and Pharmacokinetics, Bioanalytical Group, Rahway, NJ

MARK J. COLE, Research Fellow, Pfizer Global Research and Development, Department of Pharmacokinetics, Dynamics and Metabolism, Groton, CT

JOSEPH M. DI BUSSOLO, Senior Applications Scientist, Thermo Scientific Applications Laboratory at West Chester University of Pennsylvania, Department of Chemistry, West Chester, PA

AYMAN EL-KATTAN, Senior Principal Scientist, Pfizer Global Research and Development, Department of Pharmacokinetics, Dynamics and Metabolism, Groton, CT

PAUL H. GAMACHE, Vice President, ESA Biosciences, Inc., Chelmsford, MA

JASON S. GOBEY, Associate Director, Pfizer Global Research and Development, Clinical Research Operations, New London, CT

MICHAEL C. GRANGER, USTAR Research Scientist, University of Utah, Center for Nanobiosensors, Salt Lake City, UT

JOSEPH L. HERMAN, Technical Director, The Children's Hospital of Philadelphia, Philadelphia, PA

CHRIS HOLLIMAN, Associate Director, Pfizer Global Research and Development, Department of Pharmacokinetics, Dynamics and Metabolism, Groton, CT

YUNSHENG HSIEH, Senior Principal Scientist, Schering-Plough Research Institute, Department of Drug Metabolism and Pharmacokinetics, Kenilworth, NJ

JOHN JANISZEWSKI, Associate Research Fellow, Pfizer Global Research and Development, Department of Pharmacokinetics, Dynamics and Metabolism, Groton, CT

ELLIOTT B. JONES, Senior LC-MS Laboratory Manager, Applied Biosystems, Foster City, CA

JONATHAN L. JOSEPHS, Principal Scientist, Bristol-Myers Squibb Pharmaceutical Research Institute, Department of Biotransformation, Pharmaceutical Candidate Optimization, Pennington, NJ

WALTER A. KORFMACHER, Distinguished Research Fellow, Schering-Plough Research Institute, Department of Drug Metabolism and Pharmacokinetics, Kenilworth, NJ

WING W. LAM, Principal Scientist, Johnson and Johnson Pharmaceutical Research and Development, Raritan, NJ

RICHARD M. LELACHEUR, Laboratory Director, Taylor Technology, Inc., Princeton, NJ

CHO-MING LOI, Associate Research Fellow, Pfizer Global Research and Development, Department of Pharmacokinetics, Dynamics and Metabolism, San Diego, CA

DAVID F. MEYER, Scientist, Amesbury, MA

ANGUS NEDDERMAN, Director, Pfizer Global Research and Development, Department of Pharmacokinetics, Dynamics and Metabolism, Kent, UK

NATALIA A. PENNER, Associate Principal Scientist, Schering-Plough Research Institute, Department of Drug Metabolism and Pharmacokinetics, Kenilworth, NJ

DIL M. RAMANATHAN, Assistant Professor, Kean University, New Jersey Centre for Science, Technology & Mathematics Education, Union, NJ

RAGU RAMANATHAN, Associate Director, Bristol-Myers Squibb Pharmaceutical Research Institute, Department of Biotransformation, Pharmaceutical Candidate Optimization, Princeton, NJ

KENNETH L. RAY, Senior Scientist, Novatia, LLC, Monmouth Junction, NJ

MARK SANDERS, Manager, Thermo Fisher Scientific, Somerset, NJ

PETIA A. SHIPKOVA, Senior Research Investigator II, Bristol-Myers Squibb Pharmaceutical Research Institute, Department of Bioanalytical and Discovery Analytical Sciences, Pharmaceutical Candidate Optimization, Pennington, NJ

DON K. WALKER, Research Fellow Pfizer Global Research and Development, Department of Pharmacokinetics, Dynamics and Metabolism, Kent, UK

JOANNA ZGODA-POLS, Associate Principal Scientist, Schering-Plough Research Institute, Department of Drug Metabolism and Pharmacokinetics, Kenilworth, NJ

DONGLU ZHANG, Principal Scientist, Bristol-Myers Squibb Pharmaceutical Research Institute, Department of Biotransformation, Pharmaceutical Candidate Optimization, Princeton, NJ

HAIYING ZHANG, Senior Research Investigator II, Bristol-Myers Squibb Pharmaceutical Research Institute, Department of Biotransformation, Pharmaceutical Candidate Optimization, Pennington, NJ

MINGSHE ZHU, Principal Scientist, Bristol-Myers Squibb Pharmaceutical Research Institute, Department of Biotransformation, Pharmaceutical Candidate Optimization, Princeton, NJ

1

Evolving Role of Mass Spectrometry in Drug Discovery and Development

Dil M. Ramanathan

Kean University, New Jersey Centre for Science, Technology & Mathematics Education, Union, New Jersey

Richard M. LeLacheur

Taylor Technology, Princeton, New Jersey

1.1   ROUTE TO MARKET: DISCOVERY AND DEVELOPMENT OF NEW DRUGS

1.1.1   Industry Research and Development

The members of the modern biopharmaceutical industry are engaged in an on-going struggle to balance the needs of medicine and patient care with the demands of running a growing, profitable business. Moreover, new drugs must be proven to possess some combination of improved efficacy and safety compared with existing treatments. Success in drug research and development (R&D) is critical for meeting all of these objectives, and R&D efforts within the biopharmaceutical indus-try, as measured by spending, continue to grow steadily (Fig. 1.1). In recent years, the rate of annual growth in R&D spending has been between 5 and 10% in the United States, with the most recent data indicating that R&D spending in 2006 exceeded $50 billion (PhRMA, 2006).

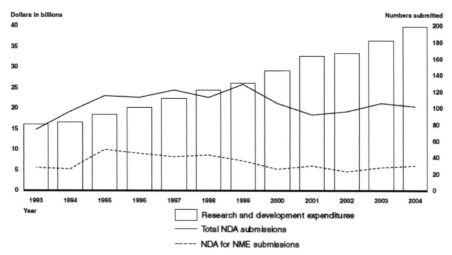

Figure 1.1. *1993–2004 Pharmaceutical R&D expenses, total new drug applications (NDAs), and NDAs for new molecular entity (NME) submission trends. [Reprinted with permission from the U.S. Government Accountability Office (GAO) 2006.]*

The many essential steps in the discovery and development of new drugs can be measured by two primary benchmarks. The first, the number of filed and approved investigational new drug (IND) applications, represents the threshold to human (clinical) testing. The second, the number of filed and approved new drug applications (NDAs), represents the threshold to marketing a drug. These numbers and their trends can represent the relative success of R&D efforts.

Given the typical 12–15 years required to discover, develop, and test a new drug (Fig. 1.2), the NDA submission and approval data will in part represent R&D

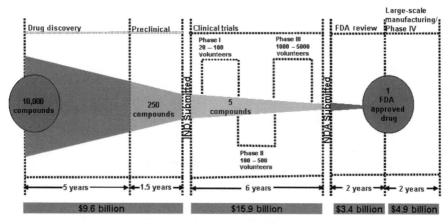

Figure 1.2. *Complex pathway of pharmaceutical R&D involved in bringing a new drug to the market. (Adapted from PhRMA, 2006.)*

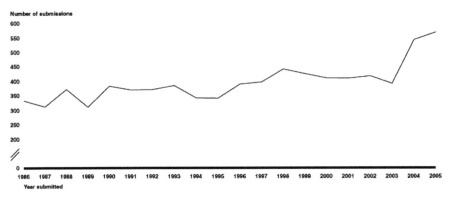

Figure 1.3. *Increase in INDs in recent years. Data are for commercial INDs. (Reprinted with permission from GAO, 2006.)*

progress from several years earlier. Since the late 1990s, the annual rate of NDA submissions and approvals has declined. A similar decline has been observed in the number of NMEs (GAO, 2006). Of the 93 NDA approvals for 2006, only 18 are considered to represent NMEs (*The Pink Sheet*, January 15, 2007). While both total NDAs and NMEs are important, the number of NMEs approved represents a particularly critical measure of overall R&D success.

The statistics of expenditure and NDA approvals can mask a major source of R&D cost and frustration in the industry: late-stage development and postmarketing failures. These types of failures attract significant unwanted publicity and only occur after hundreds of millions of dollars have been spent. Well-publicized examples have included the recent late-stage failure of torcetrapib (Tall et al., 2007) and the postmarketing withdrawals of fenfluramine-phentermine (Fen-Phen) and Vioxx (Embi et al., 2006).

Consideration of IND trends is more encouraging (Fig. 1.3). IND filings occur years before NDA filings and represent a more recent state of R&D success. The number of compounds in clinical testing has approximately doubled over the last decade to approximately 3000 compounds in 2005 in the United States alone. A recent tally of new treatments in clinical testing for various indications is summarized in Table 1.1 (PhRMA, 2006). It is encouraging to see this increase in clinical testing, but it is also important to remember that only about 8% of early-stage clinical testing drugs will produce an approved NDA (Caskey, 2007).

1.1.2   Drug Discovery and Development Process

The overall process of bringing a new drug to market is typically divided into two principal areas: drug discovery and drug development. Examples of summaries describing the entire process include the publication entitled "Drug Discovery and Development: Understanding the R&D Process" (PhRMA, February 2007) and a tutorial written by Jens Eckstein, recently available online at www.alzforurm.org/drg/tut/tutorial.asp.

TABLE 1.1. Treatments in Clinical Testing

Disease Area or Indication	Number of Compounds in Development
Oncology	682
Neurological disorders	531
Infectious diseases	341
Cardiovascular	404
Psychiatric	190
Human immunodeficiency virus/acquired immunodeficiency syndrome (HIV/AIDS)	95
Arthritis	88
Asthma	60
Alzheimer/dementia	55

Source: PhRMA, 2006.

The following description very briefly summarizes some of the steps in drug discovery and development.

1.1.2.1 Drug Discovery The first step in discovering a new medicine is to identify a therapeutic target. Drugs in today's market as well as those in recent clinical testing target less than 500 biomolecules, with more than 10 times that many potential therapeutic targets waiting to be discovered and developed (Drews, 2000). More than 50% of the newly approved drugs result from R&D involving previously clinically tested and validated targets. Once a target has been validated (proven to be related to the disease process), high-throughput screening methods may be used to determine initial structural leads. Compounds are assessed for target affinity and for their "drug-like" properties, including absorption, distribution, metabolism, and excretion (ADME) using a series of in vivo and in vitro tests. The results of these tests are used to improve the structure and therefore the properties of the next round of test compounds, until ultimately one or more acceptable compounds are advanced forward in the process. This stage of discovery, which can be lengthy and difficult to predict, is generally referred to as lead optimization. The lead selection and lead optimization studies that are used to sift out the problematic compounds are summarized in Fig. 1.4.

Mass spectrometry enters into all phases of drug discovery (Feng, 2004; Lee, 2005). Early in the discovery, target proteins are identified and characterized by MS following LC or two-dimensional gel electrophoresis separation (Kopec et al., 2005; Deng and Sanyal, 2006). The make-up of an isolated protein is determined by enzymatically digesting the protein and then analyzing the peptides by MS (Link, 1999; Kopec et al., 2005; Köpke, 2006). Once a target is validated, compounds generated from any one of the following strategies are evaluated against the target: total synthetic process (33%), derivative of natural products (23%), total synthetic product with natural product mimic (20%), biological (12%), natural product (5%), total synthetic product based on a natural product (4%), and vaccine (3%) (Newman et al., 2003; Newman and Cragg, 2007). In almost all pharmaceutical

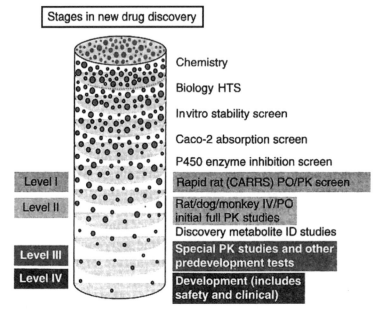

Figure 1.4. *NCE/NME progression scheme showing the various discovery stage liquid chromatography–mass spectrometry (LC–MS) and LC–tandem MS (LC–MS/MS) assays used for selecting NME/NCE to advance into development. (Reprinted with permission from Korfmacher, 2005.) (CARRS, Cassette accelerated rapid rat screening; IV, Intravenous administration; PO, Oral administration; NCE, New chemical entity)*

companies, open-access MS laboratories have been set up to allow medicinal chemists to confirm and assess the purity of their synthesis or isolated products (Chen et al., 2007). Once the compounds or compound series are confirmed, high-throughput screening (HTS) assays are used to weed out compounds that do not show any activity toward a host [protein, ribonucleic acid (RNA), deoxyribonucleic acid (DNA), etc.] (Fligge and Schuler, 2006). Mass spectrometric approaches also have been used to study noncovalent complexes involving protein–drug, DNA–drug and RNA–drug to identify structural details of the drug-binding sites (Benkestock et al., 2005; Siegel, 2005; Hofstadler and Sannes-Lowery, 2006; Jiang et al., 2007).

Compounds or compound series selected using HTS are further filtered using invitro-based solubility, chemical stability (Wilson et al., 2001), permeability (Bu et al., 2000a,b; 2001a–d; Mensch et al., 2007), and metabolic stability (Lipper, 1999; Thompson, 2000, 2005) assays before the lead selection/optimization stage (Lipper, 1999; Thompson, 2000, 2005). Most of these in vitro assays are faster, more efficient, and more sensitive due to unsurpassed involvement of the LC–MS (Thompson, 2001; Mandagere et al., 2002; Pelkonen and Raunio, 2005; Thompson, 2005). Results from such high-throughput in vitro assays are used to select compounds for additional in vitro tests and finally for in vivo testing in preclinical species (mouse, rat, dog, monkey, etc.). Similar to the early discovery stage high-throughput assays, LC–MS and LC–MS/MS assays are the methods of choice for the late-stage discovery studies (lead optimization stage, levels II and III) because they are rapid, sensitive, easy to automate, and robust.

All the discovery stage quantitative and qualitative LC–MS assays (levels I, II, and III), which are used to select drug candidates for development, are not rigorously validated and are not required to satisfy any of the good laboratory practices (GLPs) guidelines set forth by the regulatory agencies (Shah et al., 2000; Hsieh and Korfmacher, 2006; Jemal and Xia, 2006).

1.1.2.2 Drug Development

The preclinical testing represents the bridge between discovery and later clinical (human) testing. As shown (Fig. 1.2), if 10,000 compounds enter the screening stage, only about 250 will make it into the preclinical testing stage. During this stage, critical assessments of drug candidate safety are obtained in toxicology studies. Also essential understanding of the ADME, pharmacokinetic (PK), and pharmacodynamic (PD) properties of the drug is established.

1.1.2.2.1 The Drug Substance

Before starting any long-term toxicological studies in rodent (rat or mouse) and nonrodent (dog or monkey) species, it is imperative to work out all the chemical, pharmaceutical, large-scale synthesis, purification, stability, and formulation issues associated with the drug substance (Smith et al., 1996; van De Waterbeemd et al., 2001).

For a drug substance to move further in the development pipeline, its physical and salt forms have to be optimized in pharmacokinetics studies often using quantitative LC–MS/MS assays. Pharmaceuticals can exist as either a crystalline form (which has long- and short-range order in three dimensions) or an amorphous form (which lacks the long-range order present in crystalline material). In the discovery stage, usually all ADME assays (levels I, II, and III) are conducted using laboratory-grade amorphous drug substance without optimizing for physical and pharmaceutical properties of the drug (Kerns, 2001). Although the stability of an amorphous drug substance is sufficient for short-term discovery studies and for making internal recommendations, a crystalline form is the preferred form for long-term toxicological and clinical studies due to its long-term stability. However, the ability of a drug (organic molecule) to exist in more than one crystalline form leads to polymorphism. Polymorphs (same chemical composition but different internal crystal structure) of a given drug can have widely different pharmacokinetic parameters (Chapter 2), especially bioavailability due to differences in physicochemical properties such as dissolution rate, density, and melting point (Kobayashi et al., 2000; Agrawal et al., 2004; Panchagnula and Agrawal, 2004).

Changes in the method of synthesis during the large-scale manufacturing phase of drug development can also lead to changes in the crystalline form (Perng et al., 2003; Huang and Tong, 2004). A well-documented example of crystalline form change was observed with ritonavir (Norvir), a protease inhibitor approved in 1996 for treatment of HIV infections. In mid-1998, sales of ritonavir were temporarily halted due to manufacturing difficulties associated with multiple polymorphs (Bauer et al., 2001; Van Arnum, 2007). Later, in 1999, reformulation and additional LC–MS/MS-based pharmacokinetic studies allowed Abbott Laboratories to bring ritonavir back to the market. Today, the Food and Drug Administration (FDA) requires application of techniques such as X-ray diffraction and/or vibrational spectroscopic analysis [*Fourier transform infrared (FTIR)*, near infrared (NIR), Raman] to characterize polymorphic, hydrated, or amorphous forms of drug substances and for further evaluation of pharmacokinetic parameters using the final thermodynamically stable form of the drug.

Salt form selection/finalization is another crucial step in preclinical development (Engel et al., 2000; Furfine et al., 2004). Some of the common pharmaceutical salts include hydrochloride, sulfate, mesylate, succinate, tartrate, acetate, and phosphate. Similar to the changes that occur in the crystalline form, the changes that occur in the salt form also alter the oral bioavailability of a drug. When the salt form of a drug substance is changed, quantitative LC–MS/MS assays are used to reassess the key pharmacokinetic parameters as well as bridge the new parameters with the discovery stage data, if necessary. Along with physical and salt form optimization, the drug substance is also subjected to acid, base, and photostability tests, and when necessary, degradants are identified using LC–MS and nuclear magnetic resonance (NMR) techniques.

Once the salt and physical forms of a drug substance are finalized and large-scale manufacturing issues are addressed, the NCE/NMEs recommended for development and human testing is often referred to as the active pharmaceutical ingredient (API). Around this stage of the preclinical development, several kilograms of the API are manufactured under good manufacturing practices (GMP) guidelines established by the regulatory authorities (Webster et al., 2001). At this stage, LC–MS and MS/MS methods are used to fully characterize the API and to identify any major impurities and degradants present in the starting materials and/or formed during API processing (Kovaleski et al., 2007). Once all the API impurity issues are worked out, the certified API is used for toxicological studies conducted in support of first-in-human clinical studies. The International Conference on Harmonization (ICH) guidelines on the API suggest that impurities >0.15% and >0.05% respectively for ≤2 g and >2 g daily dose should be characterized and the impurity levels should be reduced if there are any known human risks.

Before the start of toxicological studies, an LC–MS/MS method to quantify the drug substance and/or its metabolites in plasma is developed using the certified API. This quantitative LC–MS/MS assay is developed under GLP guidance. Most often a stable isotope labeled form of the drug is used as the internal standard to correct for any experimental limitations. Upon completion of the rodent and nonrodent toxicological studies using the quantitative LC–MS/MS assays, safe human doses to be used in the first-in-human study come to light and the pharmaceutical company is ready to file for an IND. For perspective, the total testing regime up to this stage is estimated to consume about one-quarter of the total R&D expenditure in the industry (PhRMA, 2006). Of the 250 compounds that entered preclinical testing, only 5 on average will advance into human clinical testing.

1.1.2.2.2 Clinical Trials   Once an IND is approved, clinical trials take place typically in three sequential phases, phases 1–3. However, based on the recent FDA guidelines, traditional phase 1 studies could be preceded by "phase 0" or "exploratory

IND" studies. These studies involve the administration of a single subtherapeutic dose of a radiolabeled NME to healthy adult volunteers to assess the human pharmacokinetics and/or metabolism (Lappin and Garner, 2005; Hill, 2007). Subtherapeutic doses are defined as the smaller of either $1/100$ of the expected pharmacologically effective dose, or $100 \, \mu g$. The FDA guidelines also require animal toxicity studies to be completed using doses above the human subtherapeutic doses to show no risk of toxicity before starting phase 0 clinical studies. Phase 0 studies may allow identification of "less promising" compounds earlier and at lower cost. According to a recent presentation, phase 0 studies can shorten the drug development process by 6–12 months (Kummar et al., 2007). However, most of the phase 0 studies cannot be completed using conventional LC–MS techniques because administered doses are around $100 \, \mu g$ and require the use of accelerator mass spectrometry (AMS), the only ultrasensitive technique capable of quantifying ^{14}C-labeled compounds with attomole ($10^{-18} M$) sensitivity (Chapters 2 and 7). However, several laboratories are hard at work developing ultrasensitive LC–MS techniques capable of detecting drugs and/or metabolites from microdosing studies (Lebre et al., 2007; Seto et al., 2007; Yamane et al., 2007).

Phase 1 clinical trials are conducted on a small number (20–100) of healthy adult volunteers to determine the potential toxicity of a drug, whether severe side effects can occur, and safe dosage ranges. An assessment of pharmacokinetics and drug metabolism is also included. For obtaining all the PK parameters, quantitative LC–MS/MS assays developed under GLP guidance are used. However, metabolism studies are conducted using non-GLP-based qualitative LC–MS and LC–MS/MS methods to get a glimpse of the metabolites present in human plasma and urine (Chowdhury, 2007; Ramanathan et al., 2007c; Ramanathan et al., 2007d). In specialized cases, phase 1 trials may include subjects with the targeted disease (e.g., oncology drugs). Overall, the critical criteria for phase 1 are the safety profile of the drug and determination of a safe dosage.

Phase 2 trials involve the administration of the potential drug to 100–500 volunteer patients to demonstrate the efficacy of the drug against the targeted disease or condition. A phase 2a trial is considered a relatively small, early study with a limited number of patients and may include both efficacy testing and refinement of the dosing regime. A successful phase 2a trial could be followed by a larger phase 2b trial to expand the available data, particularly on efficacy under the defined dosing regime. The first testing of efficacy in a patient population can also be called a proof-of-concept study.

Following a successful determination of safety and efficacy in phase 2, phase 3 trials are conducted on hundreds to thousands of volunteers suffering from the target disease or condition. The large size of phase 3 trials makes this by far the most expensive stage of clinical testing. Drugs that fail in phase 3 or later represent a significant cost without return and the industry as a whole has increased efforts to identify and terminate development investments in such compounds before the expense of phase 3 is incurred.

Upon completion of successful phase 3 clinical trials, a NDA is filed with the FDA for marketing approval of the new drug against a particular disease or condition.

NDA approval leads to large-scale manufacturing and marketing of the medicine. Clinical trials may continue to assess efficacy against different diseases or assess long-term safety in a larger population than was possible under phase 3 testing. As noted in Fig. 1.2, of the 5000–10,000 compounds that entered testing, approximately 1 will emerge as an approved drug.

1.2 DRUG METABOLISM AND PHARMACOKINETICS IN DRUG DISCOVERY AND DEVELOPMENT

Prior to the 1990s, the pharmaceutical lead finding activities were mainly driven by human diseases and dominated by chemistry and pharmacology ("disease-driven method," or "old paradigm"). During the 1990s, combinatorial chemistry, parallel chemical synthesis, and HTS revolutionized the drug discovery process and put forward a vastly increased number of biologically active NME/NCE leads. The increase in leads, the 50% success rate in Phase 3 for NME (PhRMA, 2006), and the increase in time required to complete clinical trials (3.1 years in the 1960s to 8.6 years in the 1990s (DiMasi, 2001b)); resulted in shifting to a new drug discovery and development paradigm. A new paradigm was also indicated by retrospective analysis that demonstrated the unacceptable pharmacokinetic (PK) characteristics, not identified in preclinical testing, was a significant cause of clinical failure (Prentis et al., 1988; Milne, 2003; Wahlstrom et al., 2006). Under the "new paradigm," or "target-driven method," pharmaceutical companies started to incorporate PK components early in the drug discovery process to generate more promising clinical candidates. A subsequent study 10 years later showed that the incorporation of PK early in the drug discovery process helped to reduce the clinical stage drug candidate failures associated with unacceptable PK characteristics to <15% (Hopkins and Groom, 2002; Kola and Landis, 2004).

Pharmacokinetics is the science that describes the movement of a drug in the body (Jang et al., 2001). In other words, PK is concerned with the time course of a drug's concentration in the body, mainly in the blood (plasma). The PK parameters are discussed in Chapter 2. Four separate but somewhat interrelated processes influence a drug's movement in the body: absorption (A), distribution (D), metabolism (M), and excretion (E). These four major components which influence a drug's level, its kinetics of exposure to tissues, and its performance as a drug are described in the following:

- *Absorption* The process by which a drug molecule moves from the site of administration into the systemic circulation (bloodstream). When a drug is administered intravenously (IV), the drug is 100% absorbed (bioavailability is 100%). However, when a drug is administered via other routes [such as orally (by mouth, PO, *per os*), subcutaneously (under the skin), intradermal (into the skin)], its absorption (bioavailability) is influenced by many factors, including the rate of dissolution, metabolism before absorption and the ability to cross the gastrointestinal tract (Martinez and Amidon, 2002). Therefore, bioavailability, as detailed in Chapter 2, is one of the essential tools in pharmacokinetics, as bioavailability must be considered when determining dosing regimens and formulations for nonintravenous routes of administration.

- *Distribution*   The process of a drug being carried via the bloodstream to its site of action, including extracellular fluids and/or cells of tissues and organs. Factors that affect a drug's distribution include blood flow, plasma protein binding, tissue binding, lipid solubility, pH/pK_a, and membrane permeability (Vesell, 1974). Although distribution is typically not the rate-limiting step, distribution to sites such as the central nervous system, bones, joints, and placenta could be slow, inefficient, and therefore the rate-limiting step (De Buck et al., 2007).

- *Metabolism*   Metabolism or biotransformation is the process by which the body (human and animal) or a system (cell based or in vitro) breaks down and converts a drug generally via oxidation, reduction, hydrolysis, hydration, and/or conjugation reactions into an active, inactive, or toxic chemical substance. Enzymes (e.g., cytochrome P450s) present in the liver are responsible for metabolizing many drugs (Guengerich, 2006). When a drug is administered intravenously (or other nonoral routes such intramuscular and sublingual), some of these metabolism pathways are avoided.

- *Excretion/Elimination*   The irreversible removal (elimination) of a drug and/or its metabolites from the systemic circulation or from the site of measurement. The process of elimination usually happens through the kidneys (urine) or the feces. Unless excretion is complete, accumulation of drugs and/or metabolites can lead to adverse affects. Other elimination routes include the lung (through exhalation), skin (through perspiration), saliva, and mammary glands.

Phermacodynamics (PD) is the relationship between a drug's concentration at the site of action and its pharmacological, therapeutic, or toxic response at the site of action. It is often difficult to measure a drug's concentration at the site of action. Therefore, the PK/PD relationship (Chapter 2) becomes essential to understand and relate a drug's concentration in the blood (plasma) or other biological fluids with its pharmacological, therapeutic, or toxic response at the site of action (Derendorf and Meibohm, 1999). In the pharmaceutical drug discovery and development arena, the parameters that define PK and/or PD are the primary drivers in the selection of a drug candidate to move forward to the clinic and finally to the patients. Therefore, for a NME/NCE to be an effective drug, it not only must be pharmacologically active against a target but must also possess the appropriate ADME properties necessary to make it suitable for use as a drug (Thompson, 2000).

1.3   MASS SPECTROMETRY FUNDAMENTALS

The dramatic increase in the complexity of the new drug discovery and development paradigm involving an evaluation of a vast number of leads for favorable activity, selectivity, and ADME properties in turn puts more pressure on the drug discovery and early development teams. For drug metabolism and pharmacokinetics (DMPK) scientists, evaluating large numbers of compounds with limited supply meant creating high-throughput ADME assays that can provide answers quickly. The speed of analysis contributed directly to the discovery and development of optimized lead candidates, which in turn impacted the overall time required for developing new medicines. The inherent sensitivity, selectivity, and speed of MS turned out to be a superb solution for drug metabolism and pharmacokinetics applications, especially high-throughput ADME assays.

1.3.1 History

Mass spectrometry is an analytical technique that measures the mass-to-charge ratio (m/z) of gas-phase ions formed from molecules ranging from inorganic salts to proteins. The mass spectrometer is a device or instrument that measures the mass-to-charge ratio of gas-phase ions and provides a measure of the abundance of each ionic species. To measure the m/z of ions, the mass analyzer and detector must be maintained under high-vacuum conditions and calibrated using ions of known m/z. As explained in the following section, some ion sources can be maintained at atmospheric pressure, while others require vacuum conditions.

For excellent perspectives on the historical developments in MS, readers are directed to several outstanding books and reviews, including the American Society for Mass Spectrometry's 50th anniversary book (Grayson, 2002). Similar to any other field, the field of MS is laced with several Nobel laureates, including the father of modern MS, J. J. Thomson:

Scientist	Nobel Prize Year and Field	Contribution
Joseph J. Thomson	1906, Physics	Discovery of electrons
Francis W. Aston	1922, Chemistry	Stable isotopes
Wolfgang Paul	1989, Physics	Development of quadruplole and quadrupole ion trap
John B. Fenn	2002, Chemistry	Development of electrospray ionization (ESI)
Koichi Tanaka	2002, Chemistry	Development of matrix-assisted laser desorption ionization (MALDI)

The analytical capability of MS has been evolving at an astounding rate as Nobel laureates and developers push what is an inherently powerful analytical technique to even higher levels of capability. During the last decade, numerous ionization and analyzer configurations have been commercialized. Some of the most recent developments have made MS the gold standard for many pharmaceutical analyses, and has made the biopharmaceutical industry the major purchaser of mass spectrometers (Cudiamat, 2005).

1.3.2 Fundamental Concepts and Terms

For greater detail, the reader is referred to a comprehensive text on MS (Gross, 2004; Watson and Sparkman, 2007) or on terminology in MS (Sparkman, 2006). For brevity, a relatively simple list of definitions is provided here. For most mass spectrometry users, the concept of mass has been limited to the relatively simplistic integer mass level. The proliferation of high resolution and high mass accuracy instruments in the last decade, however, necessitates a brief consideration of the fundamentals of mass beyond the integer level. For beginners, the "mass" comes from protons and neutrons (and, marginally, electrons), and the "charge" comes from an excess of either protons ($+$ charge) or electrons ($-$ charge). Mass spectrometers can only detect charged species. Finally, it is worth noting that the dominant focus of this book is on small molecules, where in general only a single charge resides during MS analysis. For these types of species, $z = 1$ and mathematically, $m/z = m$. The MS user community commonly discuss mass where mass-to-charge ratio would be accurate.

1.3.2.1 Mass Terminology

- *Mass Unit* The unified atomic mass unit, or u, is the fundamental unit of mass for most mass spectrometrists. The Dalton, or Da, is also generally accepted and is commonly used in descriptions of large, biological molecules. The mass unit is defined as one-twelfth of the mass of carbon-12. Atomic mass unit, or amu, is technically incorrect but still commonly used. The unit Thomson (Th) has been used as a unit of m/z. However, Th is not accepted by most mass spectrometry journals and the International Union of Pure and Applied Chemistry (IUPAC). Therefore, m/z used for labeling the x-axis of mass spectra is unit less.

- *Average Mass* Mass calculated using the weighted average atomic mass of each element. Average mass is not measured using a mass spectrometer; rather this is calculated using the values reported on the periodic table. For example, the average mass of dextromethorphan ($C_{18}H_{25}NO$) is 271.4 [(18×12.011) + (25×1.0079) + (1×14.0067) + (1×15.9994)].

- *Nominal Mass* The whole-number (nominal) mass of a molecule (or atom) is calculated from the integer mass of the most abundant, stable isotope of each constituent atom. For example, the nominal mass of protonated dextromethorphan ($C_{18}H_{25}NO + H^+$) is 272 [(18×12) + (26×1) + (1×14) + (1×16)].

- *Exact Mass* A calculated mass, and theoretically the mass (for $z = 1$) that should be observed on the mass spectrometer; sometimes also used to refer to a measured mass (see accurate mass below). The exact mass of a molecule is determined by adding the exact mass of a particular isotope for each constituent atom in the molecule. For example, the exact mass of protonated dextromethorphan ($C_{18}H_{25}NO + H^+$) is 272.2009 [(18×12.0000) + (25×1.0078) + (1×14.0031) + (1×15.9949) + (1×1.0073). The importance of the electron mass (0.00055 u) in the calculation of exact mass has been explained in detail by Ferrer and Thurman (2007).

- *Accurate Mass*   A measured mass. Accurate mass is the observed mass to some specified number of decimal places of a molecule (or similar) as measured on the mass spectrometer. A so-called accurate mass measurement can be obtained on any mass analyzer, though it is generally assumed that the accuracy will be improved when the analysis is performed using high-resolution mass spectrometers (see below).

- *Monoisotopic Mass*   An exact mass, derived from the mass of the most abundant, stable isotope of each constituent atom in the molecule. For example, the monoisotopic mass of protonated dextromethorphan containing one ^{13}C ($^{12}C_{17}$ $^{13}C\,^1H_{25}\,^{14}N\,^{16}O + {}^1H^+$) is 273.2032 [(17 × 12.0000) + (1 × 13.0034) + (25 × 1.0078) + (1 × 1.0073) + (1 × 14.0031) + (1 × 15.9949).

- *Mass Defect*   The difference between the exact mass of an ion or molecule and the nominal (integer) mass. The mass defect can be highly characteristic of the constituent atoms and is useful in data handling (see below and Chapters 5 and 6).

1.3.2.2   Mass Calibration and Resolution

- *Mass Calibration*   The process by which the mass analyzer is calibrated such that a measured and displayed m/z is accurate. Well-characterized calibration compounds are utilized, and measured m/z values for these compounds are compared to theoretical m/z values. Calibrants commonly used include various polymeric species (such as polypropylene glyol, or PPGs; polytyrosine (poly-t)) or fluorinated species (perfluorokerosene or PFK) but can be any compound or mixture (NaI/KI) of compounds properly characterized for MS.

- *Internal Calibration*   The process by which one or more calibrant is introduced into the mass spectrometer simultaneously with the unknown sample, and the mass calibration is continuously updated during analysis. Considered the most effective means of obtaining highly accurate mass analysis (provided the calibrant does not interfere with the analysis of the unknown) (Herniman et al., 2004).

- *External Calibration*   When mass calibration is conducted in an entirely separate exercise from analysis of an unknown. External calibration can be performed infrequently, avoiding the potential problem of simultaneous analysis of calibrant and unknown (direct interferences, suppression, etc.).

- *Lock Mass*   Similar to internal calibration. The lock mass compound is monitored during analysis of the unknown, and the mass calibration is adjusted based on the comparison of the measured m/z and the theoretical m/z for the lock mass compound. If multiple lock mass compounds are used across the m/z range, the process effectively becomes internal calibration. Lock mass compound(s) can be introduced into the LC–MS source via a tee into the LC flow or sheath liquid inlet or dedicated sprayer.

- *Resolution*   The width (in u) of a mass spectral peak at a given m/z value. Also frequently used interchangeably with resolving power below. Along with mass calibration, the mass resolution is the most essential parameter to control in the mass analysis.

- *Resolving Power* (RP)   A measurement of how effectively a mass analyzer can distinguish between two peaks at different, but similar m/z. Mathematically, the formula $M/\Delta M$ is used, where M is the m/z value for one of the peaks and ΔM is the spacing, in unified atomic mass units, between the peaks. Most commonly, ΔM is the mass resolution, either via the 10% valley or FWHM definitions (see below). (Note that the definition used will affect the resolving power calculated.) Resolving power of 500–1000 approximately corresponds to unit resolution (e.g., at m/z 700 and FWHM resolution of 0.7, RP = 1000).

- *FWHM*   Full width at half-maximum. Mass resolution is often difficult to determine at or near the base of a peak due to baseline noise and peak overlap. It is more common to measure the width of the peak halfway to the peak maximum, where a clean measurement is possible. The most common alternative to FWHM was the 10% valley definition, in which the peak width at 10% of height was examined. This latter definition is common in the literature, especially for magnetic sector mass spectrometers, but is currently used much less frequently than FWHM. The choice of FWHM or 10% valley has an impact on the calculation of resolving power.

- *Unit Resolution*   Setting the resolution to produce a peak 1 mass unit wide at the base. For a Gaussian-shaped peak, the FWHM width for unit resolution is about 0.7 u.

- *High Resolution*   There is no specific definition for high resolution, but it is generally accepted that a resolving power over 5000 or 10,000 represents the beginning of high resolution. For small molecules, this typically corresponds to a mass resolution of approximately 0.1 (FWHM) or below. The acronym HRMS (high-resolution mass spectrometry) is often used to describe analysis at a high resolving power.

- *Parts Per Million*   The term parts per million (ppm) is a relative measure commonly used in discussing mass accuracy. One ppm is determined as the measured m/z divided by 10^6. For reference, accuracy within 1 ppm at m/z 500 would establish a yield of 500 ± 0.0005 u.

- mDa or mmu   One mDa is 0.001 u. The millidalton (mDa) and the equivalent milli mass unit (mmu) are also used in describing small mass differences.

1.3.3 Mass Spectrometer Components

A mass spectrometer consists of a sample inlet, an ion source, a mass analyzer, and a detector (Fig. 1.5). Each component is described below.

1.3.3.1 Sample Inlet and Source   A key component of any mass spectrometer is the mechanism of introducing the sample into the instrument. The first

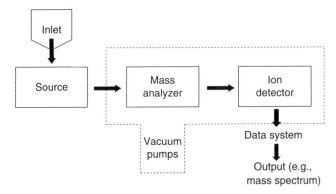

Figure 1.5. *Components of a mass spectrometer.*

component is the sample inlet. In many cases, this will be the liquid (or gas) chromatograph, which delivers the sample to the mass spectrometer source. Sources used with gas chromatography include electron impact ionization (EI) and chemical ionization (CI). Use of GC–MS has declined significantly due to improvements in LC–MS, and GC–MS sources are not described here. For MALDI systems, samples are typically "spotted" onto a surface (the target). The target is then physically placed in the source (Chapters 11 and 12). There are several common source types, as described below. For successful analysis, the sample introduced to the source must be converted from the liquid or solid phase to the gas phase and must be ionized before entering the mass analyzer.

- *API*   The atmospheric pressure ionization (API) source is the most common category of source for LC–MS analysis, in which ionization is performed outside of the high-vacuum region of the mass spectrometer. Electrospray ionization (ESI) and atmospheric pressure chemical ionization (APCI) sources are both examples of API sources.
- *ESI*   A common LC–MS source in which the effluent from a liquid chromatograph is directed through a fine capillary to which a high electric field has been applied. Ions are formed in a solution via acid–base or redox chemistry and converted to the gas phase through some combination of ion evaporation or ion ejection mechanisms (Labowsky et al., 1984; Kebarle, 2000). ESI is considered a soft ionization technique, where little fragmentation of the analyte occurs. The technique is capable of creating multiple charges on a single molecule and is highly effective for analysis of large molecules such as peptides and proteins. ESI can also lead to a profusion of different ion types, such as $[M + H]^+$, $[M + Na]^+$, and $[M + NH_4]^+$, in the positive-ion mode and $[M - H]^-$ in the negative-ion mode.
- *APCI*   Atmospheric pressure chemical ionization (APCI) is a source for LC–MS analysis in which the effluent from a liquid chromatograph is directed through a fine capillary and sprayed into a heated tube. The liquid is converted

to the gas phase through evaporation. Upon exiting the heated region, the gaseous effluent passes a high-voltage corona discharge needle, leading to the formation of reagent ions (typically derived from solvent species). Gas-phase proton transfer between reagent ions and analyte molecules ultimately leads to ionization of the analyte (Bruins, 1991; Duffin et al., 1992).

- *APPI* Atmospheric pressure photoionization (APPI) is an ionization source similar to APCI but the corona discharge needle is replaced with an irradiation source (e.g., krypton lamp). In comparison to ESI and APCI, APPI can be used to efficiently ionize broad classes of nonpolar compounds. In the bioanalytical tool box, APPI is an important complement to ESI and APCI (Hanold et al., 2004; Syage et al., 2004; Cai et al., 2005; Hsieh, 2005).

- *MALDI* A soft (gentle) method for creating gas-phase ions that utilizes energy from a laser targeted onto a mixture of analyte and a chemical matrix. Analyte ions can be formed from a combination of vaporization of existing ions and by vaporization of neutrals followed by ionization in the gas phase (Hillenkamp et al., 1990).

- *DESI* Desorption electrospray ionization (DESI) is a recently developed technique that permits formation of gas-phase ions at atmospheric pressure without requiring prior sample extraction or preparation. A solvent is electrosprayed at the surface of a condensed-phase target substance. Volatilized ions containing the electrosprayed droplets and the surface composition of the target are formed from the surface and subjected to mass analysis (Takats et al., 2005; Wiseman et al., 2005; Kauppila et al., 2006).

- *DART* Direct analysis in real time (DART) is an analogous technique to DESI that does not require the electrospray solvent (Cody et al., 2005; McEwen et al., 2005; Williams et al., 2006).

- *NSI* Nanospray ionization (NSI) is a low-flow (10–500-nL/min) ESI technique with many advantages over conventional-flow ESI ($\sim$200 µL/min) for the analysis of drugs, metabolites, peptides, proteins, and other macromolecules. Advantages of NSI over ESI include decreased sample consumption and increased sensitivity (Wilm and Mann, 1996; Corkery et al., 2005). NSI can be used for LC–MS or direct-infusion MS analysis of molecules (Wickremsinhe et al., 2006; Ramanathan et al., 2007c).

1.3.3.2 Mass Analyzers Analysis based on mass-to-charge ratio occurs within the mass analyzer of the instrument. The mass analyzer is often used as the basis for differentiating and discussing various types of mass spectrometers. Mass analyzers commonly considered to operate at a high resolving power are denoted by HRMS under the mass analyzer listing.

- *QMF* The quadrupole mass filter (QMF) or the transmission quadrupole is a mass analyzer that utilizes four parallel conducting rods arrayed such that a combination of two voltages permits the passage or filtering of only a single m/z value. Varying the amplitude of the fields permits a sequential range of

m/z ions to pass through the mass analyzer to create a mass spectrum. Low operating voltages (therefore tolerant of high operating pressures of $\sim 10^{-6}$ torr) and fast scanning capabilities make quadrupole analyzers ideal for coupling with LC systems (Dawson, 1986; Kero et al., 2005).

- *QIT* The quadrupole ion trap (QIT) utilizes a cylindrical ring and two end-cap electrodes to create a three-dimensional (3D) quadrupolar field for mass analysis. These instruments are capable of selectively trapping or ejecting ions and are often used for the sequential fragmentation and analysis experiments of product ion MS/MS. Also known as a 3D trap due to the configuration (March, 1997).

- *LIT* The linear ion trap (LIT) (also referred to as a two-dimensional, or 2D, trap) is a variation on the transmission quadrupole mass analyzer. In the LIT, the quadrupole is constructed such that either ions can be analyzed immediately or, ions can be trapped and held in the quadrupole region and then analyzed (Hager, 2002; Schwartz et al., 2002). Various types of MS/MS can be performed, as described in Chapter 3.

- *TOF* The time-of-flight (TOF) mass analyzer is conceptually the simplest of all. Ions are "gated" from the source region by an electrical field pulse and accelerated down the TOF flight tube. Low m/z ions travel at a higher velocity and reach the detector quicker than the slower ions with high m/z. Calibration of the accelerating field and resulting flight times permits mass analysis for unknowns. Hybrid instruments combining quadrupole and TOF mass analyzers (Q-TOF) have become common in recent years (Morris et al., 1997; Hopfgartner and Vilbois, 2000) (HRMS).

- *FTICR* The Fourier transform ion cyclotron resonance (FTICR) mass analyzer represents the highest performance in terms of resolving power. The FT (ICR) utilizes a strong magnetic field to store ions of various m/z in a cylindrical flight path (X and Y directions). An electric field is used to excite ions, which are detected when they pass near a detector plate. The frequency with which ions of a particular m/z pass the detector is recorded and fast Fourier transform is used to deconvolute the resulting data (Marshall et al., 1998). Hybrid FTICR often utilizes a quadrupole mass analyzer prior to the ICR cell (Patrie et al., 2004; P. O'Connor et al., 2006). Overall, these high-performance mass analyzers are the most expensive and massive of the common instruments and exist in relatively limited numbers compared to other instrument types (HRMS).

- *Orbitrap* The newest of the major mass analyzers, the Orbitrap is a hybrid MS consisting of a LIT mass analyzer, or transmission quadrupoles connected to the high-resolution Orbitrap mass analyzer. The Orbitrap utilizes electrical fields between sections of a roughly egg-shaped outer electrode and an inner (spindle) electrode (Chapter 5). Ions orbit between the inner and outer electrodes and their oscillation is recorded on detector plates (Hardman and Makarov, 2003; Hu et al., 2005). As with the FTICR, fast Fourier transform of the raw data is used to convert the data for mass analysis, making the Orbitrap the second major type of FTMS instrument. The resolving power of the Orbitrap is intermediate between the TOF and FTICR, as is the price. Ease of ownership and use versus the hybrid FTICR instruments and the higher performance versus the Q-TOF instruments have both worked in favor of the Orbitrap (HRMS).

- *Tandem Mass Spectrometer*   An instrument capable of performing multiple mass (m/z) analyses. There are two major categories: (1) tandem-in-space instruments (triple quadrupole and Q-TOF), (2) tandem-in-time instruments (QIT and FTICR).

- *Hybrid Mass Spectrometer*   A tandem mass spectrometer comprised of multiple mass analyzers of different types. A Q-TOF is a hybrid, but a triple quadrupole is not. Ideally, a hybrid instrument harnesses the best features of each mass analyzer type to produce a system perhaps greater than the sum of the parts.

- *MS/MS*   A process in which mass (m/z) selection or analysis is typically performed in two distinct serial steps. Operational examples include selected reaction monitoring or constant neutral loss scanning (see below).

- *MS^n*   A series of n steps in which m/z selection is performed. MS^n can be conducted by linking a series of mass analyzers, each of which performs one selection step, or more commonly by using ion-trapping instruments such as QITs (2D or 3D) or FTICR.

Use of Mass Analyzer: Scan Types   Depending on the configuration of the instrument, tandem and hybrid mass spectrometers are capable of far more than simply identifying the mass of a species that emerges from the source. The following is a brief list of relevant terminology and scan types that can be useful in generating additional information to support the identification of an unknown. Note that not all scan types are feasible on all types of instrument.

- *Full Scan*   The mass analysis process by which a controlled series of m/z are allowed to be detected. The m/z range over which a mass analyzer can be used (e.g., m/z 20 to 4000) is one defining characteristic of the instrument.

- *Selected Ion Monitoring (SIM)*   The mass analysis process in which only a single m/z value is selected by the mass analyzer and transmitted to the detector. Also referred to as the "single ion monitoring."

- *Precursor Ion (MS/MS)*   Generally the ion selected by the first mass analysis of an MS/MS process. Also formerly referred to as the "parent ion."

- *Product Ion (MS/MS)*   The species formed by fragmentation of the precursor ion. Also formerly referred to as the "daughter ion."

- *Product Ion Scan (MS/MS)*   Determination of all possible product ions formed from a specific precursor ion. A key step in the characterization of an unknown species that can facilitate functional group and structure identification.

- *Precursor Ion Scan (MS/MS)*   Determination of all possible precursor ions that form a specific product ion. Useful when a characteristic or significant product ion has been noted and the sources of that ion are sought. An example would be the detection of structurally similar compounds (i.e., metabolites, degradants, etc.) by identifying all species that produce a common fragment. Also referred to as the "parent ion scan."

- *Constant Neutral Loss Scan (MS/MS)* Determination of precursor/product ion combinations that exhibit a specific, characteristic loss of a portion of a molecular ion. Particularly useful when the characteristic species (loss) is neutral and cannot be detected directly by the mass spectrometer. Analysis of glutathione conjugates via neutral loss of 129 is an example. For the purposes of this book, NLS is used to describe these types of MS/MS experiments.
- *Selected Reaction Monitoring (MS/MS)* Selected reaction monitoring (SRM) is the process by which the first mass analysis selects a specific m/z (the precursor ion) to be fragmented in the collision cell and the second mass analysis selects and detects a specific product ion. Most commonly used in the quantitative analysis of well-characterized, targeted species for which optimized precursor–product pairs can be established. In SRM-based LC–MS assays no qualitative information can be obtained. However, SRM can be used to trigger product ion, neutral loss, or precursor ion scans.

1.3.3.3 Detector The detector is the last major portion of the mass spectrometer, and it detects the presence, and preferably abundance, of ions after they have exited the mass analyzer. Examples include the electron multiplier, common on quadrupole instruments, and the microchannel plate (an array of electron multipliers), which have been common on TOF instruments. For most users, the actual detector is a relatively "invisible" portion of the instrument that needs little or no regular attention.

1.4 MASS SPECTROMETRY IN QUANTITATIVE ANALYSIS

Over the past 20 years, LC–MS-based quantitative bioanalysis has grown to replace every other quantitative analytical method, including LC–UV and GC–MS. As evidenced in Chapter 2 and a number of recent reviews, today, quantitative LC–MS/MS is the most important application area of MS (Hsieh and Korfmacher, 2006; Jemal and Xia, 2006). Routine quantification of drugs and metabolites is achieved using LC–MS run times of less than 5 min. Technological advances discussed in this book provide further evidence that LC–MS run times of less than 1 min are becoming standard practice in many laboratories. Quantitative LC–MS and LC–MS/MS assays (simplified as LC–MS except where differentiation is necessary) are required not only during the journey of a drug through discovery and development stages (ADME, toxicological and clinical studies), but also during the postapproval marketing period. Although quantitative LC–MS methods developed during the drug discovery stage may not be adequate to support the drug development stage studies, discovery stage assays may be improved and validated as necessary to satisfy the regulatory and the sensitivity requirements of development preclinical and clinical studies. The fundamental parameters for LC–MS method validation include selectivity, sensitivity, linearity, precision, accuracy, matrix effects, recovery, stability, reproducibility, and dilution integrity (Jones, 2006; Shah, 2007). Components and criteria that define and/or impact a quantitative LC–MS assay are as follows:

- *Liquid Chromatography* The process by which the components of a liquid sample are physically separated based on their partitioning between a stationary phase and a moving (mobile) phase. Major modes include reverse phase, in which the stationary phase is non-polar, and normal phase, in which the stationary phase is polar. HILIC (Hydrophobic interaction chromatography) is a popular variant on the latter (Goodwin et al., 2007).

  For developing an LC method with high precision and accuracy, information about the sample/analyte such as number of possible analytes present, chemical structure of the analytes, molecular weight, concentration range, and solubility are crucial. LC separations are optimized by changing the following variables in the order listed: (1) mobile-phase composition/gradient, (2) column temperature, (3) solvent type, (4) additives, (5) pH, and (6) column type. For comprehensive description of HPLC systems, techniques, and method development, the readers are directed to specialized texts and review articles (Sadek, 2000; Tang et al., 2000; Tolley et al., 2001).

- *Mass Spectrometry* Mass spectrometer components, types of mass spectrometers, ionization sources, and scan types are described in Section 1.1.1.

  MS Dwell Time: Dwell time describes the time spent on a single step in a SRM or SIM analysis. Longer dwell time results in fewer data points but better signal-to-noise ratio, and should be optimized to produce acceptable data for each consideration. Common SRM dwell times in LC–MS would be 25–300 milliseconds (ms).

  MS Scan Time/MS Cycle Time: Scan time describes the time required to perform one complete MS data point for all targeted m/z. In SRM or SIM, this is the sum total time for each individual dwell, plus any additional time required by the system. In full scan or other scanning modes, scan time is the time required to complete one entire scan, e.g., from m/z 100 to m/z 1100 in one second.

  Run Time: The time for one complete injection and analysis, including any autosampler time required between injections. Run time is critical to determining the overall time required for analysis of a number of samples.

- *Sample Preparation/Extraction* The process of separating potentially interfering components from a sample prior to LC–MS analysis for the purposes of improving sensitivity, specificity, and/or method ruggedness. Variations include solid phase extraction (SPE), liquid–liquid extraction (LLE), and protein precipitation (PPT). Extraction may be performed off-line, in which the cleanup is completely independent from the LC–MS analysis, or on-line, in which the cleanup is integrated directly into the LC–MS analysis.

- *Method Validation* The procedure by which an LC–MS method (extraction, chromatographic separation, and MS detection) developed for quantitative measurement of an analyte, in a given biological matrix, is demonstrated to be reliable and reproducible for the intended use. For analytes present in different biological matrices (plasma vs. urine), separate methods have to be validated. Cross-validation and/or partial validation experiments are required when changes (MS or LC instrument type, extraction methods, etc.) are made to a validated assay.

- *Standard Curve/Calibration Curve*   The response from samples containing known, spiked quantities of analyte is mathematically regressed to create a calibration curve for each analyte. The response is usually peak area ratio (analyte area/internal standard area), but can be derived from area or height. The calculated curve is most commonly fit to the data using linear regression, but quadratic, power fit, and other models may be used. The variance observed across the assay range is often a function of concentration, and weighting such as $1/\text{concentration}^2$ (also known as $1/x^2$) is often used to improve the fit of the regression line to the data.

- *Internal Standard (IS)*   The internal standard (IS) is a compound added in a fixed, known amount to every quantitation sample to serve as an internal control for the analysis. Most commonly, the IS is used to normalize response through determination of peak area ratio as described above. The ideal IS will track with the analyte(s) through the extraction, chromatography, and mass spectrometry to account for variable recovery, minor spills, and changes in response over time. Stable-isotope versions of the analytes are ideal IS for LC–MS quantitation, but in many cases structural analogs exhibit sufficiently similar chemistry to be useful in this role (Jemal et al., 2003; Wieling, 2002; Stokvis et al., 2005).

- *Quality Control (QC)*   QC samples are used to check the performance of the bioanalytical method as well as to assess the precision and accuracy of the results of postdose samples. QC samples are prepared by spiking the analyte of interest and the IS into a blank/control matrix and processing similar to the postdose samples. QC samples cover the low ($3 \times$ LLOQ; LLOQ = lower limit of quantitation), medium, and high (70–85% of ULOQ; ULOQ = upper limit of quantitation) concentration ranges of the standard curve and are spaced across the standard curve and the postdose sample batch.

- *Matrix Effects*   The suppression or enhancement of LC–MS response due to the presence of biological matrix components such as salts, proteins, metabolites, coadministered drugs, degradants, additives, impurities, and phospholipids (King et al., 2000; Avery, 2003; Weaver and Riley, 2006). Matrix effects may result in shifts in analyte retention times, poor chromatographic peak shapes, and inaccurate quantitative assessments. Although APCI is less susceptible to matrix effects in comparison to ESI, most of the pharmaceutical assays require the use of ESI due to thermal instability of the analyte (Matuszewski et al., 2003). Generally, matrix effects are examined by comparing the absolute LC–MS peak area for an analyte in neat solution with that of analyte spiked post-extraction into a blank matrix at the same concentration (Matuszewski et al., 1998,

2003). Alternatively, matrix effects can be evaluated using postcolumn infusion methods described in detail in this chapter and elsewhere (King et al., 2000; Weng and Halls, 2002; Mei, 2005).

- *Carryover* Analyte or IS response transferred from a previous analysis to a sub-sequent analysis. Carryover is classically considered to occur within the LC–MS system in the autosampler (syringe, injection canula, switching valve) or LC column, but can also occur in sample handling devices such as liquid handlers (pipets, robotic pipets) used during extraction. Carryover is assessed by injecting one or more control/blank matrix extracts and/or mobile-phase mixtures after a high-concentration QC, postdose sample, or standard. The typical benchmark for carryover of an analyte is a relative measure, with a target level of less than 20% of the LLOQ response measured following analysis of a ULOQ standard (Weng and Halls, 2002).

- *Crosstalk* An unwanted contribution to a LC–MS/MS transition from a previous LC–MS/MS transition. The potential for crosstalk is higher when multiple analytes with identical product ion mass-to-charge ratios are being monitored and when sufficient time is not provided for emptying the collision cell between MRM or SRM transitions. Crosstalk leads to over- or underestimation of an analyte of interest (Tong et al., 1999).

- *Acceptance Criteria* The acceptance criteria recommended by the current guidance calls for $\leq 15\%$ for all the calibration curve standards and QCs with the exception of the LLOQ, where the acceptance criterion is increased to a 20% deviation. At least four of the six QCs must pass with $\leq 20\%$ of the nominal value. In addition, at least one QC sample per concentration range must pass with this criterion. If additional QCs are used in a batch, at least 50% of the QCs need to be within each concentration range.

- *Lower Limit of Quantification (LLOQ)* The lowest concentration of the analyte of interest in a matrix that can be quantitatively determined using the standard curve with acceptable precision and accuracy. The LLOQ is usually defined as the lowest concentration at which the assay imprecision does not exceed 20%.

- *Upper Limit of Quantification (ULOQ)* The highest concentration of an analyte in a matrix that can be quantitatively determined using the standard curve with an acceptable precision and accuracy. If the analyte concentrations in the postdose samples are higher than the ULOQ, then a dilution QC is needed to cover the highest anticipated dilution.

- *Linear Range* The concentration range where increasing concentrations of an analyte have a proportional increase in LC–MS response. Overall QqQ-type mass spectrometers (triple quadrupoles, Q-TRAPS) are superior in terms of linearity. Most common causes for nonlinear response include MS detector saturation, dimmer/adduct formation, API droplet/vapor saturation at high concentrations, and space charge effects.

- *Analyte Stability* Analyte stability experiments are carried out mimicking the sample collection, storage, and processing conditions as closely as possible. Stability experiments are conducted for the assay duration in the same matrix containing the same type of anticoagulant [Na–heparin, Li–heparin, Na$_2$–ethylenediaminetetraacetic acid (EDTA), etc]. Typical short-term stability evaluations include three freeze-and-thaw cycles, 4-hs at room temperature in matrix, and stability of final extracts (autosampler stability). Long-term stability experiments cover storage of unprocessed postdose samples at $-80°$C, $-70°$C, and/or $-20°$C for weeks and if necessary months or years.

- *Recovery/Extraction Efficiency*   A ratio between the response of an analyte spiked into a blank matrix preextraction and the response of the same analyte spiked into a blank matrix postextraction. Although the recovery of an analyte need not be close to 100%, the extent of the recovery at all QC levels should be consistent, precise, and reproducible.

- *Dilution Integrity*   To check dilution integrity, a QC sample prepared at a concentration greater than the ULOQ is analyzed using dilution in blank matrix. Acceptable assay precision and accuracy are required.

- *Inter- and Intra-Assay Precision*   Intraassay precision and accuracy are assessed within one batch (QCs, standards, etc.), whereas interassay precision and accuracy are assessed using separate batches.

1.4.1 Applications in Pharmacokinetics

Quantitative analysis to track the concentration of one or more targeted species throughout the course of the drug discovery and development processes is broadly referred to as pharmacokinetic analysis. The data obtained permit critical determination of the movement and transformation of the initial drug, as described in Chapter 2. For a number of years, the quantitation required for pharmacokinetic studies was largely performed by LC with spectrophotometeric detection such as ultraviolet/visible (UV/Vis) absorbance, or occasionally fluorescence. While the latter technique offered good specificity, UV/Vis detection generally did not. This relative lack of specificity frequently necessitated careful sample-processing/extraction techniques and relatively long run times to minimize quantitation interferences.

Over the past two decades, QMF-based quantification assays have become the technique of choice for quantification of drug candidates and their metabolites. Combining a mass spectrometer with LC provides an additional degree of selectivity and makes the combined technique the method of choice for quantitative bioanalysis of drugs and metabolites. Among the mass spectrometer types, QMF are ideal for coupling with LC and atmospheric pressure ionization sources (ESI, APCI, APPI, DART, DESI, etc.) because QMFs have the lowest voltage requirements and vacuum requirements.

With the advent of the practical API-based LC–MS interfaces, the high specificity of mass spectral analysis permitted a radical decrease in the amount of analytical time invested (sample preparation, injection, chromatography) prior to final detection (Hsieh et al., 2006; Maurer, 2007). Although SRM detection as the final step in LC–MS analysis can incorporate several stages of specificity (Chapter 3), some form of sample preparation/extraction is still performed to remove unwanted matrix components (proteins, phospholipids, salts, etc.). The extraction step can range from simple protein removal to highly specific solid-phase extraction (Kuhlenbeck et al., 2005; Chang et al., 2007a). Sample preparation is followed by chromatographic separation to resolve the analyte-like interferences from the peak of interest. In general, the combination of extraction and chromatography probably brings less specificity enhancement to LC–MS/MS analysis than it does to LC–UV analysis. But this is feasible because of the many levels of specificity in the final analytical step. To be detected in a SRM-based LC–MS/MS assay, an analyte must be eluting from the chromatography system at the correct retention time and be vaporized and ionized to the desired polarity under the conditions employed in the source. It must then have the correct m/z to transit Q1 successfully. In Q2 (sometimes the notation q2 is used because this set of quadrupoles cannot function as a mass analyzer and sometimes hexapoles and octapoles are used instead of quadrupole collision cells), the compound must fragment under the optimized conditions of gas pressure (argon, nitrogen, or helium) and energy employed, and only a fragment at the correct m/z can be transmitted through Q3 to reach the detector. The various drug discovery and development stages that require pharmacokinetic analysis are listed in Fig. 1.4.

1.4.2 LC–MS/MS in Pharmacokinetics: Example

In a recent example, a sensitive LC–MS/MS method was successfully applied to assay for fexofenadine in plasma following a single oral administration of a microdose (100-μg solution) and a clinical dose (60-mg dose) to eight healthy volunteers (Yamane et al., 2007). Fexofenadine and terfenadine (IS) eluted at 0.95 and 2.07 min, respectively, and the correct m/z for the protonated precursor ions were observed at m/z 502.2 and 472.2. For SRM (or MRM) experiments, both precursor ions were fragmented separately in the collision cell and the fragment ions of m/z 466 and 436, respectively, were monitored for fexofenadine and terfenadine. The details of the fexofenadine assay are given in the following:

Analyte (drug)

Fexofenadine, [M + H]$^+$ at m/z 502.2

Internal standard

Terfenadine, [M + H]$^+$ at m/z 472.2

Mass spectrometer	Sciex API 5000, triple quadrupole
Ionization source/mode	Turbo IonSpray/positive
Scan type	SRM (MRM); transitions = 502.2 → 466.1 and 472.2 → 436.1
Sample preparation	SPE (Waters Oasis HLB)
LC system	Waters Acquity
Column	Waters XBridge C18 (2.1 x 100 mm, 3.5 μm)
Mobile phase	A: 2 m*M* ammonium acetate; B: Acetonitrile
LC flow rate	0.6 mL/min
Test system (species)	Human
Postdose blood sampling	0.5, 1, 2, 3, 4, 6, 8, and 12 h

Using the sensitive quantitative LC–MS/MS method described above, linear PK profiles between clinical dosing and microdosing were obtained. Furthermore, Yamane et al. (2007) demonstrated that concentrations in human plasma after an oral dose of 100 μg is quantifiable using LC–ESI–MS/MS (Fig. 1.6), similar to what can be achieved using AMS (Chapter 2).

1.4.3 Focus: Matrix Effects

In the above example, successful quantification of fexofenadine in the concentration ranges of 10–1000 pg/mL and 1–500 ng/mL required two standard curves because

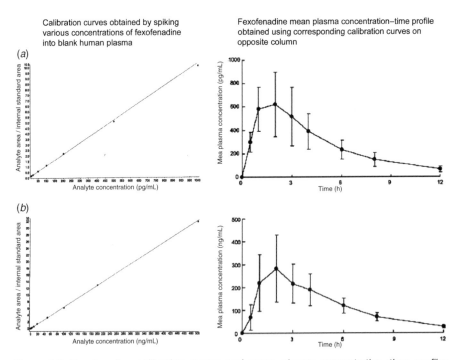

Figure 1.6. *Fexofenadine calibration curves and mean plasma concentration–time profiles following a single oral administration of (a) 100 μg (microdosing) or (b) 60 mg (clinical dosing) fexofenadine to healthy volunteers. (Reprinted with permission from Yamane et al., 2007.)*

often the linear dynamic range of an LC–MS or LC–MS/MS assay is limited due to calibration curve nonlinearity over wide concentration ranges. Calibration curve nonlinearity occurs due to detector saturation, adduct formation (dimers/multimer), and chromatographic carryover at higher concentrations as well as matrix effects (Matuszewski, 2006). Reduction (ion suppression) or enhancement of a MS signal caused by chromatographically coeluting matrix components was noted as a major issue in the 1990s (Matuszewski et al., 1998) and remains a significant issue in quantitation (Mei et al., 2003; Mei, 2005; Viswanathan et al., 2007). As far as the regulatory guidance is concerned, matrix effects are not required to be considered during a validation of a GLP assay. However, matrix effects can hamper assay reproducibility and/or linearity. Therefore, prior to validation and qualification of a quantitative LC–MS/MS method, matrix effects should be addressed. Figure 1.7 illustrates the steps necessary to evaluate a matrix effect (Bonfiglio et al., 1999; King et al., 2000).

As shown in Fig. 1.7, the method for evaluating ion suppression/enhancement encountered during a bioanalytical assay involves injection of a processed blank matrix sample on the column with continuous postcolumn infusion of a mixture of an analyte and an internal standard into the LC stream. The analyte and the internal standard are monitored (MRM or SRM scan) throughout the entire LC run time while the matrix components are eluting from the column. Data from a matrix effect experiment obtained using the postcolumn addition method are given in Fig. 1.8.

Extensive studies performed by several leading quantitative bioanalytical laboratories indicate that matrix effects can be limited by selecting the appropriate sample preparation techniques (Muller et al., 2002) and selecting the appropriate internal standard (Matuszewski, 2006).

Occasionally, interfering peaks are observed from metabolites, dosing vehicles, or the sample matrix itself. Suppression and interfering peaks can often be eliminated by changing the MS conditions, including the source type and resolving power

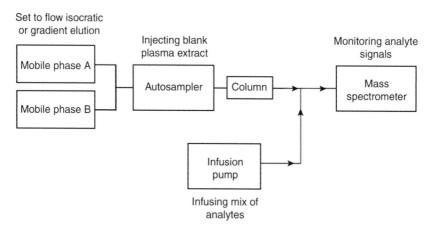

Figure 1.7. *Postcolumn infusion method to evaluate matrix effect originally described by Bonfiglio et al. (1999) and King et al. (2000). (Reprinted with permission from Bakhtiar and Majumdar, 2007.)*

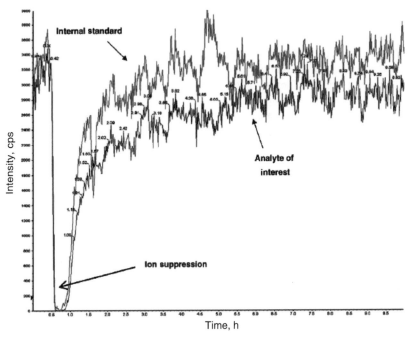

Figure 1.8. *MRM scans for analyte and internal standard obtained using ion suppression experiment described by Bonfiglio et al. (1999); and King et al. (2000). (Reprinted with permission from Bakhtiar and Majumdar, 2007.)*

(Xu, 2005), or by improving the sample preparation/extraction or chromatography. The latter approach, while powerful, is often seen as time-consuming in both the development of the required technique and the operation during sample analysis.

1.4.4 Applications in Toxicokinetics

The goal of toxicology experiments in drug discovery and development is to find out a drug's human health risks from the results in toxicological species. A recent survey (Lasser et al., 2002) reports that, among the drugs approved from 1975 through 2000, 45 drugs received one or more black-box warnings, and 16 were withdrawn from the market. Although a limited patient population and incomplete patient representation (ethnic, age, gender differences, etc.) are the main reasons for the failures of early clinical detection of drug toxicity, pharmaceutical scientists strive to conduct a well-designed preclinical toxicological study to avoid any human ethical issues. Toxicokinetic (TK) studies are required in the course of conducting toxicological experiments. In TK studies, PK parameters (Chapter 2) are applied to understand the relationship between a drug's exposure and its toxicity. Since a drug's exposure is a function of dose and time, historically, toxicology studies have been conducted using much higher doses than those which are pharmacologically relevant for a drug's action. However, over the years, the science of toxicology has changed from very high doses and adverse events such as death and changes in organ size/weight to more relevant doses and more sensitive endpoints such as biochemical and functional changes in the immune system, endocrine system, and neurological system. As a result, analytical techniques used for TK studies were also shifted from less sensitive LC–UV and GC–MS assays to more sensitive LC–MS/MS assays.

Regulatory guidelines issued by the FDA and the ICH dictate the required drug toxicity studies in preclinical species for supporting the start of phase I, II, and III clinical studies. Drug TK assessments may be performed in the following preclinical toxicity studies: (1) safety pharmacology, (2) single dose and rising single dose (1–2 weeks), (3) repeat dose (1–4 weeks), (4) longer repeat dose (6–12 months), (5) reproduction, (6) genotoxicity, and (7) carcinogenicity. In all preclinical in vivo studies, blood draw (sample size) is limited to less than 10% of the circulating blood in rodents (rat/mouse) and nonrodents (dog/monkey). Regular blood draws, over several days, for clinical chemistry and hematology changes limit the blood draws for TK analysis. Therefore, TK studies involve a sparse sampling of blood with five or six sampling time points. Sample limitations in TK studies further require the application of rugged and sensitive LC–MS methodologies for quantitative monitoring of drugs in blood and plasma. All development stage TK study samples are analyzed using quantitative LC–MS or LC–MS/MS bioanalytical methods developed under GLP guidance. Most often, GLP bioanalytical assays developed for TK studies are further validated and used in clinical trials and the postmarketing period (Srinivas, 2007).

1.4.5 Special Techniques in LC–MS/MS Quantitation

1.4.5.1 Quantitative Bioanalysis with High Mass Resolution
Prior to the introduction of the API sources for LC–MS, GC–MS was the dominant format for mass spectrometry. Within GC–MS, mass analysis at high resolution using magnetic sector instruments was relatively common, especially in the central mass spectrometry facilities of major corporations and universities. Uses of these instruments included quantitation by GC-HRMS for improved specificity and sensitivity.

Quantitation using high resolution mass spectrometry faded with the shift to MS/MS, and was particularly driven by the combination of the electrospray source and triple quadrupole mass spectrometer. Significant efforts continue however to reintroduce HRMS for quantitation using LC–MS on time of flight, quadrupole, or Orbitrap mass analyzers. As described above in Section 1.3.2.2, high resolution is a broad term, and the instrumentation cited here reflects that. The typical maximum resolving power at m/z 400 is about 4000 for the TSQ Quantum (high resolution triple quadrupole; HRMS on both mass analyzers), about 10–20,000 for a modern time of flight (HRMS on TOF only if Q-TOF hybrid), and 15–60,000 for the Orbitrap (HRMS on Orbitrap only if LTQ-Orbitrap hybrid).

A number of authors have discussed the utility of a high resolution triple quadrupole such as the TSQ Quantum (Fig. 1.9) (Jemal and Ouyang, 2003; Xu et al., 2003; Hughes et al., 2003; Paul et al., 2003). As shown in Fig. 1.10, nominally isobaric

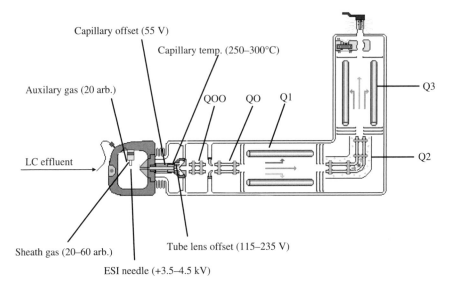

Figure 1.9. *Triple-stage quadrupole Quantum mass spectrometer capable of operating under enhanced resolution conditions. To reduce the chemical noise and to improve the sensitivity, the mass analyzers were oriented in an "L" shape rather than the conventional "straight" design. An additional benefit of the "L" shape orientation is a smaller foot print. (Courtesy of ThermoFisher Scientific.)*

PPG ions were resolved from both the 35Cl and 37Cl isotopic peaks of mometasone through the use of high resolution in Q1, Q3, or in both mass analyzers (Yang et al., 2002). Transmission losses on increased resolution, which are traditionally high for most quadrupole mass analyzers, were examined by Yang et al. (2002) for SRM quantitation. These results (Fig. 1.11 and Fig. 1.12) demonstrated that while absolute signal was reduced by approximately a factor of three, the true sensitivity as signal : noise ratio was maintained or possibly improved. In cases where interfering peaks or high background noise levels are problematic in SRM quantitation, a quick examination of the utility of enhanced resolution would potentially be far more time efficient than redevelopment of chromatography or extraction conditions.

Time of flight instruments, and perhaps especially hybrid Q-TOF systems (Fig. 1.13), have also been examined as quantitation systems (Zhang and Henion 2001; Yang et al., 2001b; O'Connor and Mortishire-Smith, 2006; D. O'Connor et al., 2006). While most testing has successfully demonstrated the concept, comparisons of sensitivity between conventional unit mass triple quadrupole and Q-TOF systems (Fig. 1.14) have shown that the quadrupoles generally produce greater sensitivity (Yang et al., 2001b). However, the rate of change and improvement in TOF systems has been extremely rapid in the last ten years, and such platform comparisons need to be revisited frequently to reflect the state of current commercial systems (Hashimoto et al., 2005; Weaver et al., 2007; De vlieger et al., 2007; Hopfgartner et al., 2007; Inohana et al., 2007).

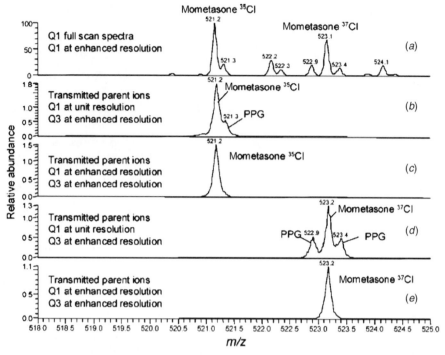

Figure 1.10. *LC–MS spectra of mometasone in the presence of PPG interferences obtained under unit- and enhanced-resolution settings showing the minimum loss in ion transmission under enhanced-resolution settings. (Reprinted with permission from Yang et al., 2002.)*

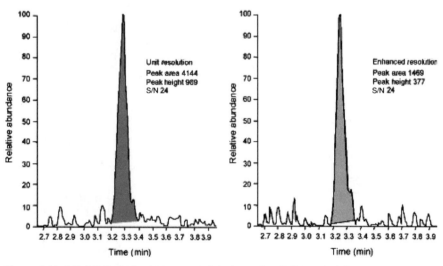

Figure 1.11. *LC–MS/MS chromatograms of desloratadine (SCH 34117) (fortified into plasma) obtained under unit- and enhanced-mass-resolution conditions. (Reprinted with permission from Yang et al., 2002.)*

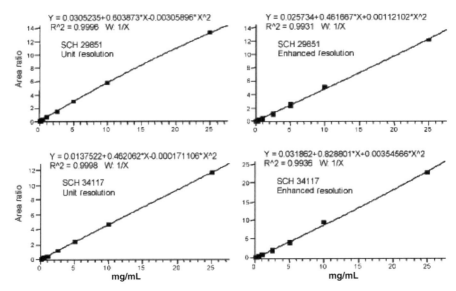

Figure 1.12. *Calibration curves for loratadine (SCH 29851) and desloratadine (SCH 34117) obtained under unit- and enhanced-resolution conditions. The precision and accuracy under both unit- and enhanced-resolution conditions met the assay acceptance criteria, correlation coefficients at enhanced resolution (0.993) were lower than those obtained at unit resolution (0.999). The lower correlation coefficients under enhanced-resolution conditions might have resulted from a slight mass window shift during the long overnight 17-h run. (Reprinted with permission from Yang et al., 2002.)*

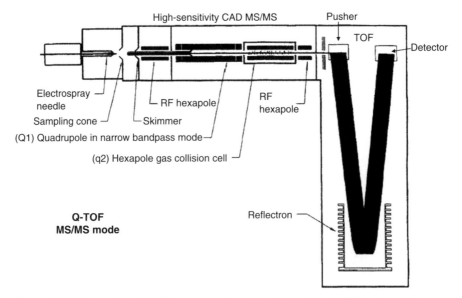

Figure 1.13. *Schematic of Q-TOF mass spectrometer. An updated Q-TOF schematic is presented in Chapter 4. (Reprinted with permission from Morris et al., 1996.)*

Triple-stage quadrupole (TSQ) data   Q-TOF data

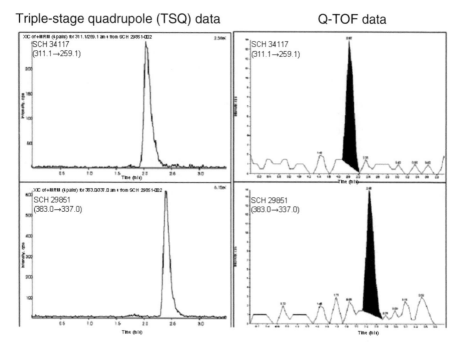

Figure 1.14. *MRM chromatograms of SCH 29851 (383.0→337.0) and SCH 34117 (311.1→259.1) obtained using Sciex API 3000 (triple-stage quadrupole) and Sciex QSTAR pulsar (Q-TOF). Comparison of MRM chromatograms of SCH 29851 and SCH 34117 obtained at the LOQ (1 ng/mL) using the API 3000 mass spectrometer with those from the Q-TOF mass spectrometer indicated that the S/N ratio is at least 10–20 times better on the API 3000 mass spectrometer. However, the MRM chromatograms from the API 3000 mass spectrometer do not provide the option to further examine the MS/MS spectra whereas the full-scan MS/MS spectra from a Q-TOF based quantitative bioanalysis assay allows one to easily eliminate any questions about false-positive data. (Reprinted with permission from Yang et al., 2001b.)*

The Orbitrap-based systems have emerged as the newest option for LC-HRMS. When configured as hybrid linear trap-Orbitrap (LTQ-Orbitrap), the systems are conceptually similar to Q-TOF in that mass analyzer 1 is nominally a unit mass analyzer, and mass analyzer 2 is capable of high resolution. These systems are capable of either LC-HRMS or LC–MS/HRMS operation. A new variant on the commercial Orbitrap, the Exactive, is expected to be released in late 2008. This system, which consists only of the single mass analyzer, has shown promising results in early assessment of quantitation by LC-HRMS (Bateman et al., 2008).

Finally, one concept that must be included in assessing quantitation by HRMS is the effective scan rate of the system. Quadrupole and time of flight mass analyzer are capable of rapid scan rates for SRM-type quantitation, with individual dwell times (quad) or scans (TOF) at 10–50 milliseconds possible. This permits acquisition of numerous data points across a chromatographic peak, which is critical for accurate and precise quantitation. Mass resolution is unaffected by changes in dwell time/scan rate, though signal : noise usually decreases with faster scanning. Resolution on Fourier-transform based mass analyzers is linked to scan rate however. The early LTQ-Orbitrap instruments could achieve resolving power 60,000 with a scan rate of about 1 second/scan. Resolving power dropped to 15,000 when a scan rate of say 300 milliseconds was used. This trade-off must be considered when matching chromatographic performance with mass analysis, and is critical if ultra-high performance liquid chromatography (Section 1.4.5.2 below and Chapter 4) are considered (De vlieger et al., 2007; Hopfgartner et al., 2007; Inohana et al., 2007).

1.4.5.2 Quantitative Bioanalysis with Enhanced Chromatographic Resolution
The majority of quantitative bioanalytical assays today involve the use of reverse-phase HPLC separation before MS detection. To improve upon conventional HPLC with respect to sample throughput/run time, chromatographic resolution, analyte sensitivity, and solvent usage, several laboratories are evaluating ultrahigh-pressure liquid chromatography (UHPLC) (Swartz, 2005a,b; Dong, 2007; Messina et al., 2007) or, as one vendor calls it, ultraperformance liquid chromatography (UPLC). For the purpose of this chapter, UHPLC and its variants are referred to as the UPLC. As detailed in Chapter 4, both HPLC and UPLC are governed by the van Deemter equation, which describes the relationship between plate height (N) and linear velocity (van Deemter et al., 1956; Wren, 2005; Wang et al., 2006). Based on this equation, as the particle size (d_p) decreases to less than 2 μm, there is significant gain in efficiency that does not diminish significantly at higher flow rates (Jerkovich et al., 2005; Swartz, 2005a,b; Wang et al., 2006). This creates an opportunity to optimize time efficiency while simultaneously improving chromatographic resolution and sensitivity (Jerkovich et al., 2005; Swartz, 2005a,b; Wang et al., 2006; Gritti et al., 2005; Martin and Guiochon, 2005; Plumb et al., 2004).

Most UPLC setups utilize columns of conventional LC–MS dimensions (length 30–50 mm, diameter about 2 mm), and operate at flow rates from 0.2–0.6 mL/min. As the particle size of the packing material decreases towards and below 2 μm, as is common in UPLC, the backpressure generated by resistance to flow increases rapidly to 10–15,000 psi (Plumb et al., 2004; Gritti et al., 2005; Martin and Guiochon, 2005; Swartz, 2005a; Swartz, 2005b). As most conventional HPLC systems are not designed for pressures greater than approximately 5000 psi, UPLC necessitated the development of new hardware prior to widespread commercialization.

The utility of UPLC has been demonstrated for both qualitative and quantitative analyses. In 2005, Castro-Perez et al. (2005) compared the performance of a HPLC with that of a UPLC and showed that improved chromatographic resolution and peak capacity attained with UPLC lead to reduction in ion suppression and increased MS sensitivity. Comparison of the mass spectrum obtained using HPLC with that from UPLC (Fig. 1.15) revealed that the higher resolving power of the UPLC–MS system resulted in a much cleaner mass spectrum than that obtained using the HPLC–MS system. The sensitivity improvement directly resulted in a higher ion count in the UPLC mass spectrum (855 vs. 176).

The additional sensitivity attainable with the UPLC approach is again demonstrated in Fig 1.16. While HPLC–MS resulted in signal-to-noise (S : N)

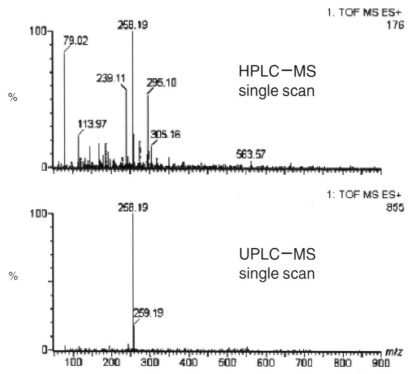

Figure 1.15. *LC–MS spectra of N-desmethyl metabolite of dextromethorphan (m/z 258.19) following incubation of dextromethorphan with rat liver microsomes. (Reprinted with permission from Castro-Perez et al., 2005.)*

ratio of $25 : 1$ for desmethyl-dextromethrophan-glucuronide metabolite $(m/z\ 434)$, UPLC–MS provided a $S : N$ ratio of $115 : 1$ for the same peak. The increased $S : N$ ratio achieved using the UPLC was attributed to improved peak resolution (sharper peak) and a reduction in ion suppression resulting from other co-eluting metabolites and endogenous compounds. The sensitivity gain was utilized by Pedraglio et al. (2007) in re-validation of a quantitative bioanalysis assay for NiK-12192, an antitumor candidate, using UPLC–MS. The UPLC–MS assay provided an LOQ of 0.1 ng/mL, whereas the previous HPLC–MS assay resulted in an LOQ of 0.5 ng/mL. Sensitivity improvements achieved with the UPLC–MS assay allowed the quantification of the 24 hour plasma samples that was previously not possible using the HPLC–MS assay (Fig. 1.17).

To highlight the advantages of UPLC for quantitative bioanalysis, Yu et al. (2006) enriched rat plasma with alprazolam, ibuprofen, diphenhydramine, naproxen, and prednisolone and compared HPLC–MS/MS and UPLC–MS/MS approaches for quantification of all five compounds. Apart from the particles that were used to pack the columns, all other separation and mass spectrometry methods were kept as similar as possible.

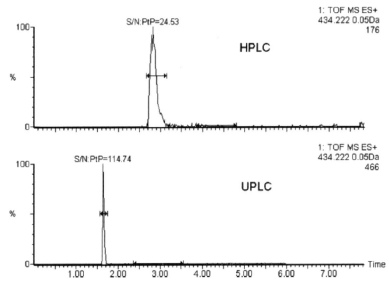

Figure 1.16. *Comparison of peak-to-peak (PtP) S/N ratio for the desmethyl-dextromethorphan-glucuronide (m/z 434.222) obtained using HPLC–MS (top trace) and UPLC–MS (bottom trace). (Reprinted with permission from Castro-Perez et al., 2005.)*

The results demonstrated some of the characteristics of UPLC–MS. Shorter retention times were observed, along with reduced chromatographic peak width. This in turn leads to less analyte dilution and improved S : N. With the HPLC approach, the alprazolam peak is about 4.8 seconds wide (Fig. 1.18) and the same peak

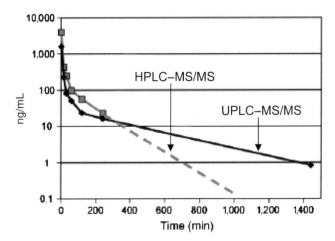

Figure 1.17. *Concentration–time plot following 3-mg/kg IV administration of NiK-12192 to mice. The dashed line is an extrapolation of the plasma sample concentration. The elimination phase determined using HPLC–MS is clearly different from that obtained using UPLC–MS. (Reprinted with permission from Pedraglio et al., 2007.)*

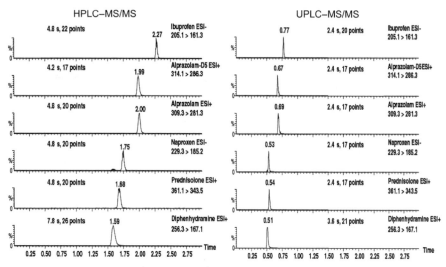

Figure 1.18. *Comparison of MRM chromatograms following HPLC–MS/MS and UPLC–MS/ MS analysis of a mixture containing alprazolam, ibuprofen, d5-alprazolam, diphenhydramine, naproxen, and prednisolone. Each set of chromatograms was obtained from a single 100-ng/ mL injection of rat plasma. d5-Alprazolam was used as the internal standard for quantification of alprazolam. (Reprinted with permission from Yu et al., 2006.)*

under the UPLC approach is about 2.4 seconds wide. In order to achieve the same number of data points (~20) across the peak, faster scanning using shorter dwell times was required for the UPLC method (Churchwell et al., 2005).

The overall benefits of going from HPLC to UPLC include increased separation efficiency, improved chromatographic resolution, and reduced analysis time (Shen et al., 2006; New et al., 2007). Although some reports suggest that the analysis time is reduced by ten-fold, a realistic, average estimate in reduction in analysis time is anywhere from three- to ten-fold. Overall, the higher chromatographic resolving power and increased separation efficiency of the UPLC result in improved MS sensitivity and a reduction in ion suppression. UPLC has proven to be one of the most promising developments in the area of LC–MS.

1.4.5.3 Quantitative Bioanalysis with Increased Selectivity: Application of FAIMS

High-field assymetric waveform ion mobility spectrometry (FAIMS) is an atmospheric pressure ion separation technique introduced to the MS community in 1998 (Purves et al., 1998). A recent review by Guevremont (2004) introduces the reader to fundamentals of FAIMS and describes its application to small- and large-molecule separation and detection. Typically, FAIMS is used in conjunction with ESI to improve the analytical selectivity of the conventional ESI–MS quantifications assays. As shown in Fig. 1.19, FAIMS is placed in between an ESI source and a mass spectrometer skimmer/orifice entrance. During the operation of FAIMS, a high-voltage asymmetric waveform is applied to the inner and outer electrodes. A time-dependent voltage difference between the

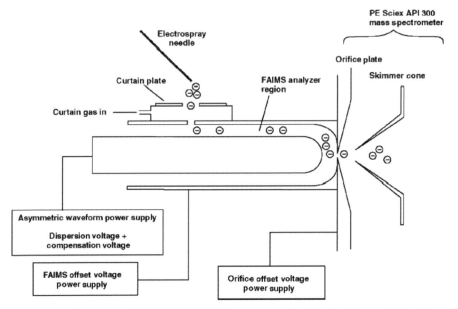

Figure 1.19. *Schematic of ESI–FAIMS–MS instrument. (Reprinted with permission from Guevremont, 2004.)*

electrodes causes the ions introduced via a carrier gas (helium, nitrogen, oxygen, carbon dioxide, or a mixture of gases) to oscillate and drift (ion mobility) toward one of the two electrodes. A compensation voltage (CV) with a correct magnitude and polarity is required to successfully transmit an ion through the electrodes. Since the CV is analyte (m/z) and temperature (Purves et al., 1998; Purves and Guevremont, 1999; Wu et al., 2007) dependent, it can be used to selectively transmit an ion of interest in the presence of other matrix or interfering ions. This unique CV-dependent selectivity feature of FAIMS allows the separation of isobaric drugs and/or metabolites as well as the separation of components that are difficult to separate under the fast LC conditions used for quantification of drugs.

Kapron et al. (2005) showed that FAIMS can be used to selectively quantify an amine compound in the presence of an interfering N-oxide metabolite (Fig. 1.20). Under the conventional LC–MS/MS settings, a SRM based precursor/fragment

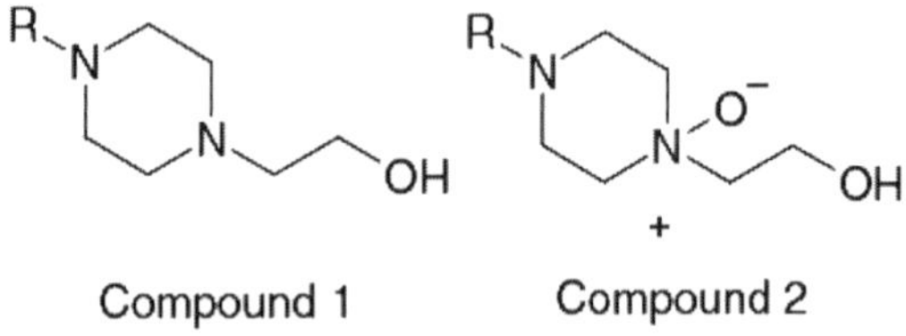

Compound 1   Compound 2

Figure 1.20. *Partial structures of the amine compound (Compound 1) and its N-oxide (Compound 2) analyzed using LC–FAIMS–MS/MS. (Reprinted with permission from Kapron et al., 2005.)*

ion pair (488/401) assay resulted in over estimation of the amine compound (compound A) (Fig. 1.21*b*) due to the N-oxide metabolite (m/z 506) undergoing fragmentation in the ESI source. In-source fragmentation of the N-oxide (compound B) resulted in the formation of m/z 488 ions through loss of an oxygen atom, which in turn contributed to Compound A transition of 488 → 401. The on-line FAIMS set-up allowed the metabolite interference to be removed before the entrance to the mass spectrometer.

Hatsis et al. (2007) showed that FAIMS can be used to increase quantitation throughput by eliminating chromatography altogether. To limit the impact from ion suppression, FAIMS was used in conjunction with a nano-flow ESI source rather than the conventional flow ESI source. As shown in Fig 1.22, the three minute LC–ESI–MS run time required to separate compound MLN A from the endogenous interference was reduced to 30 seconds by using FAIMS, although manual loading of the sample syringe added to the effective analysis time. Overall data quality was considered appropriate for the targeted discovery quantitation application.

A number of additional examples of quantitation using FAIMS have been published (McCooeye et al., 2002; McCooeye et al., 2003; Kolakowski et al., 2004; McCooeye and Mester, 2006). FAIMS also has been used to separate analytes of interest from endogenous matrices, metabolites, and other sample components (Ells et al., 2000; Venne et al., 2004; Venne et al., 2005). However, FAIMS is yet to be developed as a routine technique for the separation of complex biological samples.

1.4.5.4 Quantitative Bioanalysis with Ion Traps (3D versus 2D)

Although the pharmaceutical industry has long recognized the conventional three-dimensional (3D) ion trap as a powerful tool for structural elucidation of metabolites,

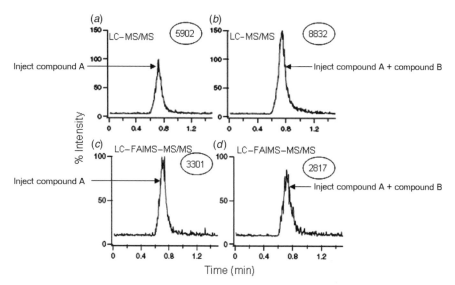

Figure 1.21. *Representative SRM (488→401) chromatograms obtained using LC–MS/MS (a,b) without and (c,d) with FAIMS. (Reprinted with permission from Kapron et al., 2005.)*

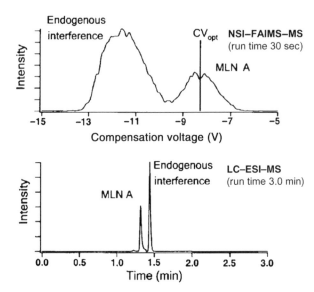

Figure 1.22. *FAIMS is used to increase the analytical throughput by eliminating the LC component altogether from quantification assays. The 3-min LC–ESI–MS run time required to separate compound MLN A from the endogenous interference was reduced to 30 s by using FAIMS. (Reprinted with permission from Hatsis et al., 2007.)*

unknowns, and degradants, due to inherent limitations, 3D ion traps (Fig. 1.23) were never accepted as the analytical technique of choice for quantification of drugs and metabolites in biological matrices. The limitations of the 3D trap as the desired quantification tool stems from its limited linear dynamic range and the extensive sample-processing requirements for achieving acceptable precision and accuracy.

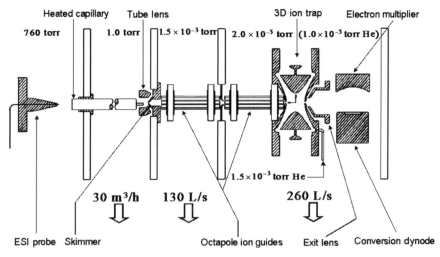

Figure 1.23. *Ion path and differentially pumped regions of LCQ mass spectrometer. (Courtesy of ThermoFisher.)*

Nevertheless, several groups demonstrated successful use of quadrupole ion trap mass spectrometers for quantification of drugs (Wieboldt et al., 1998; Chavez-Eng et al., 2000; Abdel-Hamid et al., 2001; Werner et al., 2001; Yang et al., 2001a; Naidong et al., 2002a; Werner et al. 2002; Yang et al., 2003; Salem et al., 2004; Sun et al., 2005; Vlase et al., 2007). More commonly however, QIT were the mass spectrometers of choice in the 1990s and early 2000s for performing quantification of drugs and simultaneous identification of their metabolites (Fig. 1.24) (Cai et al., 2000; Decaestecker et al., 2000; Cai et al., 2002; Kantharaj et al., 2003; Kantharaj et al., 2005a; Kantharaj et al., 2005b).

The two-dimensional (2-D) or linear ion trap (LIT) emerged in the 2000s as an effective alternative to the 3-D trap. Before 1995, linear traps were used primarily as ion storage/transfer/ion-molecule reaction devices in combination with FTICR (Senko et al., 1997; Belov et al., 2001), TOF (Collings et al., 2001), 3D ion trap (Cha et al., 2000), and triple-quadrupole (Dolnikowski et al., 1988) mass spectrometers because LITs offer better ion storage efficiencies in comparison to 3D quadrupole ion traps of the same dimensions (Hager, 2002; Schwartz et al., 2002). In 2002, commercial LITs were introduced as either stand-alone mass spectrometers (Schwartz et al., 2002) or as part of a triple quadrupole mass spectrometer (Hager, 2002).

The commercially available stand-alone LITs, marketed under the name LTQ, are made of four hyperbolic cross-sectional rods (Fig. 1.25). Since ions are trapped in an axial mode as opposed to central trapping on 3D ion traps, LTQs have been successfully coupled with Orbitrap and FTICR for achieving high-resolution capabilities (Peterman et al., 2005; Sanders et al., 2006) (Chapter 5). Functional improvements in 2D traps over 3D traps include 15 times increase in ion storage capacity, 3 times faster scanning, and over 50% improvement in detection efficiency and trapping efficiency.

The LIT introduced as part of a triple-quadrupole mass spectrometer is marketed under the name QTRAP. As shown in Fig. 1.26, the ion path and the differentially pumped region of QTRAP are similar to a triple quadrupole (API 3000, API 4000, and API 5000), except the Q3 is capable of functioning as a linear trap. QTRAP and its capabilities are described in detail in Chapter 3. Table 1.2 compares some of the advantages and limitations of QTRAP and LTQ mass spectrometers.

Most often simultaneous parent drug quantification and metabolite identification experiments are performed during early stages of drug discovery to conserve resources and expedite the lead selection process. To highlight the utility of the QTRAP mass spectrometer in this endeavor, King et al. (2003) demonstrated that information about circulating metabolites, dosing vehicle, interfering matrix components, and coeluting metabolites can be obtained by collecting full-scan data during quantification of drugs and metabolites in biological matrices. The QTRAP software controls were set up to perform a combination of SRM transitions and full scans (QTRAP scan functions are explained in detail in Chapter 3). While SRM transitions for urapidil (analyte) and labetalol (IS) were performed in the triple-quadrupole mode, full-scan spectra were acquired using the LIT mode. The LIT was operated in the enhanced MS (EMS) mode with a scan speed of 4000 daltons/sec. The cycle time for the combined scan function was 0.31 s. To

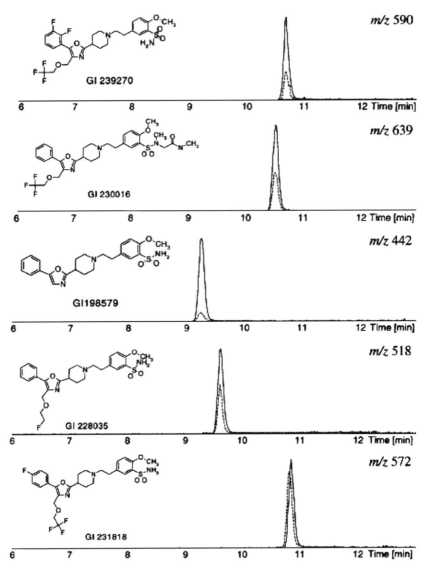

Figure 1.24. *Comparison of parent ion extracted ion chromatograms at t = 0 min (solid line) with those at t = 60 min (dotted line) following incubation of a five-compound cassette with dog microsomes. The high sensitivity of the QIT (3D) in the full-scan mode prompted Cai et al. (2000, 2001) to push the limits of QIT by simultaneously quantifying multiple drugs from cassette incubations and quantitatively studying metabolits from each drug. In an effort to increase the analytical throughput of the metabolic stability assays, compounds were individually incubated with dog microsomes for 60 min, quenched using acetonitrile, and the supernatants from four to five compounds were pooled to form cassette groups before LC–MS analysis. The percent metabolized for each compound in a cassette was calculated by measuring the extracted ion chromatogram (XIC) peak intensities for the parent analyte from each 0- and 60-min incubation. (Reprinted with permission from Cai et al., 2000.)*

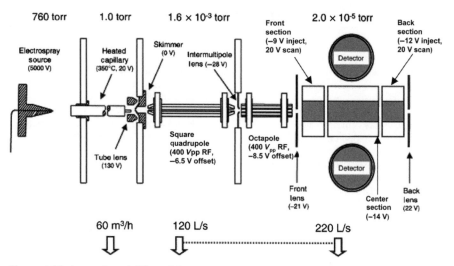

Figure 1.25. Ion path and differentially pumped regions of LTQ mass spectrometer. (Reprinted with permission from Schwartz et al., 2002.)

validate the quantification data from the alternating full-scan and SRM analysis (SRM/EMS), the data from the SRM-only analysis were compared with those acquired using the alternating method. The quantification data from both methods were generally acceptable in terms of sensitivity, accuracy, and precision. However, some loss of precision was observed at the lowest concentrations (Table 1.3).

The SRM/EMS mode of operation provided information on the coeluting glucuronide conjugates and the PEG dosing vehicle. Information about the coeluting

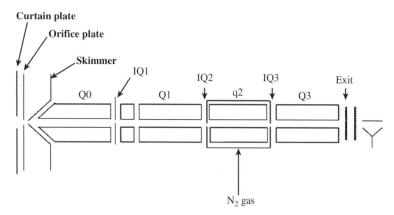

Figure 1.26. Ion path and differentially pumped regions of QTRAP mass spectrometer. (Reprinted with permission from Hager, 2002.)

TABLE 1.2. Comparison of LTQ and QTRAP Functionalities

LTQ	QTRAP
Coupled to Orbitrap, FTICR, etc.	Not practical to couple with other MS
Quantification possible with some limitations	Quantification similar to a triple quadrupole
MS^n ($n = 1, \ldots, 9$)	MS2 and MS3
Helium is used as the collision gas	Nitrogen is used as the collision gas
Inherent low-mass cutoff limits structural information form MS/MS experiments (pulsed-Q capability, however, allows one to get around this limitation)	Information rich tandem mass spectra similar to that from triple quadrupole are available
Two detectors lead to improved sensitivity	One detector
Radial ejection of ions to detectors	Axial ejection of ions to the detector
Neutral loss scan (NLS) and precursor ion scan (PIS) data obtained postacquisition by filtering MS/MS data	True NLS and PIS data can be acquired on the fly

material allowed the modification of the quantification method to improve the quality of the assay. Additionally, application of the SRM/EMS method to the in vivo PK samples provided information about circulating metabolites from early PK studies. Such information from early PK studies provided insight into the metabolic hot spots and allowed the medicinal chemist to modify the structure to optimize the PK of lead compound.

Since King et al. (2003) used SRM-triggered EMS for acquiring quantitative and qualitative data, characterization of the metabolites involved separate MS/MS acquisitions. To avoid analyzing the samples for the second time and to improve upon King et al. (2003), Li et al. (2005b) demonstrated the possibility of acquiring both parent drug quantification data and qualitative metabolite MS/MS data using the SRM-triggered information-dependent acquisition (IDA). Li et al. (2005b) tested the SRM-triggered IDA MS/MS experiments in both the conventional triple-quadrupole mode and the ion trap mode and showed that the cycle time decreased from 2.78 to 1.14 s with the latter technique. The longer cycle time in the triple-quadrupole mode of operation would have resulted in possibly missing some of the metabolites.

As shown in Fig. 1.27, a concentration–time profile for the parent molecule of compound A determined by the SRM-only method correlated very well with the SRM-triggered IDA method. Furthermore, the SRM-triggered IDA approach not only allowed Li et al. (2005b) to quantify the drug molecules, but also metabolites were quantified relative to the parent drug. As shown in Fig. 1.27, the authors were able to generate peak area–time profiles for two dioxy (M1A, M3A) metabolites and compare the PK parameters (e.g., half-life) of the metabolites with those of the parent compound. Understanding the PK parameters of the metabolites allowed the group to redesign the molecules with desired PK and metabolism profiles.

TABLE 1.3. Comparison of Accuracy and Precision from SRM-Only Mode with Those from Alternating SRM/EMS Mode (Reprinted with permission from King et al., 2003)

	SRM[a]					SRM/EMS[b]			
Expected Value	Avg.	Std. Dev.	%CV	%Accuracy	Expected Value	Avg.	Std. Dev.	%CV	%Accuracy
5.0	5.1	0.2	4.4	101.5	5.0	5.1	0.8	15.2	102.7
10.0	9.9	0.5	5.0	98.7	10.0	9.5	1.4	14.3	94.7
20.0	19.6	1.6	8.1	98.0	20.0	20.5	1.6	8.0	102.5
50.0	49.7	3.1	6.3	99.3	50.0	47.7	4.7	9.9	95.4
100.0	95.2	4.8	5.0	95.2	100.0	97.4	5.9	6.1	97.4
200.0	197.3	8.2	4.2	98.6	200.0	188.2	15.0	8.0	94.1
500.0	517.3	41.6	8.0	103.5	500.0	509.7	52.3	10.3	101.9
1000.0	1020.3	61.2	6.0	102.0	1000.0	1081.6	88.0	8.1	108.2
2500.0	2581.2	222.5	8.6	103.2	2500.0	2576.9	98.7	3.8	103.1

Note: $n = 5$ replicates. Urapidil standard curve was prepared in rat plasma and processed before analysis by QTRAP mass spectrometer.
[a]$R^2 = 0.99592$.
[b]$R^2 = 0.99042$.

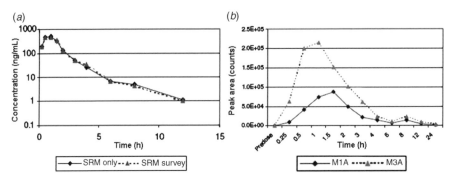

Figure 1.27. *(a) Comparison of concentration–time profile of compound A acquired using SRM-only and SRM-triggered IDA methods and (b) relative peak area–time profiles of dioxy metabolites (M1A, M3A) of compound A. (Reprinted with permission from Li et al., 2005b.)*

1.5 ADVANCES IN SAMPLE PREPARATION/CLEANUP AND COLUMN TECHNOLOGY

As summarized in Table 1.4 by Chang et al. (2007a), advances in sample preparation/clean-up and column technologies have been synonymous with improvements in analyte detection technologies over the past several decades. Improperly prepared samples for LC–MS analysis can very easily clog the LC system, associated tubing, or an API source as well as diminish the sensitivity of a mass spectrometer. On the other hand, sample preparation can not take hours and days in a high throughput bioanalytical laboratory where thousands of samples are analyzed per week. Therefore, to develop fast, sensitive and robust LC–MS and/or LC–MS/MS assays, one needs to select an appropriate and efficient sample preparation/clean-up technique and a suitable LC column to achieve adequate chromatographic separation.

1.5.1 Sample Preparation/Cleanup

Regardless of whether an assay is being developed to support a regulated (GLP) or discovery (non-GLP) study, common goals of sample preparation/cleanup include the following: (1) obtain a representative sample in solution, (2) remove coeluting/interfering matrix components with minimum analyte/drug-derived material losses, (3) achieve/maintain sufficient concentration for MS detection, (4) limit the number of steps, and (5) maintain ruggedness and reproducibility.

Some of the sample preparation/clean-up strategies used in high throughput LC–MS and LC–MS/MS quantitative bioanalytical analyses have been described in detail (Weng and Halls, 2001; Weng and Halls, 2002; Souverain et al., 2004a; Souverain et al., 2004b; Chang et al., 2007a), and a general approach is described in Fig. 1.28. While the 96-well format is now common, use of the 384-well format may permit further efficiency gains, provided that the most time-consuming steps of selecting, thawing, and aliquoting samples can be streamlined sufficiently (Chang et al., 2007a; Chang et al., 2007b).

TABLE 1.4. Advances in Sample Preparation Techniques Compared with Advances in Detection Technologies

Era	PK Requirement	Detection Technologies	New Goals of Sample Preparation	Major Sample Preparation Technology
1950– 1975	Detection of metabolites, estimate exposure	Colorimetry, radioimmunoassay (RIA), GC	Bring the analyte concentration to assay range; remove interference; make analyte volatile	Dilution, LLE, PPE, TLC, GC (normal phase and ion exchange); derivatization
1975– 1985	Determination of exposure	RIA, enzyme-linked immunoassay (ELISA), HPLC with visible UV detection	Bring analyte concentration to assay range; protein removal; remove interferences	Dilution, use of internal standard. LLE with back extraction; silica-based reverse chromatography with intention for fractional; commercial SPE cartridge
1985– 1995	GLP bioanalytical	RIA, ELISA, HPLC, GC, GC–MS, capillary zone electrophoresis	Reliability of quantitative data; validated assay with proven sample history and stability	Automation, online elution of SPE, online SPE, use of analogue internal standard
1995– 2000	Guidance for industry	ELISA, HPLC, GC, GC–MS, and HPLC–MS	Validated assay with proven specificity; cost reduction to compete with contract research organization	Commercial automation, high-throughput (high-density) assay based on 96-well SBS format; pre- and postextraction techniques in SBS format
2000 to	current	Biomarker and large-molecule determination	HPLC–MS/MS, HPLC–MS, Biacore, Mesoscare	Reduce matrix effect and improve incurred sample repeatability; reduce manual labor to compete with off-shoring Integrated process in SBS format; time sharing of MS by multiplex of HPLCs using multiple sprayers or stream selection valves; online SPE

Source: From Chang et al., 2007a.

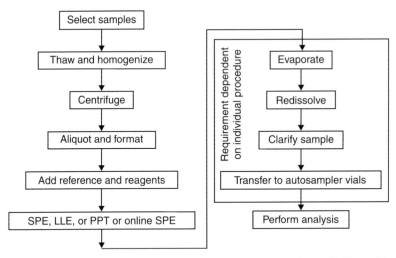

Figure 1.28. *Generic quantitative bioanalytical sample preparation scheme. (SPE = solid phase extraction; LLE = liquid-liquid extraction; PPT = protein precipitation). (Reprinted with permission from Chang et al., 2007a.)*

An alternative approach links the traditionally off-line sample preparation directly to the LC–MS system to utilize on-line SPE-type extraction. These systems have the advantages of unattended operation, and can use the LC-MS analysis time of sample one to prepare sample two, thereby saving time. Examples of such systems include the Symbiosis™ system from Spark Holland and the systems utilizing turbulent flow from Cohesive Systems/ThermoFisher (Ayrton et al., 1997; Alnouti et al., 2005; Kuhlenbeck et al., 2005; Koal et al., 2006; Smalley et al., 2007; Xu et al., 2007). The latter technique typically utilizes direct injection of the plasma sample, extraction of the analyte(s) onto a trapping column, and elution to an analytical column (Ayrton et al., 1997). The trapping column is engineered with relatively large diameter particles, and is operated at high flow rate. The theory and practice of this type of system are considered in greater detail in Chapter 10 of this book, both for quantitation and for metabolite identification.

Using a Symbiosis system, Alnouti et al. (2005) and Li et al. (2005) developed online SPE–LC–MS/MS methods for analysis of rat plasma without any prior sample processing. Direct plasma injection resulted in accuracy of 88–111% and 41–108% with and without on-line SPE, respectively. The precision was improved from 3–81% without SPE to 0.5–14% with SPE. Furthermore, Alnouti et al. (2005) demonstrated that the cost of quantitative bioanalysis can be reduced by reusing the on-line SPE cartridges up to 20 times without loss of accuracy, precision or analyte recovery.

Today, in the pharmaceutical industry, there exists a variety of technologies to which either partial or full automation of quantitative bioanalytical steps can make the process higher throughout. However, choosing the appropriate technology to automate requires an evaluation of several parameters, including number of samples, type of samples, and time required to automate in comparison to nonautomated work flow.

1.5.2 Improvements in Column Technology

Similar to MS, prior to the 1970s, only selected LC methods were available to the pharmaceutical scientists and most laboratories around the country packed their own columns, and HPLC was considered somewhat of a specialist tool. It was not until the 1980s that HPLC became a most practical analytical tool across the industry and the pharmaceutical scientist's tool of choice for separation, identification, purification, and quantification of drugs and metabolites. The past two decades have seen a vast undertaking in the development of column technology, especially in support of high-throughput bioanalysis. Today, more than 2 million analytical columns are sold per year (Unger, 2008).

To achieve the optimum reversed-phase LC separation, one needs to explore variables such as the analyte chemistry, mobile-phase composition (solvent type, solvent composition, pH, and additives), column composition, column particle size, and column temperature. For pharmaceutical analysis using mass spectrometry, the chemistry of an analyte is rarely changed beyond manipulation of the mobile phase pH, and even there options are limited. Volatile pH modifiers (buffers) are still preferred for LC–MS, and concentrations of these modifiers are kept low. Relatively simply mobile phases consisting of water, acetonitrile, and either formic acid (0.1% v/v), ammonium acetate (1–20 mM), or both have been common.

The column chemistry can be altered through changes to either or both the packing material and the bonded phase. The packing material used in LC–MS experiments is often based on 3–5-μm silica microspheres with a single pore-size (80–100 Å) distribution, with the smaller particle delivering higher efficiency separations. UPLC, as described above and in Chapter 4, extends this approach for both chromatographic and throughput efficiency gains. The common bonded phases used in LC–MS include the non-polar or "reversed phase" C18, C8, or phenyl, while less common options include the polar cyano and diol phases. Most of the separation issues that cannot be achieved by changing the mobile phase composition can be optimized by changing the bonded phase. In the example shown in Fig. 1.29, alternate selectivity for the test compounds was achieved by changing the bonded phase from either C18 or C8 to either cyano or phenyl. This type of alternate selectivity comes in handy when one is trying to separate a coeluting matrix or metabolite component from an analyte of interest.

Monolith-based column packing material emerged in the 1990s as potentially important in high throughput quantitative bioanalysis. High permeability, low pressure drop, and good separation efficiency are some of the attributes of monolithic columns. These attributes result from monolithic columns being designed to have a single piece of biporous solid material with interconnected skeletons and interconnected flow paths through the skeletons (Tanaka et al., 2002). While the larger through-pores (typically 2 μm) lead to reduced flow resistance, the smaller mesopores (12 nm) located on the silica skeleton lead to increased surface area needed to achieve separation efficiency. Data suggest that silica- and polymer-based monoliths are ideal for small- and large-molecules, respectively (Tennikova et al., 1990).

In a recent study Alnouti et al. (2005) evaluated conventional C18 and monolithic columns for online SPE–LC–MS/MS quantification of propranolol (ketoconazole

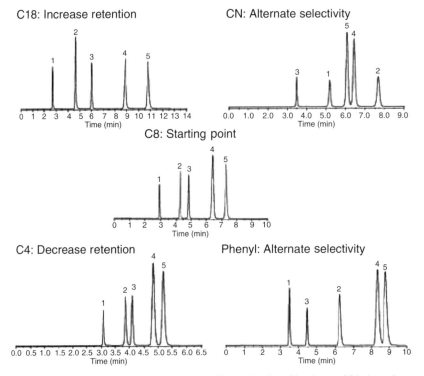

Figure 1.29. *LC separation of (1) norephedrine, (2) nortriptyline, (3) toluene, (4) imipramine, and (5) amitriptyline using columns with various bonded phases [all-ACE column: 250 mm × 4.6 mm, 5 μm; mobile phase 80/20 (v/v): methanol–25 mM phosphate buffer; usually, phosphate buffers are not preferred for MS applications]. (Reprinted with permission from Dolan, 2007.)*

was used as the IS) and diclofenac (ibuprofen was used as the IS) directly in rat plasma. As shown in Fig. 1.30, the LC–MS/MS run time was reduced from 4 to 2 min in going from a conventional Luna C18 to a monolithic column, while accuracy and precision of the method were maintained. The HPLC flow rates were 0.8 mL/min and 3.5 mL/min (split to deliver 1.5 mL/min to the MS) respectively for the C18 and the monolithic column approaches. The monolithic column approach demonstrates that high separation efficiencies can be achieved at a significantly increased HPLC flow rate. Several other groups also have evaluated the advantages and limitations of using monolithic columns for high throughput quantitative bioanalysis (Wu et al., 2001; Hsieh et al., 2002; Hsieh et al., 2003; Huang et al., 2006).

One weakness of the dominant reverse phase separations mechanism has been the poor retention of highly polar analytes, and hydrophilic interaction liquid chromatography (HILIC) has emerged as an alternative. In HILIC, a polar stationary phase such as silica gel is used to retain highly polar analytes. Mobile phases components similar to those described above for reverse phase separations are used, but the proportions of aqueous vs. organic are changed. Analytes are retained under conditions of relatively low water content, and eluted using increased water content.

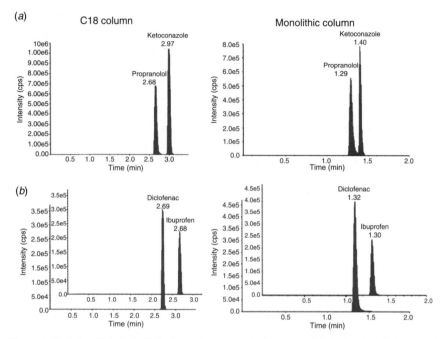

Figure 1.30. *LC–MS/MS (MRM) chromatograms of (a) propranolol (260→116)/ketoconazole (531→489) and (b) diclofenac (294→250)/ibuprofen (205→161) in rat plasma obtained with online SPE and either a Luna C18 column (left panels) or a Chromolith monolithic column (right panels) combination. (Reprinted with permission from Alnouti et al., 2005.)*

Conversely, in conventional reverse-phase HPLC, very high water content is required to retain polar analytes. The high water content in turn hinders the ionization and desolvation process during LC–MS (Hsieh and Chen, 2005; Xue et al., 2006). Therefore, HILIC allows one to elute highly polar analytes with small amounts water and maintain good LC–MS sensitivity (Hsieh and Chen, 2005). In a recent

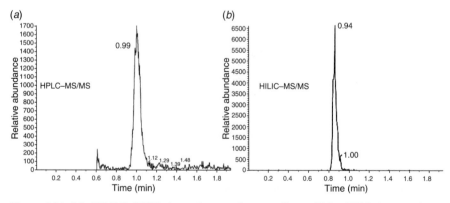

Figure 1.31. *LC–MS/MS (SRM) chromatograms for muraglitazar (517→186) in human plasma obtained using (a) HPLC and (b) HILIC. (Reprinted with permission from Xue et al., 2006)*

study, Xue et al. (2006) compared HPLC–MS/MS and HILIC–MS/MS for quantification of muraglitazar in human plasma and showed that the sensitivity of the LC–MS/MS assay can be improved by using HILIC (Fig. 1.31).

Under HILIC conditions, similar to polar analytes, polar endogenous matrix components such as phospholipids, peptides, and sugars also get retained longer. These polar matrix components can only be disrupted by using high buffer concentrations in the mobile phase. If high buffer concentrations are not an option, then the samples have to be appropriately processed to remove most of the endogenous matrix components before LC–MS/MS. Incomplete removal of these endogenous components is known to cause ion suppression during LC–MS/MS.

Table 1.5 compares various fast chromatographic approaches available for LC–MS/MS based quantitative bioanalysis.

TABLE 1.5. Comparison of Selected Fast Chromatographic Approaches

Approaches	Advantages	Limitations
Monoliths	Low back pressure, possibility to obtain high efficiency with $L_{col} > 1$ m; Compatible with conventional instruments Several monoliths available: organic and inorganic (e.g., carbon, zirconia, silica) Green chemistry (low organic modifier proportion at high temperature)	No straightforward method transfer Low resistance at high pH (pH > 7) and high pressure (>200 bars) Narrow-bore column not yet available (high solvent consumption, split with MS) Heating and cooling requirements (dedicated system) Stability of stationary phases
High-temperature liquid chromatography (HTLC)	Peak shape improvement of basic compunds (pK_a modification) Temperature an additional parameter for method development Possibility to couple HTLC with other fast-LC approaches (sub-2 μm, UPLC) Significant decrease in analysis time (up to 10 times)	Compound stability needs to be evaluated prior to analysis Not straightforward method transfer (selectivity changes with T) High back pressure with small particle size, limited efficiency
Sub-2-μm particles	Easy methiod transfer Many commercially available sub-2-μm particles (e.g., C_4, C_8, C_{18}, HILIC)	Limited compatibility with conventional instrumentation
Sub-2-μm particles at 1000 bars (UPLC)	Large decrease in analysis time (up to 20 times) Possibility to obtain high efficiency with $L_{col} \geq 15$ cm Easy method transfer	Few available stable stationary phases Dedicated instrumentation needed Solvent compressibility and frictional heating could exist at $\Delta P = 1000$ bars

Abbreviations: Lcol, length of column; T, temperature.
Source: From Guillarme et al., 2007.

1.6 SERIAL AND PARALLEL LC–MS APPROACHES FOR INCREASING QUANTITATIVE BIOANALYTICAL THROUGHPUT

The traditional, serial-mode, quantitative bioanalytical operation involves a single auto-sampler/HPLC column and a MS system. In this mode of operation, samples are injected one at a time. The serial-mode strategies used to increase the throughput include fast chromatography (Romanyshyn et al., 2000, 2001; Hop et al., 2002), automated data processing (Whalen et al., 2000; Fung et al., 2004; Briem et al., 2007), and pooling strategies (cassette dosing, pooling after individual dosing, simple sample screens, etc.) (Korfmacher et al., 2001; Kassel, 2004, 2005). Some of these strategies have resulted in 60 samples per hour. Today, fast chromatography is routinely achieved by utilizing UPLC instead of HPLC (Yu et al., 2006, 2007).

In most traditional LC–MS quantitation, the mass spectrometer is utilized for analyte and IS detection for only a brief period in the overall analysis. Significant time is lost to the autosampler, gradient delay and re-equilibration, or isocratic flushing of late-eluting materials. Throughput can be improved by establishing additional chromatographic systems in parallel. Some systems utilized simultaneous, parallel LC separations followed by detection using a multiplexed "MUX" MS interface (Fig. 1.32) (Deng et al., 2001, 2002; Jemal et al., 2001; Rudewicz and Yang, 2001; Yang et al.,

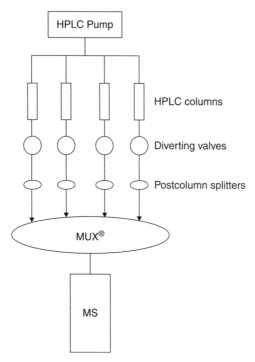

Figure 1.32. *Schematic of parallel LC–MS system consisting of binary pump, autosampler, and mass spectrometer used in combination with four LC columns connected to four-sprayer MUX interface. (Reprinted with permission from Fung et al., 2003.)*

2001a,b; Fang et al., 2002, 2003; Morrison et al., 2002; Fung et al., 2003). Other systems used parallel LC separations with a time offset such that elution from each column occurred in a discrete time period (Zweigenbaum and Henion, 2000; van Pelt et al., 2001; Xia et al., 2001). A single LC could be coupled to multiple columns, and varying lengths of "delay" tubing used to create offset elution from the columns. Alternatively, multiple independent autosamplers, pumps, and columns can be connected to a common mass spectrometer through a switching valve. Injection times are offset across the LC systems (typically two or four parallel systems), and the injection/elution/detection process is tracked through the software. This latter technique has been successfully commercialized through the Cohesive/ThermoFisher system mentioned above with or without the turbulent flow component (Chapter 10).

1.7   HIGHER THROUGHPUT QUANTITATIVE ANALYSIS WITHOUT LIQUID CHROMATOGRAPHY

A faster option is to eliminate the chromatography step altogether. In the early years of API–LC–MS, many users hoped that the techniques of ESI and APCI would facilitate just that. However, it was quickly recognized that ion suppression or metabolite coelution caused irreproducible or inaccurate results for many analytes (Cohen and Gusev, 2002). More recently, there has been considerable renewed interest in the use of desorption ionization techniques for quantitation (Notari et al., 2006). By eliminating the necessity for chromatographic separation, desorption-based techniques can achieve order-of-magnitude increases in throughput compared with high-speed LC–MS techniques. Recent examples include the commercial LDTD (laser diode thermal desorption, Phytronix Technologies) (Edge et al., 2008; Koers, 2008; Tremblay et al., 2008) and DESI (Prosolia, Inc.) systems. Use of desorption techniques for quantitative analysis is considered in Chapter 11 of this book.

1.8   MASS SPECTROMETRY IN QUALITATIVE ANALYSIS

The qualitative applications of MS in drug metabolism and pharmacokinetic studies are generally focused on addressing at least one of the two fundamental questions in DMPK. As summarized by Ma et al. (2006) and Prakash et al. (2007), we are looking to determine what drug-derived analytes are in a sample and how much there is. The first part of the question centers on identification of an unknown, a challenge for which MS is eminently suited. The combination of chromatographic separation, identification of molecular weight (and possibly elemental composition), and structural information via MS^n fragmentation is currently the strongest technique available to the drug metabolism group for the analysis of biological samples.

From the perspective of the pharmaceutical industry, the attention given to metabolites started to increase when the Pharmaceutical Research and Manufacturers of America (PhRMA) commissioned a review of the role of metabolites in drug induced toxicity. This group, called the metabolites in safety testing (MIST) committee, partnered with the U.S. Food and Drug Administration, convened several workshops, and in 2002 published a proposal outlining the types of drug metabolism studies relevant to safety assessment at all major phases of drug development. This publication, commonly referred to as the MIST document, very carefully addressed some of the contemporary issues in the safety evaluation of drug candidates and discussed how best to use the metabolite data. A brief timeline of recent publications relevant to metabolite issues is presented in Table 1.6.

TABLE 1.6. Recent Publications/Events on Metabolite in Safety Testing

Year	Organization/Reference	Outcome/Recommendations
1999	PhRMA—Commissioned a survey of current practices in dealing with metabolites in safety testing	• Several workshops held to discuss metabolites in safety testing approaches across pharmaceutical industry and the expectations from FDA.
2002	Baillie et al.—MIST Document Published	• Perform radiolabeled human ADME study as early as possible to define a major circulating metabolite.
		• A major circulating metabolite is defined to be a drug derived component(s) accounting for 25% or more of the AUC of total circulating drug derived components (relative abundance).
		• Any major human circulating metabolites should be considered for monitoring in toxicological and/or clinical studies.
		• If a metabolite is pharmacologically active or structural alert (i.e., contains a reactive functional group, glutathione conjugates) then it should be considered for monitoring in toxicological and/or clinical studies.
		• If a unique or human specific metabolite is observed during in vitro cross-species comparison studies, additional toxicological studies are warranted.
		• If a unique or human specific circulating metabolite is absent or present at relatively low concentration in toxicological species, separate studies to evaluate the toxicity of the human specific metabolite are warranted.
		• Carcinogenicity studies on a major metabolite are not recommended.
2003	Hastings et al.—Letter to the Editor	• Raised some concerns about a major metabolite being defined as a metabolite representing 25% of the systemic exposure compared to the parent drug.
		• Provided clear evidence that a minor metabolite could very easily produce toxicity.
		• The authors stated that the agency reserves the right to request carcinogenicity studies on major metabolites where required.

(Continued)

TABLE 1.6 *Continued*

Year	Organization/Reference	Outcome/Recommendations
		• Mentions that a formal guidance on the issue of drug metabolites in safety testing is warranted.
2003	Baillie et al.— Response to Hastings et al.	• Welcomed a formal guidance from the U.S. FDA
		• Additional clarification provided to distinguish between *major* and *unique* metabolites.
		• Emphasized a case-by-case approach to metabolite safety studies rather than formally outlining a set of studies for all circumstances.
2005	Smith and Obach—Commentary on MIST	• Proposed using the absolute abundance of a metabolite rather than relative abundance value as suggested in the MIST document.
2005	FDA—Draft Guidance on Safety Testing on Drug Metabolites is Published	• As early as possible, assess species differences in metabolism of a drug (in vitro studies, nonclinical animal studies).
		• Perform metabolic evaluations in humans as early as possible (radiolabeled or non-radiolabeled).
		• Early identification of unique or major human metabolites can provide clear directions for testing in animals, assist in interpreting and planning of clinical studies, and prevent delays in drug development/approval.
		• All human circulating metabolites that account for >10% of the administered dose or systemic exposure (whichever is less) and that were not present at sufficient levels to permit adequate evaluation during nonclinical animal studies should be considered for additional safety/toxicological testings.
		• If the systemic exposure for a major human circulating metabolite is equivalent to that observed in nonclinical toxicological species, then the metabolite levels may be sufficient to limit additional toxicity testing using the major human metabolite.
		• As needed, perform carcinogenicity studies on major human metabolites.
2006	Prueksaritanont, Lin and Baillie— Publication	• Highlighted the kinetic, metabolic, exposure, and toxicity differences of a preformed or synthetic metabolite(s) compared to that of a metabolite(s) generated endogenously from a parent drug.
		• Safety evaluations involving preformed or synthetic metabolite(s) of a drug should take metabolite kinetics into considerations.

(Continued)

TABLE 1.6 *Continued*

Year	Organization/Reference	Outcome/Recommendations
2008	FDA—Guidance on Safety Testing on Drug Metabolites is Published	• No major changes to recommendations provided in the Draft Guidance. • Focus changed from metabolites accounting for >10% of the administered dose (which could potentially apply to circulating as well as metabolites in excreta) to circulating metabolites accounting for >10% of parent drug's systemic exposure. • A new term, disproportionate drug metabolite, was introduced. Disproportionate drug metabolite was defined as "a metabolite present only in humans or present at higher plasma concentrations in humans than in the animals used in nonclinical studies. In general, these metabolites are of interest if they account for plasma levels greater than 10 percent of parent systemic exposure, measured as area under the curve (AUC) at **steady state**." • An updated decision tree flow diagram summarizing some of the steps required to assess safety of a major human drug metabolite is provided (Fig. 1.33).

Abbreviations: PhRMA, The **P**harmaceutical **R**esearch and **M**anufacturers of **A**merica; MIST, **M**etabolites in **S**afety **T**esting; FDA, United States **F**ood and **D**rug **A**dministration.

DECISION TREE FLOW DIAGRAM

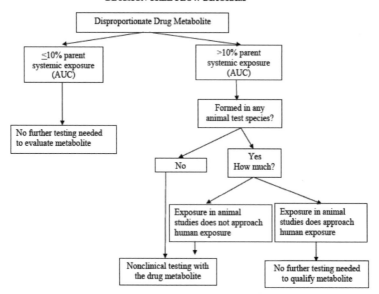

Figure 1.33. *A decision tree flow diagram describing some of the studies needed to determine safety of a human drug metabolite. (Food and Drug Administration (2008). Guidance for Industry: Safety Testing of Drug Metabolites.)*

1.8.1 Common Phase I and Phase II Biotransformation Pathways

Upon exposure to living systems, drugs undergo metabolism or biotransformation as a detoxification process to form more polar forms (metabolites) of the drug that can be readily eliminated. Undesired consequences of biotransformation include rapid clearance, formation of active metabolites, formation of reactive metabolites, and drug–drug interactions due to enzyme induction or inhibition. As described previously, metabolism can occur in different parts of the body, but the gut and liver are the major sites of metabolism. Several enzymes are involved in detoxification of drugs. These enzymes and their role in metabolism have been discussed in detail by Ghosal et al. (2005) and Johnson (2008a). Briefly, the drug detoxification pathways involve phase I and phase II metabolic reactions. As shown in Table 1.7, the phase I processes involve reactions such as oxidation, reduction, and hydrolysis. Among the phase I reactions, oxidative reactions can be catalyzed by either cytochrome P450s (CYP450) or nonmicrosomal enzymes such as monoamine oxidases (MAOs), peroxidases, and flavin-containing monooxygenases (FMOs). Among the drug-metabolizing enzymes, the CYP superfamily has

TABLE 1.7. Examples of Common Drug Biotransformations

Type of Biotransformation	Net Transformation	Nominal m/z Shift	Exact m/z Shift
	Phase I		
Hydroxylation/N-oxidation/S-oxidation	$+O$	$+16$	$+15.9949$
Dihydroxylation	$+2O$	$+32$	$+31.9898$
Dehydrogenation or reduction	$-H_2$	-2	-2.0156
Demethylation	$-CH_2$	-14	-14.0156
Deethylation	$-C_2H_4$	-28	-28.0312
Depropylation	$-C_3H_6$	-42	-42.0468
Oxidative deamination	$-NH_3, +O$	-1	-1.0316
Oxidative dechlorination	$-Cl, +OH$	-18	-17.9662
Oxidative defluorination	$-F, +OH$	-2	-1.9957
Hydration	$+H_2O$	$+18$	$+18.0105$
Methyl to an acid	$-H_2, +O_2$	$+30$	$+29.9742$
	Phase II		
Glucuronidation	$+C_6H_8O_6$	$+176$	$+176.0321$
Sulfation	$+SO_3$	$+80$	$+79.9568$
Glutathione conjugation	$+C_{10}H_{15}N_3O_6S$	$+305$	$+305.0681$
	$+C_{10}H_{17}N_3O_6S$	$+307$	$+307.0837$
Cysteine–glycine conjugation	$+C_5H_{10}N_2O_3S$	$+178$	$+178.0410$
Cysteine conjugation	$+C_3H_7NO_2S$	$+121$	$+121.0196$
N-acetyl-cysteine conjugation	$+C_5H_9NO_3S$	$+163$	$+163.0301$

Note: Exact masses of elements: $C = 12.000000$; $N = 14.003074$; $O = 15.994915$; $H = 1.007825$; $F = 18.998403$; $Cl = 34.968853$; $S = 31.972072$.

been extensively studied because they account for more than 90% of oxidative bio-transformation of drugs and xenobiotics (Ramanathan et al., 2005). To date, 750 CYP450s or polymorphs have been sequenced and 55 of them have been character-ized as human isoforms (Ramanathan et al., 2005). As listed in Table 1.7, micro-somal oxidations include aromatic and side-chain hydroxylation, N-oxidation, S-oxidation (sulfoxidation and sulfonation), N-hydroxylation, N-, O-, S-dealkylation, deamination, dehalogenation, and desulfation. Although MAOs, FMOs, and peroxi-dases have been associated with several of the above-mentioned biotransformation processes, their involvement is of lesser importance. According to the FDA's gui-dance on safety testing on drug metabolites (Food and Drug Administration, 2008), "metabolites formed from Phase I reactions are more likely to be chemically reactive or pharmacologically active and, therefore, more likely to need safety evaluation."

Phase II processes lead to ultimate detoxification reactions involving modification of a functional group ($-OH$, $-NH_2$, $-SH$, or $-COOH$) by a bulky and polar groups such as glucuronides, sulfates, amino acids, and/or glutathiones. Among these reac-tions, glucuronidation is the most common phase II reaction, and it is catalyzed by the enzymes called uridine diphosphate glucuronosyltransferases (UGTs). Most often, phase II reactions terminate the pharmacological activity of a drug. According to the FDA's guidance on safety testing on drug metabolites (Food and Drug Administration, 2008), "Phase II conjugation reactions generally render a compound more water soluble and pharmacologically inactive, thereby eliminating the need for further evaluation. However, if the conjugate forms a toxic compound such as acyl-glucuronide (Faed, 2003), additional safety assessment may be needed." Only in a few cases Phase II metabolites have been found to be pharmacologically active, for example, the phenolic glucuronide conjugate of Ezetimibe (Patrick et al., 2002) and morphine-6-glucuronide (Ishii et al., 1997).

The reasons for identification of metabolites are manyfold but all boil down to human safety of drugs under clinical investigation. Initially metabolites of a drug are characterized with in vitro systems (microsomes, hepatocytes, S9 fractions, etc.) and later lead compounds are assessed using mouse, rat, rabbit, dog, and/or monkey. Subsequently, metabolites in humans are identified following drug admin-istration to assure that the nonclinical species undergoing safety assessment are adequately exposed to human metabolites of the drug (Smith and Obach, 2006).

1.8.2 Metabolite Profiling, Detection, and Characterization Process Flow

Metabolite identification can be conducted on several levels, ranging from straightfor-ward analyses for targeted species to more complex analyses utilizing radiometric detection, MS, and possibly other detectors. A process flow for profiling and charac-terization of metabolites by LC–MS is presented in Fig. 1.34. There is an increasing emphasis on quality metabolite data from relatively early in discovery (Fernandez-Metzler et al., 1999), and some of the fundamental approaches that can be used have been summarized by Anari and Baillie (2005).

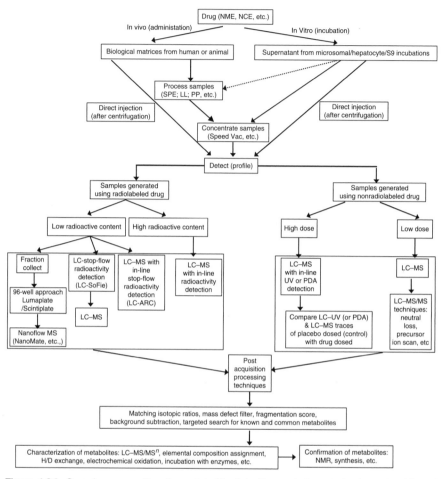

Figure 1.34. *Generic process flow for metabolite detection and characterization using LC–MS and other auxiliary techniques.*

At early stages of drug development, in the absence of radiolabeled drug, the major challenge in identification of metabolites is the presence of large amounts of mass spectral background from matrix components. To facilitate detection, initial screening for metabolites may utilize a list of targeted transformations based on the starting drug mass. Common transformations are converted to a list of target precursor ion m/z using the known mass shift of the transformation and the starting mass of the drug. The targeted list may be generic or can be modified based on known or predicted transformations expected for the compound/compound class being analyzed (Anari et al., 2004; Tiller et al., 2008). A few common example transformations are shown in Table 1.7. Reactive metabolites, discussed in detail by Johnson (2008b) and Obach et al. (2008), may be targets of especially high interest if such a structure is anticipated (Ma and Subramanian, 2006; see also Chapter 4). According to the

FDA's guidance on safety testing on drug metabolites (Food and Drug Administration, 2008), "Metabolites that form chemically reactive intermediates can be difficult to detect and measure because of their short half-lives. However, they can form stable products (e.g., glutathione conjugates) that can be measured and, therefore, may eliminate the need for further evaluation."

Once a targeted list is assembled, the appropriate LC–MS instrument can be set up to acquire both MS and MS/MS data (or MS^n data for traps) in an automated fashion. The MS/MS acquisitions would only be triggered by detection of a targeted precursor ion (from the list) at a minimum specified intensity. Linear ion trap quadrupole instruments are increasingly popular for this type of work (Hopfgartner and Zell, 2005) and are discussed in Chapter 3 of this book.

1.8.3 New Opportunities with Hybrid Mass Spectrometers

While much of the MS and MS/MS data acquisition has been performed using triple-quadrupole instruments, hybrid instruments capable of high resolution/high mass accuracy are increasingly being used for routine metabolite screening. Such systems include the Q-TOF, IT–TOF, LTQ–FTICR, and LTQ–Orbitrap configurations (Erve et al., 2008; Tiller et al., 2008). While foregoing none of the traditional MS and MS/MS data typically available, these instruments offer the added advantage of m/z measurements with high accuracy. For targeted structures with known elemental composition, the observed m/z would either be consistent with or refute the proposed structure.

When a targeted list is not used to drive the process, analysis may be conducted to assess all metabolites present. The observed m/z is compared with all possible elemental compositions that could produce the observed value, within constraints imposed by the user as described below. Accuracy and precision in the m/z measurement become critical in limiting the possible compositions to the greatest extent possible, greatly simplifying data reduction (Grange and Sovocool, 2008; Ruan et al., 2008). These techniques are discussed in Chapters 4 and 5.

1.8.4 Auxiliary Techniques to Facilitate Metabolite Profiling and Identification

Studies utilizing dosage of drug labeled with radioisotopes such as ^{14}C or ^3H are frequently used to create a complete mass balance, including all metabolites (Ramanathan et al., 2007a,b). Analysis of both eliminated and circulating metabolites is accomplished by simultaneous use of radio flow detection and mass spectrometric analysis. While targeted analysis may facilitate identification of most major metabolites, radiotracer studies are still considered the benchmark for complete metabolite profiling. An update on this technique is provided in Chapter 7, in which methods for determining low levels of circulating metabolites are assessed.

Identification of a metabolite mass shift such as $+16$ clearly implies addition of oxygen, but does little else. The nature of the functional group and its location on the molecule may be determined through MS/MS analysis, but more stringent experimental techniques may be needed. Derivatization techniques optimized for different functional groups may be useful (Hop and Prakash, 2005) but can be relatively cumbersome. A common probe for determining the nature of the functional group is hydrogen/deuterium (H/D) exchange. Depending on the nature of the functional group, different adjacent hydrogen atoms may be easily exchanged with deuterium. Analysis using D_2O instead of H_2O in the mobile phase can be sufficient to facilitate such exchange, and these techniques are explored in Chapter 9 of this book.

While highly successful, the API interfaces are not optimal for all small molecules. For cases where an analyte appears to ionize and vaporize with poor efficiency, a fundamental change in the analyte can be made prior to the API source. The use of coulometric assisted ionization is described in Chapter 8. In this method, electrochemistry techniques are used to alter the analyte. Improvements in specificity and sensitivity can result.

Mass spectrometry is also being used to help assess the ability of the drug to reach the target location in the body (McLean et al., 2007; Cornett et al., 2008; Kertesz and Van Berkel, 2008). The ability to determine the presence of the drug in the target organ, and even the drug/metabolite profile within the target organ, is explored in Chapter 12.

1.8.5 Tools and Techniques for Streamlining Metabolite Detection and Characterization

With the proliferation of mass spectrometers in drug metabolism came a vast amount of data to be processed and understood. In particular, techniques and tools for identification of unknown compounds such as metabolites, degradation products, or impurities have seen continual improvement in recent years, especially with respect to data obtained under high-resolution (and presumed high-mass-accuracy) conditions. A few possible approaches are summarized here.

One common target of data reduction is determination of the elemental composition for a particular observed m/z. Unfortunately, as m/z increases, so does the number of possible combinations of atoms that could produce the observed ion. The use of high-resolution instruments allows one to limit the possibilities of the observed m/z values through greater accuracy and specificity. High resolution should improve specificity but needs to be coupled with careful management of calibration to ensure high and known accuracy. Knowledge of the effective error bars on the m/z measurement permits limiting the search for matching formulas. Most instrument data systems now include algorithms for converting measured m/z to postulated elemental composition. The user is permitted to specify types of atoms to include (e.g., only C, H, N, and O), maximum numbers for each, and an estimate of the degree of bond saturation in the molecule. The result is a list of possible formulas that fit the criteria and the relative error of each from the observed m/z value.

These techniques still are not sufficient for true specificity. As mass increases from 200 to 500, the number of formulas mathematically possible for that mass increases rapidly, and a mass accuracy yielding a single, unambiguous formula quickly becomes impossible (Kind and Fiehn, 2006; Kind et al., 2007). To reduce the possibilities, the atom types and abundance required to produce the observed isotope pattern can be included in screening for possible formulas. In addition to the isotope pattern itself, an accurate mass measurement of each isotope peak can be considered. Only a proposed formula that satisfies the constraints for the original m/z (accuracy, atom type and number limitations, etc.), the isotope pattern, and is consistent with the proposed formulas at each isotope m/z would now be acceptable (Thurman, 2006; Cuyckens et al., 2008; De Maio et al., 2008; Gallagher et al., 2008; McGibbon et al., 2008).

While highly accurate mass measurement has become more common, efforts are still made to maximize the utility of so-called low-resolution instruments. It is reported to be possible to produce highly accurate data from unit-resolution instruments using commercial products such as MassWorks™ (Gu et al., 2006), theoretically facilitating the use of a quadrupole instrument for high-mass-accuracy work. Given the proliferation of LITs, this is an attractive (and relatively inexpensive) option. However, the possibility will still exist for signal overlap at low resolution that would be resolved at high resolution.

In an effort to further simplify data, the concept of the mass defect filter was introduced (Chapter 6). The mass defect filter utilizes the well-characterized changes in mass introduced for different types of metabolite and, in conjunction with highly accurate m/z measurement, yields stringent requirements for m/z changes that must be met. For example, introduction of oxygen to a molecule produces a nominal mass increase of 16. In more exact terms the change is 15.9949. Therefore, introduction of an oxygen atom would reduce the mass defect of the resulting molecule by approximately 0.005 u.

In addition to software tools to help postacquisition processing, software tools to help mass spectral interpretation, particularly MS/MS, have taken new strides as well (Heinonen et al., 2008). One example of such a software tool is the MathSpec™ program. The details of the MathSpec approach have been explained (Sweeney, 2003). MathSpec software is used in conjunction with MS/MS spectra obtained under high-resolution conditions. The software systematically attempts to assemble possible parts (from the MS/MS fragment data) of the molecule into a rational molecule. Other examples of structure elucidation software include HighChem's Mass Frontier and ACD/Labs ACD/MS Manager (Bayliss et al., 2007). Other metabolite prediction software tools such as Meteor are also being incorporated into LC–MS software as tools to help accelerate metabolite detection and characterization (Testa et al., 2005; Ives et al., 2007).

REFERENCES

Abdel-Hamid, M., Novotny, L., and Hamza, H. (2001). Liquid chromatographic-mass spectrometric determination of celecoxib in plasma using single-ion monitoring and its use in clinical pharmacokinetics. *J. Chromatogr. B: Biomed. Sci. Appl.* **753**:401–408.

Agrawal, S., Ashokraj, Y., Bharatam, P. V., Pillai, O., and Panchagnula, R. (2004). Solid-state characterization of rifampicin samples and its biopharmaceutic relevance. *Eur. J. Pharm. Sci.* **22**:127–144.

Alnouti, Y., Srinivasan, K., Waddell, D., Bi, H., Kavetskaia, O., and Gusev, A. I. (2005). Development and application of a new on-line SPE system combined with LC–MS/MS detection for high throughput direct analysis of pharmaceutical compounds in plasma. *J. Chromatogr. A* **1080:**99–106.

Anari, M. R., and Baillie, T. A. (2005). Bridging cheminformatic metabolite prediction and tandem mass spectrometry. *Drug Discov. Today* **10:**711–717.

Anari, M. R., Sanchez, R. I., Bakhtiar, R., Franklin, R. B., and Baillie, T. A. (2004). Integration of knowledge-based metabolic predictions with liquid chromatography data-dependent tandem mass spectrometry for drug metabolism studies: Application to studies on the biotransformation of indinavir. *Anal. Chem.* **76:**823–832.

Avery, M. J. (2003). Quantitative characterization of differential ion suppression on liquid chromatography/atmospheric pressure ionization mass spectrometric bioanalytical methods. *Rapid Commun. Mass Spectrom.* **17:**197–201.

Ayrton, J., Dear, G. J., Leavens, W. J., Mallett, D. N., and Plumb, R. S. (1997). The use of turbulent flow chromatography/mass spectrometry for the rapid, direct analysis of a novel pharmaceutical compound in plasma. *Rapid Commun. Mass Spectrom.* **11:**1953–1958.

Baillie, T. A., Cayen, M. N., Fouda, H., Gerson, R. J., Green, J. D., Grossman, S. J., Klunk, L. J., LeBlanc, B., Perkins, D. G., and Shipley, L. A. (2002). Drug metabolites in safety testing. *Toxicol. Appl. Pharmacol.* **182**(3)**:**188–196.

Baillie, T. A., Cayen, M. N., Fouda, H., Gerson, R. J., Green, J. D., Grossman, S. J., Klunk, L. J., LeBlanc, B., Perkins, D. G., and Shipley, L. A. (2003). Drug metabolites in safety testing. *Toxicol. Appl. Pharmacol.* **190**(1)**:**93–94.

Bakhtiar, R., and Majumdar, T. K. (2007). Tracking problems and possible solutions in the quantitative determination of small molecule drugs and metabolites in biological fluids using liquid chromatography-mass spectrometry. *J. Pharmacol. Toxicol. Methods* **55:**227–243.

Bauer, J., Spanton, S., Henry, R., Quick, J., Dziki, W., Porter, W., and Morris, J. (2001). Ritonavir: An extraordinary example of conformational polymorphism. *Pharm. Res.* **18:**859–866.

Bayliss, M. A., Antler, M., McGibbon, G., and Lashin, V. (2007). Rapid Metabolite Identification Using Advanced Algorithms for Mass Spectral Interpretation. In *Proceedings of the 55th ASMS Conference on Mass Spectrometry and Allied Topics.* ASMS, Indianapolis, IN.

Belov, M. E., Nikolaev, E. N., Alving, K., and Smith, R. D. (2001). A new technique for unbiased external ion accumulation in a quadrupole two-dimensional ion trap for electrospray ionization Fourier transform ion cyclotron resonance mass spectrometry. *Rapid Commun. Mass Spectrom.* **15:**1172–1180.

Benkestock, K., Edlund, P. O., and Roeraade, J. (2005). Electrospray ionization mass spectrometry as a tool for determination of drug binding sites to human serum albumin by noncovalent interaction. *Rapid Commun. Mass Spectrom.* **19:**1637–1643.

Bonfiglio, R., King, R. C., Olah, T. V., and Merkle, K. (1999). The effects of sample preparation methods on the variability of the electrospray ionization response for model drug compounds. *Rapid Commun. Mass Spectrom.* **13:**1175–1185.

Briem, S., Martinsson, S., Bueters, T., and Skoglund, E. (2007). Combined approach for high-throughput preparation and analysis of plasma samples from exposure studies. *Rapid Commun. Mass Spectrom.* **21:**1965–1972.

Briscoe, C. J., Stiles, M. R., and Hage, D. S. (2007). System suitability in bioanalytical LC/MS/MS. *J. Pharm. Biomed. Anal.* **44:**484–491.

Bruins, A. P. (1991). Liquid chromatography-mass spectrometry with ionspray and electrospray interfaces in pharmaceutical and biomedical research. *J. Chromatogr.* **554:**39–46.

Bu, H. Z., Knuth, K., Magis, L., and Teitelbaum, P. (2001a). High-throughput cytochrome P450 (CYP) inhibition screening via a cassette probe-dosing strategy. V. Validation of a direct injection/on-line guard cartridge extraction-tandem mass spectrometry method for CYP1A2 inhibition assessment. *Eur. J. Pharm. Sci.* **12:**447–452.

Bu, H. Z., Knuth, K., Magis, L., and Teitelbaum, P. (2001b). High-throughput cytochrome P450 (CYP) inhibition screening via cassette probe-dosing strategy: III. Validation of a direct injection/on-line guard cartridge extraction-tandem mass spectrometry method for CYP2C19 inhibition evaluation. *J. Pharm. Biomed. Anal.* **25:**437–442.

Bu, H. Z., Magis, L., Knuth, K., and Teitelbaum, P. (2000a). High-throughput cytochrome P450 (CYP) inhibition screening via cassette probe-dosing strategy. I. Development of direct injection/on-line guard cartridge extraction/tandem mass spectrometry for the simultaneous detection of CYP probe substrates and their metabolites. *Rapid Commun. Mass Spectrom.* **14:**1619–1624.

Bu, H. Z., Magis, L., Knuth, K., and Teitelbaum, P. (2001c). High-throughput cytochrome P450 (CYP) inhibition screening via a cassette probe-dosing strategy. VI. Simultaneous evaluation of inhibition potential of drugs on human hepatic isozymes CYP2A6, 3A4, 2C9, 2D6 and 2E1. *Rapid Commun. Mass Spectrom.* **15:**741–748.

Bu, H. Z., Magis, L., Knuth, K., and Teitelbaum, P. (2001d). High-throughput cytochrome P450 (CYP) inhibition screening via cassette probe-dosing strategy II. Validation of a direct injection/on-line guard cartridge extraction-tandem mass spectrometry method for CYP2D6 inhibition assessment. *J. Chromatogr. B: Biomed. Sci. Appl.* **753:**321–326.

Bu, H. Z., Poglod, M., Micetich, R. G., and Khan, J. K. (2000b). High-throughput Caco-2 cell permeability screening by cassette dosing and sample pooling approaches using direct injection/on-line guard cartridge extraction/tandem mass spectrometry. *Rapid Commun. Mass Spectrom.* **14:**523–528.

Cai, Y., Kingery, D., McConnell, O., and Bach, A. C. 2nd (2005). Advantages of atmospheric pressure photoionization mass spectrometry in support of drug discovery. *Rapid Commun. Mass Spectrom.* **19:**1717–1724.

Cai, Z., Han, C., Harrelson, S., Fung, E., and Sinhababu, A. K. (2001). High-throughput analysis in drug discovery: Application of liquid chromatography/ion-trap mass spectrometry for simultaneous cassette analysis of alpha-1a antagonists and their metabolites in mouse plasma. *Rapid Commun. Mass Spectrom.* **15:**546–550.

Cai, Z., Sinhababu, A. K., and Harrelson, S. (2000). Simultaneous quantitative cassette analysis of drugs and detection of their metabolites by high performance liquid chromatography/ion trap mass spectrometry. *Rapid Commun. Mass Spectrom.* **14:**1637–1643.

Caskey, C. T. (2007). The drug development crisis: Efficiency and safety. *Annu. Rev. Med.* **58:**1–16.

Castro-Perez, J., Plumb, R., Granger, J. H., Beattie, I., Joncour, K., and Wright, A. (2005). Increasing throughput and information content for in vitro drug metabolism experiments using ultra-performance liquid chromatography coupled to a quadrupole time-of-flight mass spectrometer. *Rapid Commun. Mass Spectrom.* **19:**843–848.

Cha, B., Blades, M. W., and Douglas, D. J. (2000). An interface with a linear quadrupole ion guide for an electrospray-ion trap mass spectrometer system. *Anal. Chem.* **72:**5647–5654.

Chang, M. S., Ji, Q., Zhang, J., and El-Shourbagy, T. A. (2007a). Historical review of sample preparation for chromatographic bioanalysis: Pros and cons. *Drug Dev. Res.* **68:**107–133.

Chang, M. S., Kim, E. J., and El-Shourbagy, T. A. (2007b). Evaluation of 384-well formatted sample preparation technologies for regulated bioanalysis. *Rapid Commun. Mass Spectrom.* **21:**64–72.

Chavez-Eng, C. M., Constanzer, M. L., and Matuszewski, B. K. (2000). Determination of rofe-coxib (MK-0966), a cyclooxygenase-2 inhibitor, in human plasma by high-performance liquid chromatography with tandem mass spectrometric detection. *J. Chromatogr. B Biomed. Sci. Appl.* **748:**31–39.

Chen, G., Pramanik, B. N., Liu, Y. H., and Mirza, U. A. (2007). Applications of LC/MS in structure identifications of small molecules and proteins in drug discovery. *J. Mass Spectrom.* **42:**279–287.

Chernushevich, I. V., Loboda, A. V., and Thomson, B. A. (2001). An introduction to quadru-pole-time-of-flight mass spectrometry. *J. Mass Spectrom.* **36:**849–865.

Chowdhury, S. K. (2007). Early Assessment of human metabolism: Why, how, challenges and opportunities. In *Proceedings of the 55th ASMS Conference on Mass Spectrometry and Allied Topics.* ASMS, Indianapolis, IN.

Churchwell, M. I., Twaddle, N. C., Meeker, L. R., and Doerge, D. R. (2005). Improving LC–MS sensitivity through increases in chromatographic performance: Comparisons of UPLC-ES/MS/MS to HPLC-ES/MS/MS. *J. Chromatogr. B Anal. Technol. Biomed. Life Sci.* **825:**134–143.

Cody, R. B., Laramee, J. A., and Durst, H. D. (2005). Versatile new ion source for the analysis of materials in open air under ambient conditions. *Anal. Chem.* **77:**2297–2302.

Cohen, L. H., and Gusev, A. I. (2002). Small molecule analysis by MALDI mass spectrometry. *Anal. Bioanal. Chem.* **373:**571–586.

Collings, B. A., Campbell, J. M., Mao, D. M., and Douglas, D. J. (2001). *Rapid Commun. Mass Spectrom.* **15:**1777–1795.

Corkery, L. J., Pang, H., Schneider, B. B., Covey, T. R., and Siu, K. W. (2005). Automated nanospray using chip-based emitters for the quantitative analysis of pharmaceutical com-pounds. *J. Am. Soc. Mass Spectrom.* **16:**363–369.

Cornett, D. S., Frappier, S. L., and Caprioli, R. M. (2008). Imaging drugs and metabolites in tissue using Fourier transform mass spectrometry. In *Proceedings of the 56th ASMS Conference on Mass Spectrometry and Allied Topics.* ASMS, Denver, CO.

Cudiamat, G. (2005). Peaks of Interest. *LCGC N. Am.* **23:**984.

Cuyckens, F., Hurkmans, R., Leclercq, L., and Mortishire-Smith, R. (2008). Extracting rel-evant data out of the MS background. In *Proceedings of the 56th ASMS Conference on Mass Spectrometry and Allied Topics.* ASMS, Denver, CO.

Dawson, P. H. (1986). Quadrupole mass analyzers: Performance, design and some recent applications. *Mass Spectrom. Rev.* **5:**1–37.

De Buck, S. S., Sinha, V. K., Fenu, L. A., Gilissen, R. A., Mackie, C. E., and Nijsen, M. J. (2007). The prediction of drug metabolism, tissue distribution, and bioavailability of 50 structurally diverse compounds in rat using mechanism-based absorption, distribution, and metabolism prediction tools. *Drug Metab. Dispos.* **35:**649–659.

Decaestecker, T. N., Clauwaert, K. M., Van Bocxlaer, J. F., Lambert, W. E., Van den Eeckhout, E. G., Van Peteghem, C. H., and De Leenheer, A. P. (2000). Evaluation of automated single mass spectrometry to tandem mass spectrometry function switching for comprehensive drug profiling analysis using a quadrupole time-of-flight mass spectrometer. *Rapid Commun. Mass Spectrom.* **14:**1787–1792.

De Maio, W., Hoffmann, M., Carbonara, M., Moore, R., Mutlib, A., and Talaat, R. E. (2008). Identification of drug metabolites by UPLC–MS with isotope pattern directed mass chromatograms and UPLC with radioactivity flow detection. In *Proceedings of the 56th ASMS Conference on Mass Spectrometry and Allied Topics.* ASMS, Denver, CO.

Deng, G., and Sanyal, G. (2006). Applications of mass spectrometry in early stages of target based drug discovery. *J. Pharm. Biomed. Anal.* **40:**528–538.

Deng, Y., Wu, J. T., Lloyd, T. L., Chi, C. L., Olah, T. V., and Unger, S. E. (2002). High-speed gradient parallel liquid chromatography/tandem mass spectrometry with fully automated sample preparation for bioanalysis: 30 seconds per sample from plasma. *Rapid Commun. Mass Spectrom.* **16:**1116–1123.

Deng, Y., Zeng, H., Unger, S. E., and Wu, J. T. (2001). Multiple-sprayer tandem mass spectrometry with parallel high flow extraction and parallel separation for high-throughput quantitation in biological fluids. *Rapid Commun. Mass Spectrom.* **15:**1634–1640.

Derendorf, H., and Meibohm, B. (1999). Modeling of pharmacokinetic/pharmacodynamic (PK/PD) relationships: Concepts and perspectives. *Pharm. Res.* **16:**176–185.

Devine, J. W., Cline, R. R., and Farley, J. F. (2006). Follow-on biologics: Competition in the biopharmaceutical marketplace. *J. Am. Pharm. Assoc.* **46:**213–221.

De vlieger, J., Krabbe, J. G., Commandeur, J. N. M., Vermeulen, N. P. E., Niessen, W. M. A., and Loftus, N. (2007). Characterisation of metabolites generated by mutant cytochromes P450 enzymes using a 3D ion trap-time of flight mass spectrometer. In *Proceedings of the 55th ASMS Conference on Mass Spectrometry and Allied Topics.* ASMS, Indianapolis, IN.

DiMasi, J. A. (2001a). New drug development in the United States from 1963 to 1999. *Clin. Pharmacol. Ther.* **69:**286–296.

DiMasi, J. A. (2001b). New drug development in the United States from 1963 to 1999. *Clin. Pharmacol. Ther.* **69:**286–296.

DiMasi, J. A., Hansen, R. W., and Grabowski, H. G. (2003). The price of innovation: New estimates of drug development costs. *J. Health Econ.* **22:**151–185.

Dolan, J. W. (2007). The perfect method, Part V: Changing column selectivity. *LCGC N. Am.* **25:**1014–1020.

Dolnikowski, G. G., Kristo, M. J., Enke, C. G., and Watson, J. T. (1988). *Int. J. Mass Spectrom. Ion Processes* **82:**1–15.

Dong, M. W. (2007). Ultrahigh-pressure LC in pharmaceutical analysis: Performance and practical issues. *LCGC N. Am.* **25:**656–666.

Drews, J. (2000). Drug discovery: A historical perspective. *Science* **287:**1960–1964.

Duffin, K. L., Wachs, T., and Henion, J. D. (1992). Atmospheric pressure ion-sampling system for liquid chromatography/mass spectrometry analyses on a benchtop mass spectrometer. *Anal. Chem.* **64:**61–68.

Edge, T., Smith, C., Hill, S., Picard, P., Letarte, S., Wilson, I. D., and Vince, P. (2008). Comparison of laser diode thermal desorption (LDTD) source vs LC-MS for the analysis of a theraputic drug in biological extracts. In *Proceedings of the 56th ASMS Conference on Mass Spectrometry and Allied Topics*. ASMS, Denver, CO.

Ells, B., Froese, K., Hrudey, S. E., Purves, R. W., Guevremont, R., and Barnett, D. A. (2000). Detection of microcystins using electrospray ionization high-field asymmetric waveform ion mobility mass spectrometry/mass spectrometry. *Rapid Commun. Mass Spectrom.* **14:**1538–1542.

Embi, P. J., Acharya, P., McCuistion, M., Kishman, C. P., Haag, D., and Marine, S. (2006). Responding rapidly to FDA drug withdrawals: Design and application of a new approach for a consumer health website. *J. Med. Internet Res.* **8:**e16.

Engel, G. L., Farid, N. A., Faul, M. M., Richardson, L. A., and Winneroski, L. L. (2000). Salt form selection and characterization of LY333531 mesylate monohydrate. *Int. J. Pharm.* **198:**239–247.

Erve, J. C. L., DeMaio, W., and Talaat, R. E. (2008). Rapid metabolite identification with sub parts-per million mass accuracy from biological matrices by direct infusion nanoelectrospray ionization after clean-up on a ZipTip and LTQ/Orbitrap mass spectrometry. *Rapid Commun. Mass Spectrom.* **22:**3015–3026.

Faed, E. M. (1984). Properties of acyl glucuronides. Implications for studies of the pharmacokinetics and metabolism of acidic drugs. *Drug Metab. Rev.* **15:**1213–1249.

Fang, L., Cournoyer, J., Demee, M., Zhao, J., Tokushige, D., and Yan, B. (2002). High-throughput liquid chromatography ultraviolet/mass spectrometric analysis of combinatorial libraries using an eight-channel multiplexed electrospray time-of-flight mass spectrometer. *Rapid Commun. Mass Spectrom.* **16:**1440–1447.

Fang, L., Demee, M., Cournoyer, J., Sierra, T., Young, C., and Yan, B. (2003). Parallel high-throughput accurate mass measurement using a nine-channel multiplexed electrospray liquid chromatography ultraviolet time-of-flight mass spectrometry system. *Rapid Commun. Mass Spectrom.* **17:**1425–1432.

Feng, W. Y. (2004). Mass spectrometry in drug discovery: A current review. *Curr. Drug Discov. Technol.* **1:**295–312.

Fernandez-Metzler, C. L., Owens, K. G., Baillie, T. A., and King, R. C. (1999). Rapid liquid chromatography with tandem mass spectrometry-based screening procedures for studies on the biotransformation of drug candidates. *Drug Metab. Dispos.* **27:**32–40.

Ferrer, I., and Thurman, E. M. (2007). Importance of the electron mass in the calculations of exact mass by time-of-flight mass spectrometry. *Rapid Commun. Mass Spectrom.* **21:**2538–2539.

Fligge, T. A., and Schuler, A. (2006). Integration of a rapid automated solubility classification into early validation of hits obtained by high throughput screening. *J. Pharm. Biomed. Anal.* **42:**449–454.

Food and Drug Administration (2005). Draft Guidance for Industry: Safety Testing of Drug Metabolites. http://www.fda.gov/ohrms/dockets/98fr/2005d-0203-gdl0001.pdf (June, 2005).

Food and Drug Administration (2008). Guidance for Industry: Safety Testing of Drug Metabolites. http://www.fda.gov/cder/guidance/6897fnl.pdf (February, 2008).

Frank, R. G. (2007). Regulation of follow-on biologics. *N. Engl. J. Med.* **357:**841–843.

Fung, E. N., Chu, I., Li, C., Liu, T., Soares, A., Morrison, R., and Nomeir, A. A. (2003). Higher-throughput screening for Caco-2 permeability utilizing a multiple sprayer liquid chromatography/tandem mass spectrometry system. *Rapid Commun. Mass Spectrom.* **17:**2147–2152.

Fung, E. N., Chu, I., and Nomeir, A. A. (2004). A computer program for automated data evaluation to support in vitro higher-throughput screening for drug metabolism and pharmacokinetics attributes. *Rapid Commun. Mass Spectrom.* **18:**2046–2052.

Furfine, E. S., Baker, C. T., Hale, M. R., Reynolds, D. J., Salisbury, J. A., Searle, A. D., Studenberg, S. D., Todd, D., Tung, R. D., and Spaltenstein, A. (2004). Preclinical pharmacology and pharmacokinetics of GW433908, a water-soluble prodrug of the human immunodeficiency virus protease inhibitor amprenavir. *Antimicrob. Agents Chemother.* **48:**791–798.

Gallagher, R. T., Wilson, I. D., and Hobby, K. (2008). New approach for identification of metabolites of a model drug: Partial isotope-enrichment combined with novel mass spectral modeling software. In *Proceedings of the 56th ASMS Conference on Mass Spectrometry and Allied Topics.* ASMS, Denver, CO.

General Accountability Office (GAO) (2006). New drug development: Science, business, regulatory, and intellectual property issues cited as hampering drug development Efforts, Report GAO-07-49. GAO, Washington, DC, November 2006.

Ghosal, A., Ramanathan, R., Kishnani, N., Chowdhury, S. K., and Alton, K. B. (2005). Cytochrome P450 (CYP) and UDP-glucuronosyltranferase (UGT) enzymes: Role in drug metabolism, polymorphism, and identification of their involvement in drug metabolism. In *Identification and Quantification of Drugs, Metabolites and Metabolizing Enzymes by LC–MS* (Chowdhury S. K., Ed.). Elsevier, San Diego, CA, pp. 277–336.

Goodwin, L., White, S. A., and Spooner, N. (2007). Evaluation of ultra-performance liquid chromatography in the bioanalysis of small molecule drug candidates in plasma. *J. Chromatogr. Sci.* **45:**298–304.

Grange, A. H., and Sovocool, G. W. (2008). Automated determination of precursor ion, product ion, and neutral loss compositions and deconvolution of composite mass spectra using ion correlation based on exact masses and relative isotopic abundances. *Rapid Commun. Mass Spectrom.* **22:**2375–2390.

Grayson, M. A. (2002). *Measuring Mass: From Positive Rays to Proteins.* Chemical Heritage Press, Philadelphia, PA.

Gritti, F., Martin, M., and Guiochon, G. (2005). Influence of pressure on the properties of chromatographic columns. II. The column hold-up volume. *J. Chromatogr. A* **1070:**13–22.

Gross, J. H. (2004). *Mass Spectrometry: A Textbook.* Springer-Verlag, Berlin.

Gu, M., Wang, Y., Zhao, X. G., and Gu, Z. M. (2006). Accurate mass filtering of ion chromatograms for metabolite identification using a unit mass resolution liquid chromatography/mass spectrometry system. *Rapid Commun. Mass Spectrom.* **20:**764–770.

Guengerich, F. P. (2006). Cytochrome P450s and other enzymes in drug metabolism and toxicity. *AAPS J.* **8:**E101–E111.

Guevremont, R. (2004). High-field asymmetric waveform ion mobility spectrometry: A new tool for mass spectrometry. *J. Chromatogr. A.* **1058:**3–19.

Guillarme, D., Nguyen, D. T. T., Rudaz, S., and Veuthey, J. L. (2007). Recent developments in liquid chromatography: Impact on qualitative and quantitative performance. *J. Chromatogr. A* **1149**:20–29.

Hager, J. W. (2002). A new linear ion trap mass spectrometer. *Rapid Commun. Mass Spectrom.* **16**:512–526.

Hanold, K. A., Fischer, S. M., Cormia, P. H., Miller, C. E., and Syage, J. A. (2004). Atmospheric pressure photoionization. 1. General properties for LC/MS. *Anal. Chem.* **76**:2842–2851.

Hardman, M., and Makarov, A. A. (2003). Interfacing the orbitrap mass analyzer to an electrospray ion source. *Anal. Chem.* **75**:1699–1705.

Hashimoto, Y., Waki, I., Yoshinari, K., Shishika, T., and Terui, Y. (2005). Orthogonal trap time-of-flight mass spectrometer using a collisional damping chamber. *Rapid Commun. Mass Spectrom.* **19**:221–226.

Hastings, K. L., El-Hage, J., Jacobs, A., Leighton, J., Morse, D., and Osterberg, R. E. (2003). Drug metabolites in safety testing. *Toxicol. Appl. Pharmacol.* **190**(1):91–92.

Hatsis, P., Brockman, A. H., and Wu, J. T. (2007). Evaluation of high-field asymmetric waveform ion mobility spectrometry coupled to nanoelectrospray ionization for bioanalysis in drug discovery. *Rapid Commun. Mass Spectrom.* **21**:2295–2300.

Heinonen, M., Rantanen, A., Mielkainen, T., Kokkonen, J., Kiuru, J., Ketola, R. A., and Rousu, J. (2008). FiD: A software for ab initio structural identification of product ions from tandem mass spectrometric data. *Rapid Commun. Mass Spectrom.* **22**:3043–3052.

Herniman, J. M., Bristow, T. W. T., O'Connor, G., Jarvis, J., and Langley, G. J. (2004). Improved precision and accuracy for high-performance liquid chromatography/Fourier transform ion cyclotron resonance mass spectrometric exact mass measurement of small molecules from the simultaneous and controlled introduction of internal calibrants via a second electrospray nebuliser. *Rapid Commun. Mass Spectrom.* **18**:3035–3040.

Hill, T. P. (2007). Phase 0 trials: Are they ethically challenged? *Clin. Cancer Res.* **13**:783–784.

Hillenkamp, F., Karas, M., Ingendoh, A., and Stahl, B. (1990). Matrix assisted UV-laser desorption/ionization: A new approach to mass spectrometry of large biomolecules. *Biol. Mass Spectrom. Proc. Int. Symp. Mass Spectrom. Health Life Sci.* **2**:49–60.

Hofstadler, S. A., and Sannes-Lowery, K. A. (2006). Applications of ESI-MS in drug discovery: Interrogation of noncovalent complexes. *Nat. Rev. Drug Discov.* **5**:585–595.

Hop, C. E., Tiller, P. R., and Romanyshyn, L. (2002). In vitro metabolite identification using fast gradient high performance liquid chromatography combined with tandem mass spectrometry. *Rapid Commun. Mass Spectrom.* **16**:212–219.

Hop, C. E. C. A., and Prakash, C. (2005). Metabolite identification by LC–MS: Applications in drug discovery and development. In *Identification and Quantification of Drugs, Metabolites and Metabolizing Enzymes by LC–MS* (Chowdhury, S. K., Ed.). Elsevier, Amsterdam, The Netherlands, pp. 123–158.

Hopfgartner, G., and Zell, M. (2005). *Q Trap MS: A New Tool for Metabolite Identification.* (Korfmacher, W. A., Ed.). CRC Press, Boca Raton, FL, pp. 277–304.

Hopfgartner, G., Sleno, L., Loftus, N., Warrander, J., and Ashton, S. (2007). A statistical based approach in metabolite identification by high mass accuracy MSn analysis. In *Proceedings of the 55th ASMS Conference on Mass Spectrometry and Allied Topics.* ASMS, Indianapolis, IN.

Hopfgartner, G., and Vilbois, F. (2000). The impact of accurate mass measurements using quadrupole/time-of-flight mass spectrometry on the characterisation and screening of drug metabolites: Structure elucidation by LC–MS. *Analusis* **28**:906–914.

Hopkins, A. L., and Groom, C. R. (2002). The druggable genome. *Nat. Rev. Drug Discov.* **1**:727–730.

Hsieh, Y. (2005). *APPI: A New Ionization Source for LC–MS/MS Assays.* (Korfmacher, W. A., Ed.) CRC Press, Boca Raton, FL, pp. 253–276.

Hsieh, Y., and Chen, J. (2005). Simultaneous determination of nicotinic acid and its metabolites using hydrophilic interaction chromatography with tandem mass spectrometry. *Rapid Commun. Mass Spectrom.* **19**:3031–3036.

Hsieh, Y., Fukuda, E., Wingate, J., and Korfmacher, W. A. (2006). Fast mass spectrometry-based methodologies for pharmaceutical analyses. *Comb. Chem. High Throughput Screening* **9**:3–8.

Hsieh, Y., and Korfmacher, W. A. (2006). Increasing speed and throughput when using HPLC-MS/MS systems for drug metabolism and pharmacokinetic screening. *Curr. Drug Metab.* **7**:479–489.

Hsieh, Y., Wang, G., Wang, Y., Chackalamannil, S., Brisson, J. M., Ng, K., and Korfmacher, W. A. (2002). Simultaneous determination of a drug candidate and its metabolite in rat plasma samples using ultrafast monolithic column high-performance liquid chromatography/tandem mass spectrometry. *Rapid Commun. Mass Spectrom.* **16**:944–950.

Hsieh, Y., Wang, G., Wang, Y., Chackalamannil, S., and Korfmacher, W. A. (2003). Direct plasma analysis of drug compounds using monolithic column liquid chromatography and tandem mass spectrometry. *Anal. Chem.* **75**:1812–1818.

Hu, Q., Noll, R. J., Li, H., Makarov, A., Hardman, M., and Graham Cooks, R. (2005). The Orbitrap: A new mass spectrometer. *J. Mass Spectrom.* **40**:430–443.

Huang, L. F., and Tong, W. Q. (2004). Impact of solid state properties on developability assessment of drug candidates. *Adv. Drug Deliv. Rev.* **56**:321–334.

Huang, M. Q., Mao, Y., Jemal, M., and Arnold, M. (2006). Increased productivity in quantitative bioanalysis using a monolithic column coupled with high-flow direct-injection liquid chromatography/tandem mass spectrometry. *Rapid Commun. Mass Spectrom.* **20**:1709–1714.

Hughes, N., Winnik, W., Dunyach, J. J., Amad, M., Splendore, M., and Paul, G. (2003). High-sensitivity quantitation of cabergoline and pergolide using a triple-quadrupole mass spectrometer with enhanced mass-resolution capabilities. *J. Mass Spectrom.* **38**:743–751.

Inohana, Y., Hirano, I., Yamaguchi, S., Arakawa, K., Ashton, S., Loftus, N., and Warrander, J. (2007). Structure elucidation of Sildanafil analogues by MSn and accurate mass measurement. In *Proceedings of the 55th ASMS Conference on Mass Spectrometry and Allied Topics.* ASMS, Indianapolis, IN.

Ishii, Y., Takami, A., Tsuruda, K., Kurogi, A., Yamada, H., and Oguri, K. (1997). Induction of two UDP-glucuronosyltransferase isoforms sensitive to phenobarbital that are involved in morphine glucuronidation. Production of isoform-selective antipeptide antibodies toward UGT1.1r and UGT2B1. *Drug Metab. Dispos.* **25**:163–167.

Ishizuka, N., Minakuchi, H., Nakanishi, K., Soga, N., Nagayama, H., Hosoya, K., and Tanaka, N. (2000). Performance of a monolithic silica column in a capillary under pressure-driven and electrodriven conditions. *Anal. Chem.* **72**:1275–1280.

Ives, S., Gjervig-Jensen, K., and McSweeney, N. (2007). Speed up of the identification of the metabolites of Citalopram using LC/MS/MS and in silico prediction. In *Proceedings of the 55th ASMS Conference on Mass Spectrometry and Allied Topics*. ASMS, Indianapolis, IN.

Jang, G. R., Harris, R. Z., and Lau, D. T. (2001). Pharmacokinetics and its role in small molecule drug discovery research. *Med. Res. Rev.* **21:**382–396.

Jemal, M., Huang, M., Mao, Y., Whigan, D., and Powell, M. L. (2001). Increased throughput in quantitative bioanalysis using parallel-column liquid chromatography with mass spectrometric detection. *Rapid Commun. Mass Spectrom.* **15:**994–999.

Jemal, M., and Ouyang, Z. (2003). Enhanced resolution triple-quadrupole mass spectrometry for fast quantitative bioanalysis using liquid chromatography/tandem mass spectrometry: Investigations of parameters that affect ruggedness. *Rapid Commun. Mass Spectrom.* **17:**24–38.

Jemal, M., Schuster, A., and Whigan, D. B. (2003). Liquid chromatography/tandem mass spectrometry methods for quantitation of mevalonic acid in human plasma and urine: Method validation, demonstration of using a surrogate analyte, and demonstration of unacceptable matrix effect in spite of use of a stable isotope analog internal standard. *Rapid Commun. Mass Spectrom.* **17:**1723–1734.

Jemal, M., and Xia, Y. Q. (2006). LC-MS development strategies for quantitative bioanalysis. *Curr. Drug. Metab.* **7:**491–502.

Jerkovich, A. D., LoBrutto, R., and Vivilecchia, R. V. (2005). The use of ACQUITY UPLC in pharmaceutical development. *LCGC N. Am.*15–21.

Jiang, Y., Hall, T. A., Hofstadler, S. A., and Naviaux, R. K. (2007). Mitochondrial DNA mutation detection by electrospray mass spectrometry. *Clin. Chem.* **53:**195–203.

Jones, A. B. (2006). Bioanalytical quality assurance: Concepts and concerns. *Quality Assur. J.* **10:**101–106.

Johnson, W. W. (2008a). Cytochrome P450 inactivation by pharmaceuticals and phytochemicals: Therapeutic relevance. *Drug Metab. Rev.* **40:**101–147.

Johnson, W. W. (2008b). Many drugs and phytochemicals can be activated to biological reactive intermediates. *Curr. Drug Metab.* **9:**344–351.

Kantharaj, E., Ehmer, P. B., De Wagter, K., Tuytelaars, A., Proost, P. E., Mackie, C., and Gilissen, R. A. (2005a). The use of liquid chromatography-atmospheric pressure chemical ionization mass spectrometry to explore the in vitro metabolism of cyanoalkyl piperidine derivatives. *Biomed. Chromatogr.* **19:**245–249.

Kantharaj, E., Ehmer, P. B., Tuytelaars, A., Van Vlaslaer, A., Mackie, C., and Gilissen, R. A. (2005b). Simultaneous measurement of metabolic stability and metabolite identification of 7-methoxymethylthiazolo[3,2-a]pyrimidin-5-one derivatives in human liver microsomes using liquid chromatography/ion-trap mass spectrometry. *Rapid Commun. Mass Spectrom.* **19:**1069–1074.

Kantharaj, E., Tuytelaars, A., Proost, P. E., Ongel, Z., Van Assouw, H. P., and Gilissen, R. A. (2003). Simultaneous measurement of drug metabolic stability and identification of metabolites using ion-trap mass spectrometry. *Rapid Commun. Mass Spectrom.* **17:**2661–2668.

Kapron, J. T., Jemal, M., Duncan, G. F., Kolakowski, B. M., and Purves, R. (2005). Removal of metabolite interference during liquid chromatography/tandem mass spectrometry using high-field asymmetric waveform ion mobility spectrometry. *Rapid Commun. Mass Spectrom.* **19:**1979–1983.

Kassel, D. B. (2004). Applications of high-throughput ADME in drug discovery. *Curr. Opin. Chem. Biol.* **8:**339–345.

Kassel, D. B. (2005). High throughput strategies for in vitro ADME assays: How fast we go?. In *Using Mass Spectrometry for Drug Metabolism Studies* (Korfmacher, W. A., Ed.). CRC Press, Boca Raton, FL, pp. 83–102.

Kauppila, T. J., Wiseman, J. M., Ketola, R. A., Kotiaho, T., Cooks, R. G., and Kostiainen, R. (2006). Desorption electrospray ionization mass spectrometry for the analysis of pharmaceuticals and metabolites. *Rapid Commun. Mass Spectrom.* **20:**387–392.

Kebarle, P. (2000). A brief overview of the present status of the mechanisms involved in electrospray mass spectrometry. *J. Mass Spectrom.* **35:**804–817.

Kerns, E. H. (2001). High throughput physicochemical profiling for drug discovery. *J. Pharm. Sci.* **90:**1838–1858.

Kero, F. H., Pedder, R. E., and Yost, R. A. (2005). Quadrupole mass analyzers: Theoretical and practical considerations. In *Encyclopedia of Genetics, Genomics, Proteomics and Bioinformatics, Part 3: Proteomics*. Wiley, Hoboken, NJ.

Kertesz, V., and Van Berkel, G. J. (2008). Improved imaging resolution in desorption electrospray ionization mass spectrometry. *Rapid Commun. Mass Spectrom.* **22:**2639–2644.

Kind, T., and Fiehn, O. (2006). Metabolomic database annotations via query of elemental compositions: Mass accuracy is insufficient even at less than 1 ppm. *BMC Bioinformatics* **7**.

Kind, T., Tolstikov, V., Fiehn, O., and Weiss, R. H. (2007). A comprehensive urinary metabolomic approach for identifying kidney cancerr. *Anal. Biochem.* **363:**185–195.

King, R. C., Miller-Stein, C., Magiera, D. J., and Brann, J. (2002). Description and validation of a staggered parallel high performance liquid chromatography system for good laboratory practice level quantitative analysis by liquid chromatography/tandem mass spectrometry. *Rapid Commun. Mass Spectrom.* **16:**43–52.

King, R., Bonfiglio, R., Fernandez-Metzler, C., Miller-Stein, C., and Olah, T. (2000). Mechanistic investigation of ionization suppression in electrospray ionization. *J. Am. Soc. Mass. Spectrom.* **11:**942–950.

King, R. C., Gundersdorf, R., and Fernandez-Metzler, C. L. (2003). Collection of selected reaction monitoring and full scan data on a time scale suitable for target compound quantitative analysis by liquid chromatography/tandem mass spectrometry. *Rapid Commun. Mass Spectrom.* **17:**2413–2422.

Koal, T., Sibum, M., Koster, E., Resch, K., and Kaever, V. (2006). Direct and fast determination of antiretroviral drugs by automated online solid-phase extraction-liquid chromatography-tandem mass spectrometry in human plasma. *Clin. Chem. Lab. Med.* **44:**299–305.

Kobayashi, Y., Ito, S., Itai, S., and Yamamoto, K. (2000). Physicochemical properties and bioavailability of carbamazepine polymorphs and dihydrate. *Int. J. Pharm.* **193:**137–146.

Koers, J. M. (2008). Selected ion monitoring (SIM) mode data collection using the laser diode thermal desorption (LDTD) source to increase sensitivity. In *Proceedings of the 56th ASMS Conference on Mass Spectrometry and Allied Topics*. ASMS, Denver, CO.

Kola, I., and Landis, J. (2004). Can the pharmaceutical industry reduce attrition rates? *Nature Rev.* **3:**711–715.

Kolakowski, B. M., Lustig, D., and Purves, R. W. (2004). Separation and quantitation of caffeine metabolites by high-field asymmetric waveform ion mobility spectrometry (FAIMS).

In *Proceedings of the 52nd ASMS Conference on Mass Spectrometry and Allied Topics.* ASMS, Nashville, TN.

Kopec, K. K., Bozyczko-Coyne, D., and Williams, M. (2005). Target identification and validation in drug discovery: The role of proteomics. *Biochem. Pharmacol.* **69:**1133–1139.

Köpke, A. (2006). The proteomics toolbox—A review of the newest drug discovery methods. *Business Briefing: Future Drug Discovery.* Touch Briefings, London, UK.

Korfmacher, W. A. (2005). Bioanalytical assays in a drug discovery environment. In *Using Mass Spectrometry for Drug Metabolism Studies* (Korfmacher, W. A., Ed.). CRC Press, Boca Raton, FL, pp. 1–34.

Korfmacher, W. A., Cox, K. A., Ng, K. J., Veals, J., Hsieh, Y., Wainhaus, S., Broske, L., Prelusky, D., Nomeir, A., and White, R. E. (2001). Cassette-accelerated rapid rat screen: A systematic procedure for the dosing and liquid chromatography/atmospheric pressure ionization tandem mass spectrometric analysis of new chemical entities as part of new drug discovery. *Rapid Commun. Mass Spectrom.* **15:**335–340.

Kovaleski, J., Kraut, B., Mattiuz, A., Giangiulio, M., Brobst, G., Cagno, W., Kulkarni, P., and Rauch, T. (2007). Impurities in generic pharmaceutical development. *Adv. Drug Deliv. Rev.* **59:**56–63.

Kuhlenbeck, D. L., Eichold, T. H., Hoke, S. H., 2nd, Baker, T. R., Mensen, R., and Wehmeyer, K. R. (2005). On-line solid phase extraction using the Prospekt-2 coupled with a liquid chromatography/tandem mass spectrometer for the determination of dextromethorphan, dextrorphan and guaifenesin in human plasma. *Eur. J. Mass Spectrom. (Chichester, Engl.)* **11:**199–208.

Kummar, S., Kinder, R., Gutierrez, M., Rubinstein, L., Parchment, R., Phillips, L., Low, J., Murgo, A., Tomaszewski, J., and Doroshow, J. (2007). Inhibition of poly (ADP-ribose) polymerase (PARP) by ABT-888 in patients with advanced malignancies: Results of a phase 0 trial. In *American Society of Clinical Oncology Meeting.* Chicago, IL.

Labowsky, M. J., Fenn, J. B., and Yamashita, M. (1984). Method and apparatus for the mass spectrometric analysis of solutions. US patent 84-302751.

Lappin, G., and Garner, R. C. (2005). The use of accelerator mass spectrometry to obtain early human ADME/PK data. *Expert Opin. Drug Metab. Toxicol.* **1:**23–31.

Lasser, K. E., Allen, P. D., Woolhandler, S. J., Himmelstein, D. U., Wolfe, S. M., and Bor, D. H. (2002). Timing of new black box warnings and withdrawals for prescription medications. *JAMA* **287:**2215–2220.

Lebre, D. T., Aiello, M., Impey, G., and Bonelli, F. (2007). Microdosing study in rat plasma using high sensitive LC-ESI/MS/MS technique. In *Proceedings of the 55th ASMS Conference on Mass Spectrometry and Allied Topics.* ASMS, Indianapolis, IN.

Lee, M. S. (2005). *Integrated Strategies for Drug Discovery Using Mass Spectrometry.* Wiley, Hoboken, NJ.

Lewin, M., Guilhaus, M., Wildgoose, J., Hoyes, J., and Bateman, B. (2002). Ion dispersion near parallel wire grids in orthogonal acceleration time-of-flight mass spectrometry: Predicting the effect of the approach angle on resolution. *Rapid Commun. Mass Spectrom.* **16:**609–615.

Li, M., Alnouti, Y., Leverence, R., Bi, H., and Gusev, A. I. (2005a). Increase of the LC-MS/MS sensitivity and detection limits using on-line sample preparation with large volume plasma injection. *J. Chromatogr. B Anal. Technol. Biomed. Life Sci.* **825:**152–160.

Li, et al. (2005b). *Rapid Commun. Mass Spectrom.* **19**:1943–1950.

Link, A. J. (1999). Direct analysis of protein complexes using mass spectrometry. *Nat. Biotechnol.* **17**:676–682.

Lipper, R. A. (1999). How can we optimize selection of drug development candidates from many compounds at the discovery stage? *Modern Drug Discov.* **2**:55.

Liu, Y. M., Akervik, K., and Maljers, L. (2006). Optimized high resolution SRM quantitative analysis using a calibration correction method on triple quadrupole system. In *Proceedings of the 54th ASMS Conference on Mass Spectrometry and Allied Topics.* ASMS, Seattle, WA.

Ma, S., Chowdhury, S. K., and Alton, K. B. (2006). Application of mass spectrometry for metabolite identification. *Curr. Drug. Metab.* **7**:503–523.

Ma, S., and Subramanian, R. (2006). Detecting and characterizing reactive metabolites by liquid chromatography/tandem mass spectrometry. *J. Mass Spectrom.* **41**:1121–1139.

Mandagere, A. K., Thompson, T. N., and Hwang, K. K. (2002). Graphical model for estimating oral bioavailability of drugs in humans and other species from their Caco-2 permeability and in vitro liver enzyme metabolic stability rates. *J. Med. Chem.* **45**:304–311.

March, R. E. (1997). An introduction to quadrupole ion trap mass spectrometery. *J. Mass Spectrom.* **32**:351–369.

Marshall, A. G., Hendrickson, C. L., and Jackson, G. S. (1998). Fourier transform ion cyclotron resonance mass spectrometry: A primer. *Mass Spectrom. Rev.* **17**:1–35.

Martin, M., and Guiochon, G. (2005). Effects of high pressure in liquid chromatography. *J. Chromatogr. A* **1090**:16–38.

Martinez, M. N., and Amidon, G. L. (2002). A mechanistic approach to understanding the factors affecting drug absorption: A review of fundamentals. *J. Clin. Pharmacol.* **42**:620–643.

Matuszewski, B. K. (2006). Standard line slopes as a measure of a relative matrix effect in quantitative HPLC-MS bioanalysis. *J. Chromatogr. B Anal. Technol. Biomed. Life Sci.* **830**:293–300.

Matuszewski, B. K., Constanzer, M. L., and Chavez-Eng, C. M. (1998). Matrix effect in quantitative LC/MS/MS analyses of biological fluids: A method for determination of finasteride in human plasma at picogram per milliliter concentrations. *Anal. Chem.* **70**:882–889.

Matuszewski, B. K., Constanzer, M. L., and Chavez-Eng, C. M. (2003). Strategies for the assessment of matrix effect in quantitative bioanalytical methods based on HPLC-MS/MS. *Anal. Chem.* **75**:3019–3030.

Maurer, H. H. (2007). Current role of liquid chromatography-mass spectrometry in clinical and forensic toxicology. *Anal. Bioanal. Chem.* **388**:1315–1325.

McCooeye, M., Ding, L., Gardner, G. J., Fraser, C. A., Lam, J., Sturgeon, R. E., and Mester, Z. (2003). Separation and quantitation of the stereoisomers of ephedra alkaloids in natural health products using flow injection-electrospray ionization-high field asymmetric waveform ion mobility spectrometry-mass spectrometry. *Anal. Chem.* **75**:2538–2542.

McCooeye, M., and Mester, Z. (2006). Comparison of flow injection analysis electrospray mass spectrometry and tandem mass spectrometry and electrospray high-field asymmetric waveform ion mobility mass spectrometry and tandem mass spectrometry for the determination of underivatized amino acids. *Rapid Commun. Mass Spectrom.* **20**:1801–1808.

McCooeye, M. A., Ells, B., Barnett, D. A., Purves, R. W., and Guevremont, R. (2001). Quantitation of morphine and codeine in human urine using high-field asymmetric waveform ion mobility spectrometry (FAIMS) with mass spectrometric detection. *J. Anal. Toxicol.* **25**:81–87.

McCooeye, M. A., Mester, Z., Ells, B., Barnett, D. A., Purves, R. W., and Guevremont, R. (2002). Quantitation of amphetamine, methamphetamine, and their methylenedioxy derivatives in urine by solid-phase microextraction coupled with electrospray ionization-high-field asymmetric waveform ion mobility spectrometry-mass spectrometry. *Anal. Chem.* **74**:3071–3075.

McEwen, C. N., McKay, R. G., and Larsen, B. S. (2005). Analysis of solids, liquids, and biological tissues using solids probe introduction at atmospheric pressure on commercial LC/MS instruments. *Anal. Chem.* **77**:7826–7831.

McGibbon, G. A., Bayliss, M. A., Antler, M., and Lashin, V. (2008). Automated software analysis of isotpe cluster mass differences for components in LC-MS datasets. In *Proceedings of the 56th ASMS Conference on Mass Spectrometry and Allied Topics.* ASMS, Denver, CO.

McLean, J. A., Ridenour, W. B., and Caprioli, R. M. (2007). Profiling and imaging of tissues by imaging ion mobility-mass spectrometry. *J. Mass Spectrom.* **42**:1099–1105.

Mei, H. (2005). Matrix effects: Causes and solutions. In *Using Mass Spectrometry for Drug Metabolism Studies* (Korfmacher, W. A., Ed.). CRC Press, Boca Raton, FL, pp. 103–150.

Mei, H., Hsieh, Y., Nardo, C., Xu, X., Wang, S., Ng, K., and Korfmacher, W. A. (2003). Investigation of matrix effects in bioanalytical high-performance liquid chromatography/tandem mass spectrometric assays: Application to drug discovery. *Rapid Commun. Mass Spectrom.* **17**:97–103.

Mensch, J., Noppe, M., Adriaensen, J., Melis, A., Mackie, C., Augustijns, P., and Brewster, M. E. (2007). Novel generic UPLC/MS/MS method for high throughput analysis applied to permeability assessment in early drug discovery. *J. Chromatogr. B Anal. Technol. Biomed. Life Sci.* **847**:182–187.

Messina, C. J., Grumbach, E. S., and Diehl, D. M. (2007). Development and validation of a UHPLC method for paroxetine hydrochloride. *LCGC N. A.* **25**:1042–1049.

Michael, S. M., Chien, M., and Lubman, D. M. (1992). An ion trap storage/time-of-flight mass spectrometer. *Rev. Sci. Instrum.* **63**:4277–4284.

Milne, G. M. (2003). Pharmaceutical productivity: The imperative for a new paradigm. In *Annual Reports in Medicinal Chemistry* (Doherty, A. M., Ed.). Elsevier, Amsterdam, pp. 383–396.

Morris, H. R., Paxton, T., Dell, A., Langhorne, J., Berg, M., Bordoli, R. S., Hoyes, J., and Bateman, R. H. (1996). High sensitivity collisionally-activated decomposition tandem mass spectrometry on a novel quadrupole/orthogonal-acceleration time-of-flight mass spectrometer. *Rapid Commun. Mass Spectrom.* **10**:889–896.

Morris, H. R., Paxton, T., Panico, M., McDowell, R., and Dell, A. (1997). A novel geometry mass spectrometer, the Q-TOF, for low-femtomole/attomole-range biopolymer sequencing. *J. Protein Chem.* **16**:469–479.

Morrison, D., Davies, A. E., and Watt, A. P. (2002). An evaluation of a four-channel multiplexed electrospray tandem mass spectrometry for higher throughput quantitative analysis. *Anal. Chem.* **74**:1896–1902.

Muller, C., Schafer, P., Stortzel, M., Vogt, S., and Weinmann, W. (2002). Ion suppression effects in liquid chromatography-electrospray-ionisation transport-region collision induced dissociation mass spectrometry with different serum extraction methods for systematic toxicological analysis with mass spectra libraries. *J. Chromatogr. B Anal. Technol. Biomed. Life Sci.* **773**:47–52.

Naidong, W., Addison, T., Schneider, T., Jiang, X., and Halls, T. D. (2003). A sensitive LC/MS/MS method using silica column and aqueous-organic mobile phase for the analysis of loratadine and descarboethoxy-loratadine in human plasma. *J. Pharm. Biomed. Anal.* **32**:609–617.

Naidong, W., Bu, H., Chen, Y. L., Shou, W. Z., Jiang, X., and Halls, T. D. (2002a). Simultaneous development of six LC-MS-MS methods for the determination of multiple analytes in human plasma. *J. Pharm. Biomed. Anal.* **28**:1115–1126.

Naidong, W., Shou, W. Z., Addison, T., Maleki, S., and Jiang, X. (2002b). Liquid chromatography/tandem mass spectrometric bioanalysis using normal-phase columns with aqueous/organic mobile phases: A novel approach of eliminating evaporation and reconstitution steps in 96-well SPE. *Rapid Commun. Mass Spectrom.* **16**:1965–1975.

New, L. S., Saha, S., Ong, M. M. K., Boelsterli, U. A., and Chan, E. C. Y. (2007). Pharmacokinetic study of intraperitoneally administered troglitazone in mice using ultra-performance liquid chromatography/tandem mass spectrometry. *Rapid Commun. Mass Spectrom.* **21**:982–988.

Newman, D. J., and Cragg, G. M. (2007). Natural products as sources of new drugs over the last 25 years. *J. Nat. Prod.* **70**:461–477.

Newman, D. J., Cragg, G. M., and Snader, K. M. (2003). Natural products as sources of new drugs over the period 1981–2002. *J. Nat. Prod.* **66**:1022–1037.

Notari, S., Mancone, C., Tripodi, M., Narciso, P., Fasano, M., and Ascenzi, P. (2006). Determination of anti-HIV drug concentration in human plasma by MALDI-TOF/TOF. *J. Chromatogr. B Anal. Technol. Biomed. Life Sci.* **833**:109–116.

Obach, R. S., Kalgutkar, A. S., Soglia, J. R., and Zhao, S. X. (2008). Can in vitro metabolism-dependent covalent binding data in liver microsomes distinguish hepatotoxic from non-hepatotoxic drugs? An analysis of 18 drugs with consideration of intrinsic clearance and daily dose. *Chem. Res. Toxicol.* **21**:1814–1822.

O'Connor, D., and Mortishire-Smith, R. (2006). High-throughput bioanalysis with simultaneous acquisition of metabolic route data using ultra performance liquid chromatography coupled with time-of-flight mass spectrometry. *Anal. Bioanal. Chem.* **385**:114–121.

O'Connor, D., Mortishire-Smith, R., Morrison, D., Davies, A., and Dominguez, M. (2006). Ultra-performance liquid chromatography coupled to time-of-flight mass spectrometry for robust, high-throughput quantitative analysis of an automated metabolic stability assay, with simultaneous determination of metabolic data. *Rapid Commun. Mass Spectrom.* **20**:851–857.

O'Connor, P. B., Pittman, J. L., Thomson, B. A., Budnik, B. A., Cournoyer, J. C., Jebanathirajah, J., Lin, C., Moyer, S., and Zhao, C. (2006). A new hybrid electrospray Fourier transform mass spectrometer: Design and performance characteristics. *Rapid Commun. Mass Spectrom.* **20**:259–266.

Panchagnula, R., and Agrawal, S. (2004). Biopharmaceutic and pharmacokinetic aspects of variable bioavailability of rifampicin. *Int. J. Pharm.* **271**:1–4.

Patrick, J. E., Kosoglou, T., Stauber, K. L., Alton, K. B., Maxwell, S. E., Zhu, Y., Statkevich, P., Iannucci, R., Chowdhury, S., Affrime, M., and Cayen, M. N. (2002). Disposition of the selective cholesterol absorption inhibitor ezetimibe in healthy male subjects. *Drug Metab. Dispos.* **30:**430–437.

Patrie, S. M., Charlebois, J. P., Whipple, D., Kelleher, N. L., Hendrickson, C. L., Quinn, J. P., Marshall, A. G., and Mukhopadhyay, B. (2004). Construction of a hybrid quadrupole/ Fourier transform ion cyclotron resonance mass spectrometer for versatile MS/MS above 10 kDa. *J. Am. Soc. Mass. Spectrom.* **15:**1099–1108.

Paul, G., Winnik, W., Hughes, N., Schweingruber, H., Heller, R., and Schoen, A. (2003). Accurate mass measurement at enhanced mass-resolution on a triple quadrupole mass-spectrometer for the identification of a reaction impurity and collisionally-induced fragment ions of cabergoline. *Rapid Commun. Mass Spectrom.* **17:**561–568.

Pedraglio, S., Rozio, M. G., Misiano, P., Reali, V., Dondio, G., and Bigogno, C. (2007). New perspectives in bio-analytical techniques for preclinical characterization of a drug candidate: UPLC-MS/MS in in vitro metabolism and pharmacokinetic studies. *J. Pharm. Biomed. Anal.* **44:**665–673.

Pelkonen, O., and Raunio, H. (2005). In vitro screening of drug metabolism during drug development: Can we trust the predictions? *Expert Opin. Drug Metab. Toxicol.* **1:**49–59.

Perng, C. Y., Kearney, A. S., Palepu, N. R., Smith, B. R., and Azzarano, L. M. (2003). Assessment of oral bioavailability enhancing approaches for SB-247083 using flow-through cell dissolution testing as one of the screens. *Int. J. Pharm.* **250:**147–156.

Peterman, S. M., Dufresne, C. P., and Horning, S. (2005). The use of a hybrid linear trap/FT-ICR mass spectrometer for on-line high resolution/high mass accuracy bottom-up sequencing. *J. Biomol. Tech.* **16:**112–124.

PhRMA (2006). *Pharmaceutical Industry Profile 2006.* Pharmaceutical Research and Manufacturers of America, Washington, DC.

Plumb, R., Castro-Perez, J., Granger, J., Beattie, I., Joncour, K., and Wright, A. (2004). Ultra-performance liquid chromatography coupled to quadrupole-orthogonal time-of-flight mass spectrometry. *Rapid Commun. Mass Spectrom.* **18:**2331–2337.

Prakash, C., Shaffer, C. L., and Nedderman, A. (2007). Analytical strategies for identifying drug metabolites. *Mass Spectrom. Rev.*

Prentis, R. A., Lis, Y., and Walker, S. R. (1988). Pharmaceutical innovation by the seven UK-owned pharmaceutical companies (1964–1985). *Br. J. Clin. Pharmacol.* **25:**387–396.

Prueksaritanont, T., Lin, J. H. et al. (2006). Complicating factors in safety testing of drug metabolites: Kinetic differences between generated and preformed metabolites. *Toxicol. Appl. Pharmacol.* **217**(2):143–152.

Purves, R. W., and Guevremont, R. (1999). Electrospray ionization high-field asymmetric waveform ion mobility spectrometry-mass spectrometry. *Anal. Chem.* **71:**2346–2357.

Purves, R. W., Guevremont, R., Day, S., Pipich, C. W., and Matyjaszczyk, M. S. (1998). Mass spectrometric characterization of a high-field asymmetric waveform ion mobility spectrometer. *Rev. Sci. Instrum.* **69:**4094–4105.

Ramanathan, R., Chowdhury, S. K., and Alton, K. B. (2005). Oxidative metabolites of drugs and xenobiotics: LC-MS methods to identify and characterize in biological matrices. In *Identification and Quantification of Drugs, Metabolites and Metabolizing*

Enzymes by LC-MS (Chowdhury, S. K., Ed.). Elsevier, Amsterdam, The Netherlands, pp. 225–276.

Ramanathan, R., Reyderman, L., Kulmatycki, K., Su, A. D., Alvarez, N., Chowdhury, S. K., Alton, K. B., Wirth, M. A., Clement, R. P., Statkevich, P., and Patrick, J. E. (2007a). Disposition of loratadine in healthy volunteers. *Xenobiotica* **37:**753–769.

Ramanathan, R., Reyderman, L., Su, A. D., Alvarez, N., Chowdhury, S. K., Alton, K. B., Wirth, M. A., Clement, R. P., Statkevich, P., and Patrick, J. E. (2007b). Disposition of desloratadine in healthy volunteers. *Xenobiotica* **37:**770–787.

Ramanathan, R., Zhong, R., Blumenkrantz, N., Chowdhury, S. K., and Alton, K. B. (2007c). Response normalized liquid chromatography nanospray ionization mass spectrometry. *J. Am. Soc. Mass. Spectrom.* **18:**1891–1899.

Ramanathan, R., Zhong, R., Blumenkrantz, N., Chowdhury, S. K., and Alton, K. B. (2007d). Quantitative assessment of metabolites early in the clinical program: Potential for using LC-Nanospray-MS. *Eastern Analytical Symposium and Exposition*, Somerset, NJ.

Romanyshyn, L., Tiller, P. R., Alvaro, R., Pereira, A., and Hop, C. E. C. A. (2001). Ultra-fast gradient vs. fast isocratic chromatography in bioanalytical quantification by liquid chromatography/tandem mass spectrometry. *Rapid Commun. Mass Spectrom.* **15:**313–319.

Romanyshyn, L., Tiller, P. R., and Hop, C. E. (2000). Bioanalytical applications of "fast chromatography" to high-throughput liquid chromatography/tandem mass spectrometric quantitation. *Rapid Commun. Mass Spectrom.* **14:**1662–1668.

Ruan, Q., Peterman, S., Szewc, M. A., Ma, L., Cui, D., Humphreys, W. G., and Zhu, M. (2008). An integrated method for metabolite detection and identification using a linear ion trap/Orbitrap mass spectrometer and multiple data processing techniques: Application to indinavir metabolite detection. *J. Mass Spectrom.* **43:**251–261.

Rudewicz, P. J., and Yang, L. (2001). Novel approaches to high throughput quantitative LC-MS/MS in a regulated environment. *Am. Pharm. Rev.* **4:**64, 66, 68, 70.

Sadek, P. C. (2000). *Troubleshooting HPLC Systems: A Bench Manual*. Wiley, New York.

Salem, I. I., Idrees, J., and Al Tamimi, J. I. (2004). Determination of glimepiride in human plasma by liquid chromatography-electrospray ionization tandem mass spectrometry. *J. Chromatogr. B Anal. Technol. Biomed. Life Sci.* **799:**103–109.

Sanders, M., Shipkova, P. A., Zhang, H., and Warrack, B. M. (2006). Utility of the hybrid LTQ-FTMS for drug metabolism applications. *Curr. Drug Metab.* **7:**547–555.

Schwartz, J. C., Senko, M. W., and Syka, J. E. P. (2002). A two-dimensional quadrupole ion trap mass spectrometer. *J. Am. Soc. Mass. Spectrom.* **13:**659–669.

Senko, M. W., Hendrickson, C. L., Emmett, M. R., Shi, S. D. H., and Marshall, A. G. (1997). *J. Am. Soc. Mass Spectrom.* **8:**970–976.

Seto, C., Ni, J., Ouyang, F., Ellis, R., Aiello, M., Jones, E. B., Welty, D., and Acheampong, A. (2007). Detection of circulating metabolites of carbamazepine in microdosing studies in rats using LC-MS/MS. In *Proceedings of the 55th ASMS Conference on Mass Spectrometry and Allied Topics*. ASMS, Indianapolis, IN.

Shah, V. P. (2007). The history of bioanalytical method validation and regulation: Evolution of a guidance document on bioanalytical methods validation. *AAPS J.* **9:**E43–E47.

Shah, V. P., Midha, K. K., Findlay, J. W., Hill, H. M., Hulse, J. D., McGilveray, I. J., McKay, G., Miller, K. J., Patnaik, R. N., Powell, M. L., Tonelli, A., Viswanathan, C. T., and Yacobi,

A. (2000). Bioanalytical method validation: A revisit with a decade of progress. *Pharm. Res.* **17:**1551–1557.

Shen, J. X., Wang, H., Tadros, S., and Hayes, R. N. (2006). Orthogonal extraction/chromatography and UPLC, two powerful new techniques for bioanalytical quantitation of desloratadine and 3-hydroxydesloratadine at 25 pg/mL. *J. Pharm. Biomed. Anal.* **40:**689–706.

Siegel, M. M. (2005). *Mass-Spectrometry-Based Drug Screening Assays for Early Phases in Drug Discovery.* Wiley, Hoboken, NJ.

Smalley, J., Marino, A. M., Xin, B., Olah, T., and Balimane, P. V. (2007). Development of a quantitative LC-MS/MS analytical method coupled with turbulent flow chromatography for digoxin for the in vitro P-gp inhibition assay. *J. Chromatogr. B Anal. Technol. Biomed. Life Sci.* **854:**260–267.

Smith, D. A., Jones, B. C., and Walker, D. K. (1996). Design of drugs involving the concepts and theories of drug metabolism and pharmacokinetics. *Med. Res. Rev.* **16:**243–266.

Smith, D. A. and Obach, R. S. (2005). Seeing through the mist: abundance versus percentage. Commentary on metabolites in safety testing. *Drug. Metab. Dispos.* **33**(10):1409–1417.

Smith, D. A., and Obach, R. S. (2006). Metabolites and safety: what are the concerns, and how should we address them? *Chem. Res. Toxicol.* **19:**1570–1579.

Sobott, F., Hernandez, H., McCammon, M. G., Tito, M. A., and Robinson, C. V. (2002). A tandem mass spectrometer for improved transmission and analysis of large macromolecular assemblies. *Anal. Chem.* **74:**1402–1407.

Souverain, S., Rudaz, S., and Veuthey, J. L. (2004a). Matrix effect in LC-ESI-MS and LC-APCI-MS with off-line and on-line extraction procedures. *J. Chromatogr. A* **1058:**61–66.

Souverain, S., Rudaz, S., and Veuthey, J. L. (2004b). Restricted access materials and large particle supports for on-line sample preparation: An attractive approach for biological fluids analysis. *J. Chromatogr. B Anal. Technol. Biomed. Life Sci.* **801:**141–156.

Sparkman, O. D. (2006). *Mass Spec Desk Reference.* Global View Publishing, Pittsburgh, PA.

Srinivas, N. R. (2007). Changing need for bioanalysis during drug development. *Biomed. Chromatogr.* **22:**235–243.

Stokvis, E., Rosing, H., and Beijnen, J. H. (2005). Stable isotopically labeled internal standards in quantitative bioanalysis using liquid chromatography/mass spectrometry: Necessity or not? *Rapid Commun. Mass Spectrom.* **19:**401–407.

Sun, J., Wang, G., Wang, W., Zhao, S., Gu, Y., Zhang, J., Huang, M., Shao, F., Li, H., Zhang, Q., and Xie, H. (2005). Simultaneous determination of loratadine and pseudoephedrine sulfate in human plasma by liquid chromatography-electrospray mass spectrometry for pharmacokinetic studies. *J. Pharm. Biomed. Anal.* **39:**217–224.

Swartz, M. E. (2005a). Ultra performance liquid chromatography (UPLC): An introduction. *LCGC N. Am.* 8–14.

Swartz, M. E. (2005b). UPLC. An introduction and review. *J. Liq. Chromatogr. Relat. Technol.* **28:**1253–1263.

Sweeney, D. L. (2003). Small molecules as mathematical partitions. *Anal. Chem.* **75:**5362–5373.

Syage, J. A., Hanold, K. A., Lynn, T. C., Horner, J. A., and Thakur, R. A. (2004). Atmospheric pressure photoionization. II. Dual source ionization. *J. Chromatogr. A* **1050:**137–149.

Takats, Z., Wiseman, J. M., and Cooks, R. G. (2005). Ambient mass spectrometry using desorption electrospray ionization (DESI): Instrumentation, mechanisms and applications in forensics, chemistry, and biology. *J. Mass Spectrom.* **40:**1261–1275.

Tall, A. R., Yvan-Charvet, L., and Wang, N. (2007). The failure of torcetrapib: Was it the molecule or the mechanism? *Arterioscler. Thromb. Vasc. Biol.* **27:**257–260.

Tanaka, et al. (2002). *J. Chromatogr. A* **965:**35–49.

Tang, L., Fitch, W. L., Alexander, M. S., and Dolan, J. W. (2000). Expediting the method development and quality control of reversed-phase liquid chromatography electrospray ionization mass spectrometry for pharmaceutical analysis by using an LC/MS performance test mix. *Anal. Chem.* **72:**5211–5218.

Tennikova, T. B., Svec, F., and Belenkii, B. G. (1990). High-performance membrane chromatography. A novel method of protein separation. *J. Liquid Chromatogr. Related Technol.* **13:**63–70.

Tiller, P. R., Yu, S., Castro-Perez, J., Fillgrove, K. L., and Baillie, T. A. (2008). High-throughput, accurate mass liquid chromatography/tandem mass spectrometry on a quadrupole time-of-flight system as a 'first-line' approach for metabolite identification studies. *Rapid Commun. Mass Spectrom.* **22:**1053–1061.

Testa, B., Balmat, A. L., Long, A., and Judson, P. (2005). Predicting drug metabolism: An evaluation of the expert system METEOR. *Chem Biodivers* **2:**872–885.

Thompson, T. N. (2000). Early ADME in support of drug discovery: The role of metabolic stability studies. *Curr. Drug. Metab.* **1:**215–241.

Thompson, T. N. (2001). Optimization of metabolic stability as a goal of modern drug design. *Med. Res. Rev.* **21:**412–449.

Thompson, T. N. (2005). Drug metabolism in vitro and in vivo results: How do these data support drug discovery? In *Using Mass Spectrometry for Drug Metabolism Studies* (Korfmacher, W. A., Ed.). CRC Press, Boca Raton, FL, pp. 35–81.

Thurman, E. M. (2006). Accurate-mass identification of chlorinated and brominated products of 4-nonylphenol, nonylphenol dimers, and other endocrine disrupters. *J. Mass Spectrom.* **41:**1287–1297.

Tolley, L., Jorgenson, J. W., and Moseley, M. A. (2001). Very high pressure gradient LC/MS/MS. *Anal. Chem.* **73:**2985–2991.

Tong, X., Ita, I. E., Wang, J., and Pivnichny, J. V. (1999). Characterization of a technique for rapid pharmacokinetic studies of multiple co-eluting compounds by LC/MS/MS. *J. Pharm. Biomed. Anal.* **20:**773–784.

Tremblay, P., Groleau, P. E., Ayotte, C., Picard, P., and Viel, E. (2008). High-throughput screening and quantification of doping agents in urine using LDTD-APCI-MS/MS. In *Proceedings of the 56th ASMS Conference on Mass Spectrometry and Allied Topics.* ASMS, Denver, CO.

Unger, K. K., Skudas, R., and Schult, M. M. (2008). Particle packed columns and monolithic columns in high-performance liquid chromatography-comparison and critical appraisal. *J. Chromatogr. A.* **1184:**393–415.

Van Arnum, P. (2007). Advancing approaches in detecting polymorphism. *Pharm. Technol.* **16:**S18–S23.

van Deemter, J. J., Zuiderweg, F. J., and Klinkenbergm, A. (1956). Longitudinal diffusion and resistance to mass transfer as causes of nonideality in chromatography. *Chem. Eng. Sci.* **5:**271–289.

van De Waterbeemd, H., Smith, D. A., Beaumont, K., and Walker, D. K. (2001). Property-based design: Optimization of drug absorption and pharmacokinetics. *J. Med. Chem.* **44:**1313–1333.

van Pelt, C. K., Corso, T. N., Schultz, G. A., Lowes, S., and Henion, J. (2001). A four-column parallel chromatogr.aphy system for isocratic or gradient LC/MS analyses. *Anal. Chem.* **73:**582–588.

Venne, K., Bonneil, E., Eng, K., and Thibault, P. (2004). Enhancement in proteomics analyses using nano LC-MS and FAIMS. *PharmaGenomics* **4:**30–32, 34, 36, 38, 40.

Venne, K., Bonneil, E., Eng, K., and Thibault, P. (2005). Improvement in peptide detection for proteomics analyses using nano LC-MS and high-field asymmetry waveform ion mobility mass spectrometry. *Anal. Chem.* **77:**2176–2186.

Vesell, E. S. (1974). Relationship between drug distribution and therapeutic effects in man. *Annu. Rev. Pharmacol.* **14:**249–270.

Viswanathan, C. T., Bansal, S., Booth, B., Destefano, A. J., Rose, M. J., Sailstad, J., Shah, V. P., Skelly, J. P., Swann, P. G., and Weiner, R. (2007). Quantitative bioanalytical methods validation and implementation: Best practices for chromatographic and ligand binding assays. *Pharm. Res.* **24:**1962–1973.

Vlase, L., Imre, S., Muntean, D., and Leucuta, S. E. (2007). Determination of loratadine and its active metabolite in human plasma by high-performance liquid chromatography with mass spectrometry detection. *J. Pharm. Biomed. Anal.* **44:**652–657.

Wahlstrom, J. L., Rock, D. A., Slatter, J. G., and Wienkers, L. C. (2006). Advances in predicting CYP-mediated drug interactions in the drug discovery setting. *Expert Opin. Drug Discov.* **1:**677–691.

Wang, G., Hsieh, Y., Cui, X., Cheng, K. C., and Korfmacher, W. A. (2006). Ultra-performance liquid chromatography/tandem mass spectrometric determination of testosterone and its metabolites in in vitro samples. *Rapid Commun. Mass Spectrom.* **20:**2215–2221.

Watson, J. T., and Sparkman, O. D. (2007). *Introduction to Mass Spectrometry: Instrumentation, Applications, and Strategies for Data Interpretation.* John Wiley & Sons, Inc., New York, NY.

Weaver, P. J., Laures, A. M. F., and Wolff, J. C. (2007). Investigation of the advanced functionalities of a hybrid quadrupole orthogonal acceleration time-of-flight mass spectrometer. *Rapid Commun. Mass Spectrom.* **21:**2415–2421.

Weaver, R., and Riley, R. J. (2006). Identification and reduction of ion suppression effects on pharmacokinetic parameters by polyethylene glycol 400. *Rapid Commun. Mass Spectrom.* **20:**2559–2564.

Webster, G. K., Li, H., Sanders, W. J., Basel, C. L., and Huang, G. (2001). Column robustness case study for a liquid chromatographic method validated in compliance with ICH, VICH, and GMP guidelines. *J. Chromatogr. Sci.* **39:**273–279.

Weng, N., and Halls, T. D. J. (2001). Systematic troubleshooting for LC/MS/MS Part 1: Sample preparation and chromatography. *BioPharm.* **14:**28, 30, 32, 34, 36, 38.

Weng, N., and Halls, T. D. J. (2002). Systematic troubleshooting for LC/MS/MS-Part 2: Large-scale LC/MS/MS and automation. *BioPharm.* **15:**22–27.

Werner, U., Werner, D., Mundkowski, R., Gillich, M., and Brune, K. (2001). Selective and rapid liquid chromatography-mass spectrometry method for the quantification of rofecoxib in pharmacokinetic studies with humans. *J. Chromatogr. B Biomed. Sci. Appl.* **760**:83–90.

Werner, U., Werner, D., Pahl, A., Mundkowski, R., Gillich, M., and Brune, K. (2002). Investigation of the pharmacokinetics of celecoxib by liquid chromatography-mass spectrometry. *Biomed. Chromatogr.* **16**:56–60.

Whalen, K. M., Rogers, K. J., Cole, M. J., and Janiszewski, J. S. (2000). AutoScan: An automated workstation for rapid determination of mass and tandem mass spectrometry conditions for quantitative bioanalytical mass spectrometry. *Rapid Commun. Mass Spectrom.* **14**:2074–2079.

Wickremsinhe, E. R., Singh, G., Ackermann, B. L., Gillespie, T. A., and Chaudhary, A. K. (2006). A review of nanoelectrospray ionization applications for drug metabolism and pharmacokinetics. *Curr. Drug Metab.* **7**:913–928.

Wieboldt, R., Campbell, D. A., and Henion, J. (1998). Quantitative liquid chromatographic-tandem mass spectrometric determination of orlistat in plasma with a quadrupole ion trap. *J. Chromatogr. B Biomed. Sci. Appl.* **708**:121–129.

Wieling, J. (2002). LC-MS-MS experiences with internal standards. *Chromatographia* **55**:S107–S113.

Williams, J. P., Patel, V. J., Holland, R. D., and Scrivens, J. H. (2006). The use of recently described ionization techniques for the rapid analysis of some common drugs and samples of biological origin. *Rapid Commun. Mass Spectrom.* **20**:1447–1456.

Wilm, M., and Mann, M. (1996). Analytical properties of the nanoelectrospray ion source. *Anal. Chem.* **68**:1–8.

Wilson, D. M., Wang, X., Walsh, E., and Rourick, R. A. (2001). High throughput log D determination using liquid chromatography-mass spectrometry. *Comb. Chem. High Throughput Screen.* **4**:511–519.

Wiseman, J. M., Takats, Z., Gologan, B., Davisson, V. J., and Cooks, R. G. (2005). Direct characterization of enzyme-substrate complexes by using electrosonic spray ionization mass spectrometry. *Angew Chem. Int. Ed. Engl.* **44**:913–916.

Wren, S. A. C. (2005). Peak capacity in gradient ultra performance liquid chromatography (UPLC). *J. Pharm. Biomed. Anal.* **38**:337–343.

Wu, J. T., Zeng, H., Deng, Y., and Unger, S. E. (2001). High-speed liquid chromatography/ tandem mass spectrometry using a monolithic column for high-throughput bioanalysis. *Rapid Commun. Mass Spectrom.* **15**:1113–1119.

Wu, S. T., Xia, Y. Q., and Jemal, M. (2007). High-field asymmetric waveform ion mobility spectrometry coupled with liquid chromatography/electrospray ionization tandem mass spectrometry (LC/ESI-FAIMS-MS/MS) multicomponent bioanalytical method development, performance evaluation and demonstration of the constancy of the compensation voltage with change of mobile phase composition or flow rate. *Rapid Commun. Mass Spectrom.* **21**:3667–3676.

Xia, Y. Q., Hop, C. E., Liu, D. Q., Vincent, S. H., and Chiu, S. H. (2001). Parallel extraction columns and parallel analytical columns coupled with liquid chromatography/tandem mass spectrometry for on-line simultaneous quantification of a drug candidate and its six metabolites in dog plasma. *Rapid Commun. Mass Spectrom.* **15**:2135–2144.

Xu, R. N., Fan, L., Rieser, M. J., and El-Shourbagy, T. A. (2007). Recent advances in high-throughput quantitative bioanalysis by LC-MS/MS. *J. Pharm. Biomed. Anal.* **44**:342–355.

Xu, X. (2005). Utilizing higher mass resolution in quantitative assays. In *Using Mass Spectrometry for Drug Metabolism Studies* (Korfmacher, W. A., Ed.). CRC Press, Boca Raton, FL, pp. 203–228.

Xu, X., Veals, J., and Korfmacher, W. A. (2003). Comparison of conventional and enhanced mass resolution triple-quadrupole mass spectrometers for discovery bioanalytical applications. *Rapid Commun. Mass Spectrom.* **17**:832–837.

Xue, Y. J., Liu, J., Pursley, J., and Unger, S. (2006). A 96-well single-pot protein precipitation, liquid chromatography/tandem mass spectrometry (LC/MS/MS) method for the determination of muraglitazar, a novel diabetes drug, in human plasma. *J. Chromatogr. B Anal. Technol. Biomed. Life Sci.* **831**:213–222.

Yamane, N., Tozuka, Z., Sugiyama, Y., Tanimoto, T., Yamazaki, A., and Kumagai, Y. (2007). Microdose clinical trial: Quantitative determination of fexofenadine in human plasma using liquid. *J. Chromatogr. B* **858**:118–128.

Yang, L., Amad, M., Winnik, W. M., Schoen, A. E., Schweingruber, H., Mylchreest, I., and Rudewicz, P. J. (2002). Investigation of an enhanced resolution triple quadrupole mass spectrometer for high-throughput liquid chromatography/tandem mass spectrometry assays. *Rapid Commun. Mass Spectrom.* **16**:2060–2066.

Yang, L., Clement, R. P., Kantesaria, B., Reyderman, L., Beaudry, F., Grandmaison, C., Di Donato, L., Masse, R., and Rudewicz, P. J. (2003). Validation of a sensitive and automated 96-well solid-phase extraction liquid chromatography-tandem mass spectrometry method for the determination of desloratadine and 3-hydroxydesloratadine in human plasma. *J. Chromatogr. B Anal. Technol. Biomed. Life Sci.* **792**:229–240.

Yang, L., Mann, T. D., Little, D., Wu, N., Clement, R. P., and Rudewicz, P. J. (2001a). Evaluation of a four-channel multiplexed electrospray triple quadrupole mass spectrometer for the simultaneous validation of LC/MS/MS methods in four different preclinical matrixes. *Anal. Chem.* **73**:1740–1747.

Yang, L., Wu, N., and Rudewicz, P. J. (2001b). Applications of new liquid chromatography-tandem mass spectrometry technologies for drug development support. *J. Chromatogr. A* **926**:43–55.

Yao, M., Swaminathan, A., and Srinivas, N. (2007). Assessment of dose proportionality of muraglitazar after repeated oral dosing in rats via a sparse sampling methodology. *Biopharm. Drug Dispos.* **28**:35–42.

Yu, K., Di, L., Kerns, E. H., Li, S. Q., Alden, P., and Plumb, R. S. (2007). Ultra-performance liquid chromatography/tandem mass spectrometric quantification of structurally diverse drug mixtures using an ESI-APCI multimode ionization source. *Rapid Commun. Mass Spectrom.* **21**:893–902.

Yu, K., Little, D., Plumb, R., and Smith, B. (2006). High-throughput quantification for a drug mixture in rat plasma—A comparison of ultra performance liquid chromatography/tandem mass spectrometry with high-performance liquid chromatography/tandem mass spectrometry. *Rapid Commun. Mass Spectrom.* **20**:544–552.

Zhang, D., Wang, L., Raghavan, N., Zhang, H., Li, W., Cheng, P. T., Yao, M., Zhang, L., Zhu, M., Bonacorsi, S., Yeola, S., Mitroka, J., Hariharan, N., Hosagrahara, V., Chandrasena, G., Shyu, W. C., and Humphreys, W. G. (2007). Comparative metabolism of radiolabeled muraglitazar in animals and humans by quantitative and qualitative metabolite profiling. *Drug Metab. Dispos.* **35:**150–167.

Zhang, H., and Henion, J. (2001). Comparison between liquid chromatography-time-of-flight mass spectrometry and selected reaction monitoring liquid chromatography-mass spectrometry for quantitative determination of idoxifene in human plasma. *J. Chromatogr. B Biomed. Sci. Appl.* **757:**151–159.

Zhang, N., Fountain, S. T., Bi, H., and Rossi, D. T. (2000). Quantification and rapid metabolite identification in drug discovery using API time-of-flight LC/MS. *Anal. Chem.* **72:**800–806.

Zweigenbaum, J., and Henion, J. (2000). Bioanalytical high-throughput selected reaction monitoring-LC/MS determination of selected estrogen receptor modulators in human plasma: 2000 samples/day. *Anal. Chem.* **72:**2446–2454.

2

Quantitative Bioanalysis in
Drug Discovery and
Development: Principles
and Applications

Ayman El-Kattan and Chris Holliman

*Pfizer Global Research and Development, Department of Pharmacokinetics, Dynamics &
Metabolism, Groton, Connecticut*

Lucinda H. Cohen

*Merck Research Laboratories, Department of Drug Metabolism and Pharmacokinetics,
Rahway, New Jersey*

Mass Spectrometry in Drug Metabolism and Pharmacokinetics. Edited by Ragu Ramanathan
Copyright © 2009 John Wiley & Sons, Inc.

2.1 INTRODUCTION

The bioanalytical revolution continues; in fact, the drug discovery revolution continues inexorably forward. Today, not only are humans benefiting from revolutionary new medicines that have dramatically improved the quality of life, but more is expected to come. Compound collections have ballooned through combinatorial library synthesis; biological science has mapped the human genome, and society impatiently awaits the promise of genetically tailored medicines that will treat virtually every ailment.

This pharmaceutical revolution could not have been achieved without the undergirdment of advanced analytical instrumentation. Mass spectrometry (MS) combined with liquid chromatography (LC) has enabled the characterization of novel potential drugs as well as quantitative measurement in an increasingly complex milieu at an incredibly rapid throughput rate. Quantitation of drugs in biological media such as blood (usually plasma or serum) and tissue has become more or less routine. Drugs can be accurately, precisely, and very sensitively determined through validated methods developed by hundreds of scientists at companies and academic institutions worldwide. The role of MS, in enabling this straightforward process, is taken for granted. However, significant challenges to bioanalysis of potential new chemical entities (NCEs) still lurk in the background, waiting to catch the unsuspecting biologist unaware and helping to keep analytical chemists gainfully employed.

The two main purposes of this chapter are, first, to describe the fundamental process of obtaining pharmacokinetic data derived from the concentration data generated via LC–MS/MS and, second, to describe the greatest challenges facing quantitative bioanalysis of new chemical entities in biological matrixes. These include first designing the method relative to the urgency and quality necessary. When the rigors provided by good laboratory practices (GLP) regulation are not required, the most significant problem often lies in defining the scientific rigor necessary for nonregulated quantitative methods. In addition, tissue analysis during drug discovery presents a significant bioanalytical challenge and will be discussed in more detail. Once an NCE enters the regulated arena, the current greatest challenges are focused on meeting the rigor of regulatory guidance and adapting to the new paradigm of microdosing. Thus, quantitative bioanalysis, specifically for pharmacokinetics, continues to offer exciting opportunities to tackle complex and difficult scientific problems.

2.2 A PHARMACOKINETICS PRIMER

Pharmacokinetics (PK) is the science that describes the time course of a circulating drug concentration in the body resulting from administration of a certain drug dose. In comparison, pharmacodynamics (PD) is the science that describes the relationship of the time course of drug concentration and the drug effects in the body (Meibohm and Derendorf, 1997). Key determinants of the PK of a drug include absorption, distribution, metabolism, and elimination (ADME) (Lin et al., 2003), as discussed in Chapter 1.

Discovering novel therapeutic agents is an increasingly time-consuming and costly process. Most estimates indicate that approximately 10–15 years and more than $800 million are required to discover and develop a successful drug product (DiMasi et al., 2003). Prior to 1991, poor drug pharmacokinetics was one of the leading causes of compound failure in preclinical and clinical development (van De Waterbeemd et al., 2001). However, by 2001, attrition due to pharmacokinetics or bioavailability had been reduced to 10%, thereby validating the need for preclinical assessment of PK and ADME properties (van De Waterbeemd et al., 2001). In discovery settings, the main outcomes of PK and PD evaluations are as follows:

- To select compounds with the maximum potential of reaching the target
- To determine the appropriate route of administration to deliver the drug (typically oral)

- To understand how blood levels relate to efficacy or toxicity in order to choose efficacious and safe doses
- To decide on the frequency and duration of dosing in order to maintain adequate drug concentration at the target site for disease modification
- To accurately predict the human PK profile prior to clinical trials.

2.2.1   Components of Pharmacokinetic Study

Typically, a PK study is composed of three phases, namely the in-life phase, bio-analysis, and data analysis. The in-life phase includes administering the compound to animals or humans and collecting samples from an appropriate matrix of interest such as blood or urine at predetermined time intervals for bioanalysis. The bioanalytical phase involves analysis of a drug and/or its metabolite(s) concentration in blood, plasma, serum, or urine. This analysis typically involves sample extraction and detection of analytes via LC–MS/MS. The third phase is data analysis using noncompartmental or compartmental PK computational methods.

In drug discovery, preliminary PK studies are usually conducted in rodents to evaluate the extent of drug exposure in vivo. This is commonly followed by PK studies in larger animals such as dog or monkey to better characterize the PK profile of the compound and to support safety studies. Pharmacokinetic scaling (also called allometry) is a discipline that is used to predict human PK profiles using preclinical data and is widely used in predicting the drug human half-life, dose, and extent of absorption. Accurate prediction of a human PK profile is imperative to minimize drug failure in development due to poor PK attributes. A detailed description of methods in predicting human PK is beyond the scope of this chapter but can be found in many excellent reviews (Obach et al., 1997; Miners et al., 2004; Poggesi, 2004; Raunio et al., 2004; Thomas et al., 2006; Hurst et al., 2007). A more in-depth discussion of various PK concepts and their applications can be found in various references (Gibaldi and Perrier, 1982; Rowland and Tozer, 1995; Hurst et al., 2007).

2.2.2   Parameters That Define Drug Pharmacokinetic Profile

2.2.2.1   *Area Under the Curve*   The first step in a PK experiment is the in-life phase, in which animals or humans are dosed with an NCE and blood samples are collected at predefined time points. Preclinically, animals are generally dosed intravenously (IV) and/or orally (PO). After sample extraction and LC–MS/MS, a concentration–time profile is generated and is shown in Fig. 2.1. The area under the curve (AUC) is then calculated from this data set. Mathematically, area under the plasma (or blood) concentration–time curve (AUC) can be calculated as

$$\mathrm{AUC} = \int_0^\infty C\,dt \qquad (2.1)$$

where AUC is a primary measure of the extent of drug availability to the systemic circulation; that is, it reflects the total amount of unchanged drug that reaches systemic

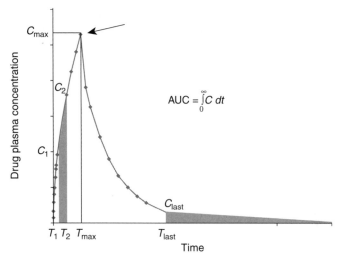

Figure 2.1. Estimation of area under the plasma concentration–time curve (AUC).

circulation following IV or extravascular administration. The unit for AUC is concentration per unit time (e.g., $ng \cdot h/mL$). The AUC is determined using simple integration methods, as shown in Equation (2.1) or a linear trapezoidal method, which is the most widely used approach to determine AUC (Fig. 2.1). The area of each trapezoid is calculated using the equation

$$\text{AUC}_{t_1 \to t_2} = \tfrac{1}{2}(C_2 + C_1) \times (t_2 - t_1) \tag{2.2}$$

The extrapolated area from t_{last} to ∞, is estimated as

$$\text{AUC}_{t_{\text{last}} \to \infty} = \frac{C_{\text{last}}}{K_e} \tag{2.3}$$

where C_{last} is the last observed concentration at t_{last} and K_e is the slope obtained from the terminal portion of the curve, representing the terminal elimination rate constant. The total AUC ($\text{AUC}_{0 \to \infty}$) is determined as

$$\text{AUC}_{0 \to \infty} = \text{AUC}_{0 \to t_{\text{last}}} + \text{AUC}_{t_{\text{last}} \to \infty} \tag{2.4}$$

The AUC is used in the calculation of clearance, apparent volume of distribution, and bioavailability (see following sections) and reflects the general extent of exposure over time.

2.2.2.2 *Clearance*   Clearance (CL) is a primary PK parameter that describes the process of irreversible elimination of a drug from the systemic circulation and is defined as the volume of blood or plasma that is totally cleared of its content of drug per unit time. Thus, CL measures the removal of drug from blood or plasma. However, CL does not indicate how much drug is being removed, but instead represents the volume of blood or plasma from which the drug is completely removed, or cleared, in a given time period. The CL unit is volume per unit time (e.g., mL/min/kg).

The most widely used approach to evaluate plasma (total) CL involves IV administration of a single dose of a drug and measuring its plasma concentration at different time points, as shown in Fig. 2.2. In this manner, the calculated clearance will not be confounded by complex absorption and distribution phenomena which commonly occur during oral dosing. Clearance is derived from the equation (Rowland and Tozer, 1995)

$$CL_{tot} = \frac{Dose_{IV}}{AUC_{IV}} \tag{2.5}$$

In general, a drug is eliminated either unchanged through excretion in the urine and/or bile or by metabolic conversion into more polar metabolite(s) that can be readily excreted in urine and/or bile. Therefore, total body clearance is a sum of all clearances by various mechanisms and can be expressed mathematically as

$$CL_{tot} = \sum CL_{hep} + CL_{ren} + CL_{bil} \tag{2.6}$$

where CL_{tot} is the total body clearance from all different organs and mechanisms, CL_{hep} is the hepatic clearance, CL_{ren} is the renal clearance, and CL_{bil} is the biliary clearance. It is interesting to note that around three-quarters of the top 200 prescribed drugs in the United States are primarily cleared by hepatic metabolism (Hosokawa et al., 1990). The hepatic extraction ratio (E_h) is a PK parameter that is widely used to assess the liver's ability to extract drug from the systemic circulation (Gibaldi and Perrier, 1982) and is defined as the fraction of a drug in the blood that is cleared (extracted) on each passage through the liver. It is calculated using

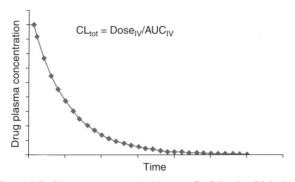

Figure 2.2. Plasma concentration time profile following IV dosing.

the equation (Gibaldi and Perrier, 1982)

$$E_h = \frac{CL_{hep}}{Q} \tag{2.7}$$

where CL_{hep} is the hepatic blood clearance and Q is the hepatic blood flow. Therefore, if the predominant clearance mechanism for a compound is via hepatic metabolism, then it is reasonable to assume that $CL_{tot} = CL_{hep}$. Thus,

$$E_h = \frac{CL_{hep}}{Q} = \frac{CL_{total}}{Q} \tag{2.8}$$

Compounds that undergo hepatic metabolism can be classified according to their E_h. Compounds with $E_h > 0.7$ are considered high-extraction drugs, whereas compounds with $E_h < 0.3$ are considered low-extraction drugs. The ratio E_h has a major impact on oral drug bioavailability (see below).

2.2.2.3 Plasma versus Blood Clearance Calculation of E_h from drug clearance in blood requires the determination of drug concentration in whole blood. Since the determination of drug concentration is usually performed in plasma or serum, knowledge of the blood/plasma concentration ratio is necessary to estimate the blood clearance. Blood clearance is calculated using the equation

$$\frac{\text{Plasma clearance}}{\text{Blood clearance}} = \frac{\text{blood concentration } (C_b)}{\text{plasma concentration } (C_p)} \tag{2.9}$$

2.2.2.4 Apparent Volume of Distribution Volume of distribution (V_d) is a proportionality factor that relates the amount of a drug in the body to its blood or plasma concentrations,

$$\text{Amount of drug in body at time } t = V_d \times C \text{ plasma at time } t \tag{2.10}$$

Following IV dosing and at $t = 0$, the amount of drug in the body is equal to the administered IV dose. At $t = 0$, V_d is termed volume of the central compartment (V_c). Similar to CL, V_d is a primary PK parameter and its unit is volume (e.g., L/kg). The V_d is used to assess the extent of drug distribution in the body. This is usually achieved by comparing the drug V_d to the total body water. If the drug has a V_d that is smaller than the total body water (human total body water $= 42\,L$ per $70\,kg$ human body weight), then this would suggest that the drug has limited tissue distribution (e.g., naproxen has $V_d = 11\,L$ per $70\,kg$ human body weight) (Wells et al., 1994). On the other hand, if a drug has a V_d larger than the total body water, then this would suggest that the drug is able to distribute to body tissues (e.g., olanzapine has $V_d = 1120\,L$ per $70\,kg$ human body weight) (Callaghan et al., 1999). In the literature, V_d ranges from 3 to more than 40,000 L per 70 kg human body weight. Therefore, the term *apparent volume of distribution* is usually used.

2.2.2.5 Apparent Volume of Distribution at Steady State

The volume of distribution that is determined when plasma concentrations are measured at steady state and in equilibrium with the drug concentration in the tissue compartment, $V_{d,ss}$, is determined as

$$V_{d,ss} = \frac{\text{amount of drug in body at equilibrium conditions}}{\text{steady-state plasma concentrations}\,(C_{ss})} \tag{2.11}$$

Although $V_{d,ss}$ is a steady-state parameter, it can be calculated using non-steady-state data as

$$V_{d,ss} = \text{CL} \times \text{MRT} \tag{2.12}$$

where MRT is the drug mean residence time (see below). Furthermore, $V_{d,ss}$ is used in the calculation of loading dose as

$$\text{Loading dose} = \frac{V_{d,ss} \times C_{ss}}{F} \tag{2.13}$$

Use of loading dose is important, especially for those drugs in which it is desirable to immediately or rapidly reach the steady-state plasma concentration (C_{ss}) (e.g., anticoagulant, antiepileptic, antiarrhythmic, and antimicrobial therapy).

2.2.2.6 Half-Life

The half-life $T_{1/2}$ is the time that is required for the amount (or plasma concentration) of a drug to decrease by one-half and is calculated by the equation

$$T_{1/2} = \frac{0.693 \times V_d}{\text{CL}} \tag{2.14}$$

where $T_{1/2}$ is a dependent PK parameter that is determined by both CL and V_d, which are independent primary PK parameters. Therefore, $T_{1/2}$ is increased by a decrease in CL or increase in V_d and vice versa. The half-life is the most widely reported PK parameter since it may constitute a major determinant of the duration of action after single and multiple dosing. In addition, $T_{1/2}$ plays a key role in determining the time that is required to reach steady state following multiple dosing and the frequency with which doses can be given. The unit for $T_{1/2}$ is time (e.g., hours).

2.2.2.7 Bioavailability According to the European Medicines Evaluation Agency (EMEA), bioavailability ($F\%$) is "the rate and extent to which an active moiety is absorbed from a pharmaceutical form, and becomes available in the systemic circulation." As a parameter, there are two types of bioavailability:

1. Absolute bioavailability, which refers to the fraction of the extravascular (or PO) dose that reaches the systemic circulation unchanged (in reference to an IV dose). Absolute bioavailability is usually determined by calculating the respective AUC after PO and IV administration as

$$\text{Absolute bioavailability} = \frac{\text{AUC}_{\text{PO}}}{\text{AUC}_{\text{IV}}} \times \frac{\text{dose}_{\text{IV}}}{\text{dose}_{\text{PO}}} \tag{2.15}$$

2. Relative bioavailability, which refers to the fraction of a dose of drug reaching the systemic circulation relative to a reference product. Relative bioavailability is usually calculated as

$$\text{Relative bioavailability} = \frac{\text{AUC}_{\text{test}}}{\text{AUC}_{\text{ref}}} \times \frac{\text{dose}_{\text{ref}}}{\text{dose}_{\text{test}}} \tag{2.16}$$

Determinants of oral bioavailability include the fraction of dose absorbed in the gastrointestinal tract and the fraction of dose that does not undergo metabolism in the intestinal tract and liver. Oral bioavailability is defined mathematically by the equation

$$F = F_a F_g F_h \tag{2.17}$$

where F is the oral drug bioavailability, F_a is the fraction of the drug that is absorbed from the intestinal lumen to the intestinal enterocytes, F_g is the fraction of the unmetabolized drug in the intestinal enterocytes, and F_h is the fraction of the unmetabolized drug in the liver calculated as

$$F_h = 1 - E_h = 1 - \frac{\text{CL}_h}{Q} \tag{2.18}$$

Thus, if a drug has a high hepatic extraction ($E_h > 0.7$), then its extent of bioavailability will be low when it is given orally ($F < 0.3$). On the other hand, if a drug has low hepatic extraction ($E_h < 0.3$), then the extent of bioavailability will be high provided that the drug is completely absorbed and not significantly metabolized by the intestine.

2.2.2.8 Mean Residence Time Mean residence time (MRT) is the average time for all drug molecules to exist in the body. MRT is another measure of drug elimination and its unit is time (e,g., hours). Following IV dosing, MRT is calculated as

$$\text{MRT} = \frac{\text{AUMC}}{\text{AUC}} = \frac{\int\limits_{0}^{\infty} Ct\,dt}{\int\limits_{0}^{\infty} C\,dt} \tag{2.19}$$

where AUMC is the curve of the area under the first moment versus time from time $t = 0$ to $t = \infty$ and is calculated using the trapezoidal rule similar to AUC.

In some cases, MRT can be a better parameter to assess drug elimination compared to half-life. This can be attributed to the greater analytical sensitivity shown with various analytical systems such as LC–MS/MS; the lower drug concentrations measured following drug administration appeared to yield longer terminal half-lives which are not related to the drug's pharmacologically relevant half-life. In a case like this, it would be recommended to measure MRT rather than half-life to assess drug elimination.

2.2.2.9 Maximum Plasma Concentration and Time of Maximum Concentration

Maximum plasma concentration ($C_{\max}$) is defined as the maximum observed drug concentration in the plasma concentration–time profile following IV or PO dosing. Most commonly, $C_{\max}$ is obtained by direct observation of the plasma concentration–time profile (Fig. 2.1). For some drugs, the pharmacological effect is dependent on the $C_{\max}$. For example, aminoglycosides, which are widely used class of antibiotics, need to achieve a $C_{\max}$ that is at least 8- to 10-fold higher than the minimum inhibitory concentration (MIC) to obtain a clinical response $\geq 90\%$ (Schentag, 1999; MacGowan, 2001). The unit of $C_{\max}$ is concentration unit (e.g., ng/mL). Time of maximum concentration ($T_{\max}$) is the time required to reach $C_{\max}$. As with $C_{\max}$, $T_{\max}$ is usually determined from direct observation of the plasma concentration–time profile. The unit of $T_{\max}$ is time (e.g., hours).

2.2.3 Noncompartmental Pharmacokinetics

Various PK parameters such as CL, V_d, F%, MRT, and $T_{1/2}$ can be determined using noncompartmental methods. These methods are based on the empirical determination of AUC and AUMC described above. Unlike compartmental models (see below), these calculation methods can be applied to any other models provided that the drug follows linear PK. However, a limitation of the noncompartmental method is that it cannot be used for the simulation of different plasma concentration–time profiles when there are alterations in dosing regimen or multiple dosing regimens are used.

2.2.4 Compartmental Pharmacokinetics

Compartmental models of PK analysis are widely used to describe drug distribution and disposition. In these models, the body is assumed to be composed of one compartment or more and the drug kinetics can be defined by differential equations

generally of first order. These compartments do not have any physiological significance. However, they may represent a group of tissues or organs with similar
distribution characteristics. For example, highly blood perfused body organs like
liver, lungs, and kidneys often have different drug distribution than fat tissue.
Compartmental models are usually arranged in a mammillary format such that
there is one or more compartments that feed from a central compartment.

2.2.4.1   One-Compartment Open Model

In the one-compartment model,
the body is assumed to be a homogenous unit where the drug is rapidly distributing
throughout the body and once eliminated the drug follows a monoexponential
decline (Fig. 2.3). Following IV dosing, the plasma drug concentration can be
calculated as

$$C = C^0 e^{-K_e t} \tag{2.20}$$

where C^0 is the plasma drug concentration immediately after IV dosing calculated as

$$C^0 = \frac{D}{V_c} \tag{2.21}$$

Unlike other compartmental models, there is only one V_d, where $V_c = V_{d,ss}$.

2.2.4.2   Two-Compartment Open Model

When the drug concentration–
time profile demonstrates a biexponential decline following IV dosing, a two-
compartment model that is the sum of two first-order processes (distribution
and elimination) best describe the data (Fig. 2.4). A drug that follows the PK of a

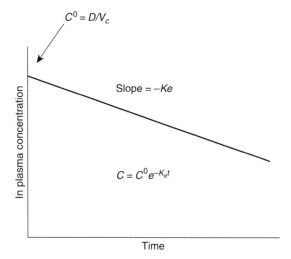

Figure 2.3. One-compartment model.

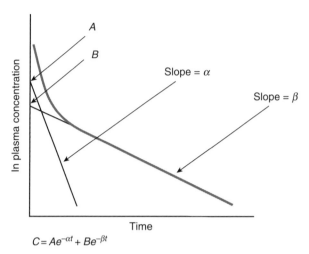

$$C = Ae^{-\alpha t} + Be^{-\beta t}$$

Figure 2.4. *Two-compartment open model.*

two-compartment model does not rapidly distribute throughout the body as evident in the one-compartment model. In the two-compartment model, the drug is assumed to distribute into two compartments, the central and the tissue compartments. The central compartment represents highly perfused body organs where the drug distributes rapidly and uniformly. On the other hand, in the tissue compartment, the drug distributes more slowly. For a drug that follows the two-compartment model, the rate of drug plasma concentration change following IV dosing can be determined as

$$C = Ae^{-\alpha t} + Be^{-\beta t} \tag{2.22}$$

where A and B are functions of the administered dose and α and β are the first-order constants for the distribution and elimination phases, respectively.

In this chapter, only the one- and two-compartment models following IV dosing were described. Other models with extravascular dosing have an additional compartment with an absorption rate constant describing input into the central compartment. Models with three or greater compartments may be used if the drug concentration versus time may be described better with additional exponential terms. However, these models present greater complexity.

2.2.5 Modeling to Predict Single- and Multiple-Dose Pharmacokinetic Profiles

As previously discussed, compartmental models can be effectively used to project plasma concentrations that would be achieved following different dosage regimens and/or multiple dosing. However, for these projections to be accurate, the drug PK profile should follow first-order kinetics where various PK parameters such as CL, V_d, $T_{1/2}$, and $F\%$ do not change with dose.

2.2.6 Linear and Nonlinear Pharmacokinetics

Drug metabolism, renal tubular secretion, and biliary secretion are usually mediated by metabolizing enzymes or transporter proteins. These protein systems usually possess good substrate selectivity with finite capacities, which are described by the Michaelis–Menten equation,

$$v = \frac{V_{max}C}{K_m + C} \tag{2.23}$$

where C is the drug plasma concentration, V_{max} is the maximum elimination or transport rate, and K_m is the Michaelis constant. The values of V_{max} and K_m are dependent on the nature of the drug and enzymatic process involved. This equation implies that when the drug plasma concentration is lower than K_m, no saturation of the enzymes or transporter proteins occurs. Therefore, various PK parameters such as CL, V_d, $T_{1/2}$, and $F\%$ remain constant with respect to dose and time and the drug is considered to follow linear PK. This is a desirable property in that prediction of the plasma exposure following various dosing regimens and over multiple dosing can be more easily achieved. However, when the drug plasma concentration is larger than K_m, saturation of the enzymes or transporter proteins occurs and the rate of elimination or transport rate is maximized and approaches that of V_{max}. As a result, drug elimination or secretion becomes a zero-order process and PK parameters such as CL, $V_{d,ss}$ and $T_{1/2}$ become a function of the administered dose or plasma concentration. Furthermore, drugs that demonstrate saturable metabolism may exhibit less than expected PO first-pass metabolism, resulting in a higher bioavailability. Consequently, there is a greater fractional increase in C_{ss} (or AUC) than the corresponding fractional increase in the rate of drug administration. Drugs with this characteristic likely will require more careful monitoring when dosage adjustment is made in order to achieve the desired therapeutic effects and to minimize the potential for adverse effects.

2.2.7 Allometric Scaling

Allometric scaling (allometry) is the discipline that predicts human PK using preclinical data (Ritschel et al., 1992). This approach is based on empirical observations that various physiological parameters are functions of body size. The most widely used equation in allometry is a one-term power function:

$$y = aB^x \tag{2.24}$$

$$\log Y = x \log B + \log a \tag{2.25}$$

where B is any independent variable such as animal body weight and y is any dependent variable. For example, in PK, y is usually CL, $V_{d,ss}$ or $T_{1/2}$. The exponent of the allometric equation (x) determines the slope of a double-logarithmic plot (Fig. 2.5). The interspecies clearance frequently scales with an exponent of 0.75, whereas the

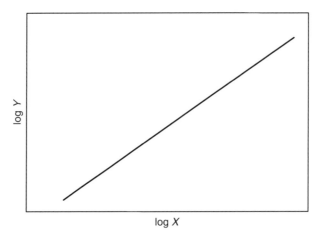

log Y

log X

Figure 2.5. *Linear transform of simple allometric equation.*

exponent for interspecies $V_{d,ss}$ frequently scales with an exponent of 1.0, and the half-life usually scales with an exponent of 0.25. Allometry sometimes fails to predict the human PK profile, in particular human clearance. This is attributed to interspecies differences in the metabolic enzyme. Therefore, it is widely accepted to use in vitro metabolic data obtained from multiple species to correct for the human PK parameters obtained from allometry.

2.2.8 Pharmacokinetic/Pharmacodynamic Modeling

Pharmacokinetics describe the drug concentration–time courses in body fluids resulting from administration of a certain drug dose, while PD describe the observed response resulting from a certain drug concentration. The rationale for PK/PD modeling is to link PK and PD so as to establish the dose–concentration–response relationships and subsequently to predict the effect–time courses following drug administration. Several relatively simple PD models, which comprise the fixed-effect model, the linear model, the long-linear model, the E_{max} model, and the sigmoid E_{max} model, can best describe concentration–response relationships (Rowland and Tozer, 1995; Derendorf et al., 1998; Beierle et al., 1999). The E_{max} model is described by a hyperbolic equation,

$$E = \frac{E_{max}C}{EC_{50} + C}$$

(2.26)

where E is the effect, intensity or response; E_{max} is the maximum effect; EC_{50} is a constant and represents the concentration which corresponds to 50% of the maximum effect; and C is the plasma drug concentration. An example of the use of the E_{max} model is shown in Fig. 2.6, which demonstrates the effect from the

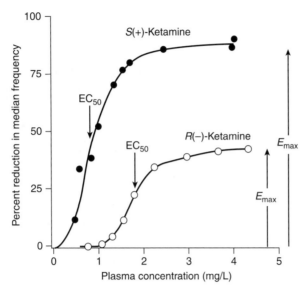

Figure 2.6. *Percent of reduction in EEG median frequency as function of ketamine plasma concentration given as the two enatiomeric forms. (Reprinted with permission from Rowland and Tozer,* Clinical Pharmacokinetics, *3rd ed., Lea & Febiger, Philadelphia, 1995, p. 341.)*

anesthetic ketamine as a function of plasma concentration for the different enantiomeric forms of the drug (Rowland and Tozer, 1995).

More complex integrated PK/PD models are necessary to link and account for a possible temporal dissociation between the plasma concentration and the observed effect. Four basic attributes may be used to characterize PK/PD models: First, the link between measured concentration and the pharmacological response mechanism that mediates the observed effect (direct versus indirect link); second, the response mechanism that mediates the observed effect (direct versus indirect response); third, the information used to establish the link between measured concentration and observed effect (hard versus soft link); and, fourth, the time dependency of the involved PD parameters (time variant versus time invariant) (Danhof et al., 1993; Steimer et al., 1993; Aarons, 1999; Lees et al., 2004). The expanded and early use of PK/PD modeling in drug discovery and development is highly beneficial for increasing the success rate of drug discovery and development and will most likely improve the current state of applied therapeutics.

2.3   QUANTITATIVE BIOANALYSIS DURING DRUG DISCOVERY

2.3.1   Engineering Your Method: Relationship of Quality to Need

As shown in Fig. 2.7, quantitative bioanalysis supports the advancement of new chemical entities throughout the continuum of drug discovery and development.

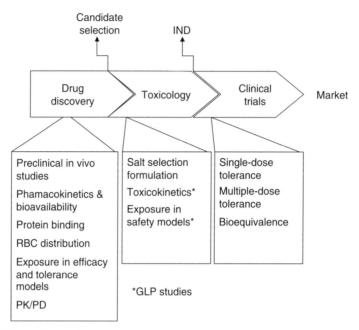

Figure 2.7. *Studies requiring bioanalytical quantitation during drug discovery and development. (GLP, Good laboratory practices; IND, Investigational new drug application.)*

Initially, when the number of compounds and samples is high, method quality can be balanced by the need for speed. The risk of a potentially incorrect decision based on low-quality bioanalysis is much more manageable in early discovery, particularly because this phase is not regulated but is for decision making and candidate selection purposes only. Expectations for method quality are generally lower for in vitro samples than those from in vivo studies. This is due in part to the amount of effort required to synthesize the relatively modest amount of compound for an in vitro experiment compared to the amount necessary to dose animals and also due to the complexity of the sample matrix, which is much simpler for in vitro samples than in vivo samples such as plasma or tissue.

As a compound advances to toxicology testing in support of first-in-human dosing and thus become regulated, the method quality required increases exponentially. During this time, the risks of erroneous bioanalytical data are significant. Consequently, appropriate time is carefully allotted towards method development to allow for sufficiently high quality methods to be validated. During the preclinical regulated phase, expectations regarding sample turnaround are (perhaps) more reasonable. However, once a compound reaches clinical development, not only must the method become rugged and high quality, but also sample turnaround times must be expedited to enable dose selection during an escalating dose clinical trial. Twenty-four-hour turnaround times from blood draw to release of data to the clinician becomes necessary to allow the patient to be dosed at a newer level the following day. This time period is probably the most intense "pressure-cooker" quantitative bioanalytical period because later clinical trials such as population PK, drug–drug interaction, or bioequivalence studies generally do not require blood-level concentrations midstudy.

Early in drug discovery when method validation is not mandatory, method quality can be increased by incorporating some of the individual elements of method validation set forth in the U.S. Food and Drug Administration (FDA, 2001) guidelines as listed in Table 2.1. Although Korfmacher (2005) has set forth "Rules of the Road" for nonregulated bioanalysis, considering this somewhat unconventional approach, the faint at heart may find more comfort in tailoring their method design according to the recommendations below. When embarking on the road to establishing best practices in the vague and ill-defined arena of "fit-for-purpose" bioanalysis, scientists are advised to exercise scientific judgment. It is generally much easier to take the lead in managing client expectations than to react to unrealistic expectations of GLP rigor in too short a time frame.

Even in regulated studies, stable-isotope- (usually deuterium-) labeled compound may not be accessible by the time the studies are conducted. Generally the stable-isotope-labeled material is considered the "gold standard" (pun intended) for quantitation. Chemically dissimilar compounds may create additional challenges to develop a chromatographic method of reasonable length. Structural analogues may still have slightly different elution times, even with rapid chromatographic methods using ballistic gradients. Due to different elution times, matrix ion suppression may affect the structural analogue internal standard (IS) to a different degree than the compound of interest. Typically, this type of problem is too subtle to observe until the moment of

TABLE 2.1. Recommendations to Engineer Quality into Fit-for-Purpose Bioanalytical Assay Characterization in Drug Discovery

Method Validation Requirement per FDA and Industry Standards	Minimum Discovery Approach	Suggestions to Engineer Quality into Nonregulated Methods
Standards and quality controls	Less than four standards, or area ratio determinations only	1. Increase number of standards 2. Use quality controls from separately weighed stock solutions
Internal standard (IS), preferably stable-isotope-labeled compound	If IS is used, may be an all-purpose compound selected for its ionizability in either polarity	Chemically similar internal standard if available in sufficient quantity; may be a lead compound currently in development or one that failed but was synthesized in large quantities
Stability studies such as stock solution, bench-top, freeze–thaw stability	Widely accepted to ignore these risks in early discovery	Freeze–thaw stability due to reasonable probability this might be necessary for repeat studies; bench-top stability assessment may be prudent for chemical series with known temperature instability

incurred sample analysis, at which point the pressure is greatest! Differential matrix ion suppression is manifested through different peak-area responses of the IS or analyte for the incurred samples versus the standards and quality controls prepared using blank plasma. Stable-isotope-labeled ISs presumably suffer less frequently from this problem (Lanckmans et al., 2007), but even so, isotope effects occasionally surface even for quantitation using stable-isotope-labeled ISs (Wang et al., 2007).

Since the introduction of the post–column infusion technique to assess matrix ionization suppression (Bonfiglio et al., 1999; King et al., 2000; Hsieh et al., 2001; Jessome and Volmer, 2006), a greater appreciation has grown for recurrent problems with ion suppression related to components from the complex biological media of plasma. Early in discovery, the problems of differential matrix effects between standards prepared from control animal plasma versus dosed animals are considered acceptable risk to ensure the speed of decisions. However, ionization suppression related to common excipients in formulations may be severe enough to affect the value of the PK parameters, particularly those derived from IV dosing (Tong et al., 2002; Larger et al., 2005; Weaver and Riley, 2006). Thus, the choice of dosing vehicle during formulation should be standardized to minimize potential problems and certain components such as Tween-80 avoided if at all possible. In the event that no other options are available, the sample extraction and chromatographic separation steps can be tailored to maximize the selectivity of sample preparation and analysis to avoid interference.

Extending chromatographic separation is also an excellent tool to minimize interference from coeluting metabolites in rapid gradient methods. When the coeluting metabolite is susceptible to in-source fragmentation in the mass spectrometer, this separation becomes even more critical (Cohen and Rossi, 2001). Generally metabolites are considered interferences to be avoided, rather than quantitated, based on the lack of suitable reference material and unknown effects on pharmacological activity. The bioanalytical scientist performing quantitation in drug discovery lives in a constant state of uncertainty with respect to common analytical problems which might, in rare circumstances, compromise data quality. Some appropriate measures can be judiciously applied to minimize these compromises, but they are balanced by the need for speed and the rate of attrition during drug discovery.

2.3.2 Tissue Preparation and Quantitation

Unlike plasma or serum, which represents complex, sometimes heterogeneous but still liquid samples, tissue quantitation represents an inherent challenge requiring significant effort to generate liquid intermediates from solid starting materials. Quantitative analysis of tissue may be utilized to assess the tissue uptake at the site of drug action, correlate drug concentration with PK and PD response, and predict toxicity and dose. However, despite the quality and importance of the desired information, preparation of biological tissue samples remains an extremely tedious and time-consuming laboratory task. From the moment the tissue is excised, great attention must be paid to how the sample is stored, processed (whether mechanical or chemical), extracted, and finally analyzed. Once this process is complete, the utility of the data obtained from tissue as well as its analytical quality (accuracy, precision, and reproducibility) is still debatable. Despite the somewhat thankless nature of tissue analysis, significant progress has been made in the last decade to explore alternative tissue sample preparation approaches as well as obtain a better understanding of the risks and benefits of conventional methods.

Tissue preparation techniques can be categorized into mechanical, digestion, or extraction instruments. Some of the techniques used successfully for other types of solid samples (Hawthorne et al., 1994; Walter et al., 1997; LeBlanc, 1999; Richter, 1999; Draisci et al., 2001; Yu and Cohen, 2003), such as soil or plant material, may also be used for tissues. Tissue samples, although solid, should be considered highly aqueous in nature, which can be exploited to rupture cells within the tissue matrix. Generally, chunks of tissue are snap frozen in liquid nitrogen immediately after sampling and are stored at very low temperatures (-20 or $-70°C$) prior to processing. Tissue samples are best processed immediately after removal from the freezer, since thawing will produce a rubbery nugget that evades slicing or dicing. If a large number of tissues must be prepared, then several small batches should be serially processed rather than allowing the entire set of samples to thaw.

A comparison of currently utilized mechanical and digestion techniques is shown in Table 2.2. Each technique has both advantages and limitations, and some may be used in combination for optimum extraction.

2.3.2.1 *Mechanical Techniques*

Homogenization, or "grinding," remains the most popular and generally practical means of preparing tissues for a range of qualitative or quantitative applications. Initially, a small stainless steel probe-style blender containing a generator and a set of blades causes vigorous mixing and turbulence as well as physically shearing of the sample into small pieces. Next, a weighed amount of sample, which may be anywhere from 10 mg to 1 g in size, is placed in a vial with a known volume of buffer solution. The pH of the buffer can be tailored to the desired extraction conditions. The resulting product, or homogenate, is semisolid in nature and can be essentially treated in the same manner as plasma. Lengthening the homogenization step or centrifuging the homogenate and decanting the supernatant will minimize large particles in the homogenate.

Homogenizers are small, compact, and relatively inexpensive and require minimal training to operate. However, extended exposure to high-velocity blending can be irritating, and ear protection should be worn. The probe should be thoroughly and repeatedly rinsed between each sample to avoid cross-contamination. The Autogizer is a parallel homogenizer that has been introduced (Wang et al., 2002) as a means to speed up the homogenization process using a multiprobe (four or six), parallel processing approach. The probes are cleaned automatically with three programmable wash stations. Fibrous material caught in the cutters may require manual intervention. The Autogizer has automated the time-consuming aspects of homogenizing and cleaning the probes and is by design four to six times both as fast and as noisy as an individual homogenizer.

An alternative to homogenization is sonication. In this technique, the tissue sample is snap frozen and then immediately ground to a fine powder using a mortar and pestle

TABLE 2.2. Preparation Techniques Used for Measuring Pharmaceutical Compounds in Tissue

Technique	Typical Vendor(s)	Handles Whole Tissue Chunks?	Analyte Stability Issues?	Automation?	Inexpensive?	High Sample Throughput?	References
			Mechanical				
Homogenizer	Polytron	Y	N	N	Y	N	Yu and Cohen, 2003
Ultrasonicator	LabCaire (North Somerset, England)	N	N	N	N	N	Yu and Cohen, 2003
Bead beater	BioSpec Products, Inc. (Bartlesville, OK)	Y	N	Y	Y	Y	Yu and Cohen, 2003
Freezer mill	SPEX CertiPrep (Metuchen, NJ)	Y	N	Y	N	Y	Yu and Cohen, 2003
Autogizer	TomTec (Hamden, CT)	Y	N	Y	N	Y	Wang et al., 2002
			Digestion				
Acid or Base	Parr (Moline, IL)	Y	Y	N	Y	Y	Posyniak et al., 2001
Enzymatic	Worthington Life Sciences (Lakewood, NJ)	Y	M	M	Y	Y	Yu et al., 2004

Key: Y, yes; N, no; M, maybe.

in a liquid nitrogen bath. The weighed powder is stored and, when ready for analysis, mixed with a known volume of buffer and sonicated using a specially designed acoustical probe placed directly into the powder–buffer mixture. This method is more straightforward than homogenization, but powderizing tissue requires significant manual labor and may lead to occupational health problems such as carpal tunnel syndrome. Similar to the homogenizer, the sonic probe should be thoroughly cleaned between each sample.

The Bead Beater represents a more hands-off approach to tissue sample preparation. Introduced several years ago, the Bead Beater is a unique but simply designed apparatus using small beads in combination with a high-speed rotor to rupture cells (Jones et al., 2002). A solid Teflon impeller rotating at high speed forces thousands of minute beads to collide in a specially shaped vessel. Cells are disrupted quickly, efficiently, and safely. Each sample is placed in a separate tube with a defined amount of beads and buffer solution and then agitated for 15–20 minutes. Homogenization inside disposable microcentrifuge vials guarantees that cross-contamination of samples is minimized. During longer agitation times, the unit is refrigerated to prevent sample heating from the beads' movement. A variety of bead types and sizes are available, including glass, stainless steel, and ceramics. A smaller cousin, the MiniBeadBeater-96™, is well suited for parallel processing.

A variation of the Bead Beater is the freezer mill, which uses small magnetic bars rather than beads to pulverize the sample. Cooling is provided by immersing the sample chambers in a liquid nitrogen bath. The freezer mill is intended for larger samples ($>500\,mg$) but may be customized for smaller samples. Both the freezer mill and the Bead Beater require that the sample be placed in secondary containers, which may or may not be disposable. Disposable secondary containers offer the very attractive advantage of reduced time and effort for cleaning.

2.3.2.2 Digestion Techniques

For nonvascularized or low-water-content tissues such as bone, cartilage, or hair, a mechanical technique may do little to disrupt cellular structure and extract analytes. Extreme measures such as digestion with strong acid (i.e., $12\,N$ HCl) are routinely used for DNA or nucleic acids, which can tolerate the harsh conditions. Alternatively, certain enzymes can be used to digest tissue samples. Commercial devices are available which contain digestion bombs fabricated from material resistant to corrosive media.

2.3.2.3 Enzymatic Digestion

Enzymatic digestion, a technique commonly used for tissue dissociation and cell harvesting of proteins and DNA, offers the advantages of unattended sample preparation, potential automation, and low cost. A range of different enzymes are available with different digestive properties and efficiencies. The choice of enzyme can be driven by the desired tissue or component, such as cartilage, to be digested. Although the enzymatic digestion technique has been utilized for decades, only a few papers (Posyniak et al., 2001; Yu et al., 2004) have been published attempting to employ enzymatic digestion in tissue sample preparation of small molecules (Fig. 2.8).

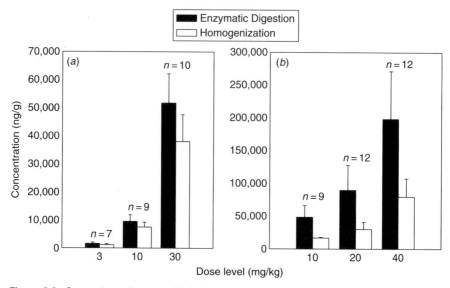

Figure 2.8. Comparison of average (a) desipramine and (b) fluoxetine concentrations in rat brain tissue extracted by enzymatic digestion or homogenization. Error bars represent standard deviation mean value and are symmetrical about the mean. (Reprinted with permission from Yu et al., 2004.)

The feasibility of enzymatic digestion as an alternate tissue preparation technique for bioanalysis of drugs has been evaluated (Yu et al., 2004). Two different enzymes, known to degrade connective tissues to allow tissue dissolution, were chosen for evaluation: collagenase and proteinase K. These enzymes were selected to represent both a more conservative digestive enzyme (collagenase) and a more aggressive digestive enzyme (proteinase K). Results indicate that enzymatic digestion has comparable extraction efficiency to homogenization, and enzymatic digestion using collagenase or proteinase K can be considered as an alternative sample preparation method for analysis of small molecules in tissue.

2.3.2.4   Considerations When Choosing Tissue Preparation Techniques

How to choose the best tissue preparation technique for every application is a difficult question to address. Sample throughput, analyte recovery, analyte thermal stability, amount of available sample, sample preparation techniques available, precision, accuracy, manual labor involved, and operator safety are only a few of the parameters that must be considered when selecting the optimum sample preparation technique.

Tissue analysis presents several unique challenges. A universal lack of reference material for any analyte in an appropriate or comparable matrix exists, regardless of the application. Questions around spatial distribution heterogeneity of the analyte within the tissue matrix are rarely answered unless an imaging technique such as autoradiography is utilized (Chapter 12). Tissue quantitation requires tedious sample weighing (slicing and dicing) for good accuracy and precision.

Thus, techniques that mandate an exact amount of tissue will require significant time just to weigh the tissue. Quantitation of the amount of a given compound in tissue is usually reported as either nanograms per milliliter of tissue homogenate (or liquid matrix generated during preparation) or nanograms per gram of equivalent tissue weight, both requiring an additional calculation correcting for dilution or sample preparation, which can complicate analysis and interpretation. Although an IS is routinely utilized, the IS is generally added as a solution to the homogenate, digest, or extract after preparation. Unfortunately no physical means has been discovered to disperse the IS into the tissue matrix, and extraction efficiency of the IS in the matrix cannot be determined. In the end, it is debatable whether the quality of end data justifies the effort necessary for sample collection, processing, and quantitative analysis.

2.4 QUANTITATIVE BIOANALYSIS FOR REGULATORY SUBMISSIONS

Parallel to most of the compounds supported, bioanalytical assays for discovery and preclinical testing have relatively short life spans within the company. Discovery bioanalytical assays are developed to support in-house decisions around target potency, in vivo pharmacology, ADME optimization, and dose range finding. As such, the degrees of accuracy, precision, and ruggedness that these assays are required to have are commensurate with the level of confidence or, conversely, the level of risk that the company is willing to accept to answer the question at hand. Once the data are obtained and the decision is made, these assays are often discarded in favor of one supporting a more interesting compound or better characterized to support a definitive study.

2.4.1 Investigational New Drug Application and Method Validation

Regulatory agencies, which for the purpose of this chapter will be focused on the FDA, are removed from the drug development process until a company requests to dose a new compound in humans and files an investigational new drug (IND) application (FDA, 1995). Within this application plasma concentrations from single-dose, repeat-dose, and safety pharmacology studies in preclinical species are presented to demonstrate that the new compound has an appropriate safety margin to support dosing in humans (FDA, 2000). Bioanalytical assays that support regulatory submissions and decisions must meet a level of confidence that the FDA or other approving body is willing to accept. To this end, the FDA and the pharmaceutical industry have maintained a dialogue around bioanalytical assay characterization or "validation" that has resulted in two workshop reports (Shah et al., 1996, 2000) and an FDA guidance for industry (FDA, 2001). To quote the latter document, "This guidance provides general recommendations for bioanalytical method validation" around the parameters of accuracy, precision, selectivity, sensitivity, reproducibility, and stability. The operative word here is *guidance*. The FDA has not mandated how an assay is characterized before it is used to support a study. Rather, the guidance is used to define the parameters that the agency recommends be characterized for "good science." Although all laboratories strive to meet the rigor prescribed by the guidance, there are variances from company to company on how this is done that are captured in each laboratory's standard operating procedures (SOPs).

The FDA guidances are "living documents" that reflect the agency's thinking at the time of document preparation and are adaptive to changing technology. Although the guidance on bioanalytical assay validation is couched in terms that address the most common biological matrices and method of analysis, that is, plasma and urine analysis by LC–triple-quadrupole MS (LC–MS/MS), the guidance is just as applicable to high-performance liquid chromatography–ultraviolet (HPLC–UV) analysis.

2.4.2   New Technology and New Guidances

New technology continues to drive both the industry and the regulatory agencies. As new technology enables bioanalytical chemists to better characterize a compound before dosing in humans, the regulatory agencies respond with updated guidance documents and submission requirements. For example, this interplay was observed as the triple-quadrupole mass spectrometer replaced traditional HPLC–UV and fluorescence detection. The new-found ability to routinely achieve nanogram-per-milliliter detection limits revolutionized the pharmaceutical industry's approach to compound development and the information it could provide in its submissions. Today LC–MS/MS sensitivity and selectivity enable the reporting of plasma exposures even at the lowest doses of studies submitted to the agencies. As LC–MS/MS became entrenched as the method of choice for the determination and submission of small-molecule plasma concentrations, the FDA issued a guidance document for the development of bioanalytical assays that specifically addressed specific LC–MS/MS issues such as ionization suppression (FDA, 2001). The industry's success with LC–MS/MS is now reflected in the agency's expectations at NDA submission. Plasma concentrations and exposures are expected to be submitted even for very potent compounds, tissue-specific compounds, or transdermally dosed compounds that result in extremely low plasma concentrations. Once again, the desire to characterize a compound as completely as possible before dosing in humans drives the development of new technology. Today, the industry strives to determine plasma concentrations at picogram-per-milliliter levels through the development of improved mass spectrometer designs or efficient sample introduction methods such as nano-LC or ultrahigh-pressure LC. A more recent example of how new technology drives both the industry and the regulatory agencies is observed with the application of accelerator mass spectrometry (AMS) to biological samples, the evolution of the human microdosing strategy, and the FDA guidance on exploratory INDs.

2.4.3   Accelerator Mass Spectrometry

Pharmaceutical analytical chemists continue to seek more sensitive techniques for the determination of drug plasma concentrations and have looked beyond triple-quadrupole technology. One of the more exciting prospects to lowering detection limits is accelerator mass spectrometry, a technology that was developed in the 1970s for the radiocarbon dating of archeological artifacts. Unlike most mass spectrometric techniques that detect the molecular ion of the analyte, AMS uses high-energy collisions to fragment and detect analytes at the atomic level. AMS analysis requires that the sample be converted to graphite. The graphite sample is then bombarded with a cesium ion beam which sputters negatively charged carbon atoms and carbon hydrides from the sample (Fig. 2.9). These species are selected according to their mass-to-charge ratio and are accelerated to undergo $1 \times 10^6 - 5 \times 10^6$ eV collisions with argon atoms. These collisions results in cleavage of the hydrides and stripping of electrons from the carbon atoms producing positively charged ions. The more abundant ^{12}C ion beam is directed for detection to a Faraday cup while the less abundant ^{14}C ion beam is directed to a more sensitive gas ionization detector. The AMS technique is capable of detecting on the order of 1000 ^{14}C ions yielding a detection limit in the range of atto to zepto (10^{-21}) moles of ^{14}C from milligrams of sample. The reader is directed to an excellent review of the AMS technique by Vogel and Love (2005).

AMS is used for the dating of archeological artifacts because the carbon composing living organisms has a $^{14}C/^{12}C$ ratio determined by the ^{14}C levels in the

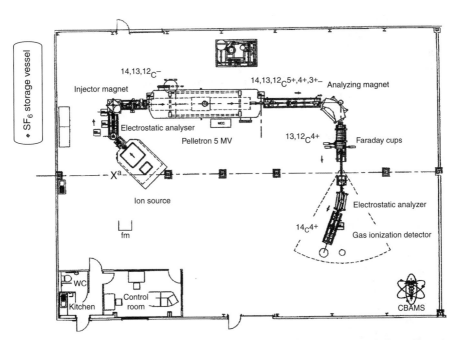

Figure 2.9. *Accelerator mass spectrometer used for bioanalytical analysis at Xceleron (formely CBAMS). Pharmaceutical companies have not adopted AMS as an in-house technique because of cost and size of instrumentation (compare the scale of the instrument to the kitchen in the lower left). Samples are outsourced to companies that specialize in the technique.*

organism's atmosphere. When the organism dies, its ^{14}C levels decrease with a half-life of 5730 years as the ^{14}C it has incorporated decays to ^{14}N. The artifact's age can therefore be determined via AMS by comparison of the artifact's current ^{14}C/^{12}C ratio to the ^{14}C/^{12}C ratio the artifact would have maintained while living.

AMS's exquisite sensitivity to detect ^{14}C was first applied to biological samples from living organisms in biomedical dosimetry experiments by Turteltaub and co-workers (Turteltaub et al., 1990; Brown et al., 2005). In these experiments mice were dosed with ^{14}C-labeled carcinogen. Isolation of mouse liver DNA followed by AMS analysis revealed an *increase* in the animal's endogenous ^{14}C/^{12}C ratio due to the formation of ^{14}C–carcinogen–DNA adducts at dose levels as low as 500 ng/kg. Further research in animals (Meirav et al., 1990; Turteltaub et al., 1990) and humans (Felton et al., 1990) established that AMS-determined plasma concentrations from very low doses of radiolabeled xenobiotics could be used to elucidate biochemical pathways and kinetics without exposing the subject to dangerous levels of radiation or the xenobiotic. In one study (Felton et al., 1990) 0.167 mg, 6.45 μCi, of ^{14}C-labeled atrazine was dermally dosed in human subjects for 24 h. Urine ^{14}C-atrazine levels were detected by AMS for the following seven days enabling the determination of the atrazine excretion half-life. Atrazine concentrations were determined from 2.5 μL of urine, yielding a 2.2-fmol/mL detection limit. Beyond the impressive limit of detection the following parameters of this experiment are significant. The subject's exposure to atrazine was significantly lower than the 104-mg/kg/day scaled no-effect level for short-term dermal atrazine exposure (Gilman et al., 1998) and the subject's exposure to radiation was also less than the level that one is exposed to during a transcontinental commercial airline flight (Vogel and Love, 2005). Finally, preclinical predictions of the human atrazine absorption rate via rat and in vitro human dermal absorption studies substantially overestimated the actual atrazine absorption rate observed in the clinic.

2.4.4 Human Microdosing Strategy

As has been previously mentioned, the failure of in vitro and preclinical in vivo studies to accurately predict clinical outcomes has impacted the entire pharmaceutical industry. A significant number of compounds identified as drug candidates from preclinical characterization are abandoned from further development after the first-in-human study indicates unacceptable human PKs (DiMasi, 2001, 2003). The costs of these failures are significant and can be divided up into three interrelated major expenses. The first expense is the completion of animal safety studies to establish that a safe dose range that includes the proposed therapeutic dose exists. The second expense is the cost and rigor surrounding GMP manufacturing of the active pharmaceutical ingredient (API) and final dosage form. A typical traditional IND application can require the manufacturing of over a kilogram of drug to complete the preclinical and clinical studies. Finally there is the overriding expense of the time and internal resources required to author protocols and reports and complete the studies. Typically the time from candidate identification to first-in-human testing is 12–15 months. This preclinical investment in the compound can be lost with the first human dose when unexpected poor human PKs terminate further development of the drug. These costs rise linearly and significantly for programs that can produce more than one quality candidate but do not have a reliable method for human PK prediction. For these types of scenarios, the pharmaceutical industry is turning toward the capabilities of AMS to select development candidates using a strategy known as the human microdosing strategy (Sandhu et al., 2004; Lappin and Garner, 2005; Lappin et al., 2006a, b; Bertino et al., 2007).

The human microdosing strategy has evolved from the previously cited advantages of AMS. In its simplest form, the subject is given a ^{14}C-labeled drug at a dose that is well below the dose expected to be therapeutic or to elicit a toxic effect (Sarapa et al., 2005). Plasma concentrations and human PK are then determined by AMS. The strategy directly addresses several of the issues faced in drug development. The microdosing strategy does not rely on animal models to obtain actual human PK data. There is a reduction in the number and costs of animal safety studies needed to support a microdosing IND application because the experiments are conducted at low doses to establish safe no-effect levels. This is unlike animal safety studies conducted to support a traditional IND. In these studies doses are escalated to high levels to set upper safety levels by eliciting the actual onset of toxicities. It follows then that the manufacturing costs to support the microdosing approach are greatly reduced. The microdosing safety studies and human doses can be conducted with 100–150 total grams of drug, versus the 1–2 kg of drug needed to support traditional IND safety studies and first-in-human trials. Finally, there is a significant savings in time and internal resources. The time from candidate identification to dosing in humans using the microdose strategy is on the order of 8 months versus the 12–15 months needed to follow the traditional IND strategy (Fig. 2.10). The value of this time saving can be dramatically significant if the reduction in development time allows the company to be first to market or adds months of patent exclusivity to a blockbuster drug.

2.4.5 Microdosing Strategy and Regulatory Agencies

Typical first-in-human studies seek not only to determine human PK parameters but also to escalate the dose to the maximum tolerated dose. Therefore the safety studies designed to support the traditional IND are designed to extensively characterize the in vivo safety profile of the drug, establishing the expected safety margin between the proposed therapeutic dose and the initial signs of drug toxicity. The first-in-human microdose study seeks only to determine human PK parameters at a single, very low dose that by design is subtherapeutic and is relatively safe for the subject. As such, the industry has sought permission from the regulatory agencies to submit abbreviated toxicology profiles for the compounds they propose to microdose into humans. The EMEA was the first regulatory agency to formally respond with the publication of a position paper on nonclinical safety studies to support clinical trials with a single microdose (EMEA, 2003). In this paper the EMEA defines an abbreviated toxicology package when the microdose meets the agency's definition that the dose is less than 1/100th of the dose calculated to yield a pharmacological effect

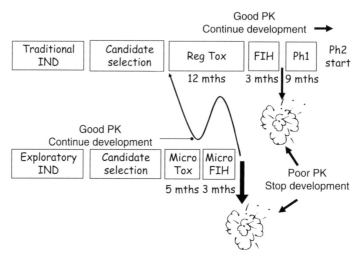

Figure 2.10. *The microdosing strategy for candidate selection offers the greatest benefit when a stop-development decision is made. This is because the exploratory IND does not forego the submission of a traditional IND to continue testing in the clinic. (IND, Investigational new drug; mths, months; FIH, First-in-human clinical study; Ph1, Phase 1 clinical studies; Ph2, Phase 2 clinical studies.)*

and that the dose is less than 100 μg. In January 2006, the FDA followed by the issuance of its guidance for industry, investigators, and reviewers on exploratory IND studies (FDA, 2006). In this document, an exploratory study is defined as a clinical trial that occurs very early in phase 1, involves very limited human exposure, and has no therapeutic intent (i.e., microdosing). Interestingly the FDA does not appear to consider its guidance for an exploratory IND a new position. Rather this document appears to be a clarification to industry that the existing regulations allow great flexibility and the data submitted in an IND application need only support the goals of the study. Indeed, it is stated in the document that because microdose studies present fewer potential risks than do traditional phase I studies, such limited exploratory IND investigations in humans can be initiated with less or different preclinical support that is required for traditional IND studies. The guidance recommends that the sponsor should demonstrate a large multiple (e.g., 100 times) of the proposed human dose does not induce adverse effects in the experimental animals. It is further recommended that scaling from animals to humans based on body surface area can be used to select the dose for the clinical trial.

Given the sensitivity of AMS, the financial advantages of the microdose strategy, and the regulatory approval to proceed with the strategy, why are not all pharmaceutical candidates microdosed? The primary reason is regulatory in nature and is described in Fig. 2.10. A microdose study in humans only provides the sponsor with human PK information. If the compound demonstrates acceptable human PK, the sponsor will decide to continue development of the compound and now must file the traditional IND to define the safety and tolerability of the drug. Therefore a microdose that demonstrates good human PK actually adds to the development time of the compound, 8 months to complete the exploratory IND submission and study followed by an additional 12–15 months to complete the traditional IND submission and study. The microdose strategy is more advantageous when poor PK is observed and the microdose study results terminate further development of the compound. This is because with the exploratory IND route poor PK is observed and development is stopped after 8 months and using approximately 200 g of compound whereas through the traditional IND route the decision would be made after 12–15 months and 1–2 kg of compound. In conclusion, if animal models and prior work with a class of compounds afford the ability to reasonably predict human PK, the traditional IND route is likely the most economical. The microdose strategy may be most advantageous when PK predictions from animal models are known to be dubious and when there are multiple candidates that are not differentiated preclinically. In this scenario microdosing of all candidates enables the use of actual human PK to select a lead and to stop development decisions for the remaining candidates with less desirable pharmacokinetics.

The other obstacle to the widespread use of the microdosing strategy is the cost and availability of AMS. As of 2008, the cost of an accelerator mass spectrometer was in the range of \$1–\$3 million. As a result, several commercial ventures have invested in AMS technology and few companies have made the investment to analyze samples in-house. It can be expected that if the microdosing strategy raises the demand for the technique, instrument and analysis costs will decrease as they have for the analysis of samples by triple-quadrupole MS (although certainly not to the levels of the latter). The introduction of instruments that fragment the carbon hydrides by multiple submegavolt collisions has already reduced the size and cost of the accelerator mass spectrometer (Hughey et al., 1997).

2.4.6 Supporting Microdosing Strategy with Triple-Quadrupole Mass Spectrometry

Although the human microdosing strategy evolved from the capabilities of AMS, the strategy can be supported using other instrumentation. For example, if the ionization efficiency of the compound is high and the extraction procedure efficient, detection limits on the order of 10 pg/mL can be achieved with triple-quadrupole MS. The support of microdosing studies with conventional mass spectrometers offers the advantages that they are available in-house, they do not require synthesis of a ^{14}C-labeled compound for detection, and the sample preparation is less complex leading to faster turnaround times. Another advantage is that the triple-quadrupole mass spectrometer can be interfaced directly to HPLC for the separation and quantitation of known metabolites. The necessity to convert the AMS sample to graphite because AMS detects ^{14}C ions and not molecular ions result in the loss of selectivity with respect to the source of the ^{14}C signal. Therefore HPLC fractionation of the sample and conversion of the fractions to graphite prior to analysis are required if AMS is to differentiate the PK of the parent from the combined PK of the parent and metabolites. Support of microdosing in animals to establish linear kinetics over the microdose through the therapeutic dose has been achieved using triple-quadrupole LC–MS/MS (Buchholz et al., 2005; Balani et al., 2006).

The exploratory IND guidance does not preclude the sponsor from supporting the microdosing strategy using triple-quadrupole MS. We have demonstrated that if there is a favorable combination of the predicted therapeutic dose, the scaled-human no-effect level, and if the ionization and extraction of the compound is sufficiently efficient, plasma concentrations from a human microdose can be determined using triple-quadrupole MS (Holliman et al., 2005; McFadden et al., 2005). However, support of the microdose strategy with LC–MS/MS is not without its caveats. While it is relatively straightforward to determine if the mass spectrometer will have the inherent sensitivity to support the microdose study using neat solutions of the analyte, it is far more difficult to achieve that level of detection after the analyte has been extracted from the biomatrix. We have found that the most difficult assay parameter to achieve is selectivity. This is because the drug must be extracted and concentrated from a relatively large volume of the biomatrix to achieve the very low limits of detection needed. Endogenous matrix species or polymers from labware that do not typically present an issue when the detection limit is in the nanogram-per-milliliter range are coconcentrated along with the analyte and can create chromatographic interferences or ionization suppression. The synthesis of stable-labeled ISs can offset ionization suppression but will not reduce the rigorous extraction and chromatographic development that are needed to isolate the compound. Given the great subject-to-subject variability in the composition of human plasma, analysts will not be certain that the assay has achieved adequate selectivity until the actual microdose samples are analyzed.

Although the lower limit of quantitation is established during assay validation and prior to microdosing, assay sensitivity remains an uncertainty until the actual analysis of the microdose samples as well. There is always the danger that plasma exposures from the microdose are lower than predicted and as a result plasma concentrations from some or all of the time points cannot be detected by the LC–MS/MS method. Reduction of this risk is achieved by collaborative communication between the bioanalytical chemist and the project team. Conservative estimates on bioavailability and clearance can be used to establish the necessary limit of detection needed to determine plasma concentrations for all time points. Updates on the progress of the assay development allow the team to decide if the achievable limit of detection will enable the determination of plasma concentrations from enough time points to make a go–no go decision. Of course, sensitivity is not an issue with AMS, which practically ensures that plasma concentrations will be determined, possibly for several days, enabling the observation of complex PK and clearance from deep compartments.

ACKNOWLEDGMENTS

The authors would like to thank Cho-Ming Loi and Scott Fountain for helpful review and discussion and Colin Garner who provided figures for this chapter.

REFERENCES

Aarons, L. (1999). Software for population pharmacokinetics and pharmacodynamics. *Clin. Pharmacokinet.* **36:**255–264.

Balani, S. K., Nagaraja, N. V., Qian, M. G., Costa, A. O., Daniels, J. S., Yang, H., Shimoga, P. R., Wu, J. T., Gan, L. S., Lee, F. W., and Miwa, G. T. (2006). Evaluation of microdosing to assess PK linearity in rats using liquid chromatography–tandem mass spectrometry. *Drug Metab. Dispos.* **34:**384–388.

Beierle, I., Meibohm, B., and Derendorf, H. (1999). Gender differences in pharmacokinetics and pharmacodynamics. *Int. J. Clin. Pharmacol. Ther.* **37:**529–547.

Bertino, J. S., Jr., Greenberg, H. E., and Reed, M. D. (2007). American College of Clinical Pharmacology position statement on the use of microdosing in the drug development process. *J. Clin. Pharmacol.* **47:**418–422.

Bonfiglio, R., King, R. C., Olah, T. V., and Merkle, K. (1999). The effects of sample preparation methods on the variability of the electrospray ionization response for model drug compounds. *Rapid Commun. Mass Spectrom.* **13:**1175–1185.

Brown, K., Dingley, K. H., and Turteltaub, K. W. (2005). Accelerator mass spectrometry for biomedical research. In *Methods in Enzymology—Biological Mass Spectrometry* (Burlingame, A. L., Ed.). Academic Press, New York, pp. 423–443.

Buchholz, L. M., McFadden, J. R., Ware, J. A., McKenzie, D. L., Weller, D. L., Holliman, C. L., and Cohen, L.H. (2005). LC-MS-MS method development in support of subtherapeutic preclinical pharmacokinetics. In *Proceedings of the 53rd ASMS Conference on Mass Spectrometry and Allied Topics*, San Antonio, TX.

Callaghan, J. T., Bergstrom, R. F., Ptak, L. R., and Beasley, C. M. (1999). Olanzapine. Pharmacokinetic and pharmacodynamic profile. *Clin. Pharmacokinet.* **37:**177–193.

Cohen, L. H., and Rossi, D. T. (2001). The LC/MS experiment. In: *Mass Spectrometry in Drug Discovery* (Rossi, D. T., Ed.). Marcel Dekker, New York.

Danhof, M., Mandema, J. W., Hoogerkamp, A., and Mathot, R. A. (1993). Pharmacokinetic-pharmacodynamic modelling in pre-clinical investigations: Principles and perspectives. *Eur. J. Drug Metab. Pharmacokinet.* **18:**41–47.

Derendorf, H., Hochhaus, G., Meibohm, B., Mollmann, H., and Barth, J. (1998). Pharmacokinetics and pharmacodynamics of inhaled corticosteroids. *J. Allergy Clin. Immunol.* **101:**S440–446.

DiMasi, J. A. (2001). Risks in new drug development: Approval success rates for investigational drugs. *Clin. Pharmacol. Ther.* **69:**297–307.

DiMasi, J. A., Hansen, R. W., and Grabowski, H. G. (2003). The price of innovation: New estimates of drug development costs. *J. Health Econ.* **22:**151–185.

Draisci, R., Marchiafava, C., Palleschi, L., Cammarata, P., and Cavalli, S. (2001). Accelerated solvent extraction and liquid chromatography–tandem mass spectrometry quantitation of corticosteroid residues in bovine liver. *J. Chromatogr. B Biomed. Sci. Appl.* **753:**217–223.

European Medicines Evaluation Agency (EMEA) (2003). Position paper on non-clinical safety studies to support clinical trials with a single microdose. EMEA, London, UK.

Felton, J. S., Turteltaub, K. W., Vogel, J. S., Balhorn, R., Glenhill, B. L., Southon, J. R., Caffee, M. W., Finkel, R. C., Nelson, D. E., Proctor, I. D., and Davis, J. C. (1990). Accelerator mass spectrometry in the biomedical sciences: Applications in low-exposure biomedical and environmental dosimetry. *Nucl. Instrum. Methods Phys. Res. Sect. B* **52:**517–523.

Food and Drug Administration (FDA) (1995). Guidance for industry: Content and format of investigational new drug applications for phase I studies of drugs, including well-characterized, therapeutic, biotechnology-derived products. FDA, Washington, DC.

Food and Drug Administration (FDA) (2000). Nonclinical safety studies for the conduct of human clinical trials for pharmaceuticals, Guideline, I. H. T. FDA, Washington, DC.

Food and Drug Administration (FDA) (2001). Guidance for industry: Bioanalytical method validation, FDA, Washington, DC.

Food and Drug Administration (FDA) (2006). Guidance for industry, investigators, and reviewers: Exploratory IND studies. FDA, Washington, DC.

Gibaldi, M., and Perrier, D. (1982). *Pharmacokinetics*. Marcel Dekker, New York.

Gilman, S. D., Gee, S. J., Hammock, B. D., Vogel, J. S., Haack, K., Buchholz, B. A., Freeman, S. P. H. T., Wester, R. C., Hui, X., and Maibach, H. I. (1998). Analytical performance of accelerator mass spectrometry and liquid scintillation counting for detection of 14C-labeled atrazine metabolites in human urine. *Anal. Chem.* **70:**3463–3469.

Hawthorne, S. B., Yang, Y., and Miller, D. J. (1994). Extraction of organic pollutants from environmental solids with sub- and supercritical water. *Anal. Chem.* **66:**2912–2920.

Holliman, C. L., Buchholz, L. M., McFadden, J. R., and Pace, G. (2005). The prediction of human pharmacokinetics at the therapeutic dose from low sub-therapeutic doses in human. Paper presented at the 11th Annual FDA Science Forum, Washington, DC.

Hosokawa, M., Maki, T., and Satoh, T. (1990). Characterization of molecular species of liver microsomal carboxylesterases of several animal species and humans. *Arch. Biochem. Biophys.* **277:**219–227.

Hsieh, Y., Chintala, M., Mei, H., Agans, J., Brisson, J. M., Ng, K., and Korfmacher, W. A. (2001). Quantitative screening and matrix effect studies of drug discovery compounds in monkey plasma [by] using fast-gradient liquid chromatography/tandem mass spectrometry. *Rapid Commun. Mass Spectrom.* **15:**2481–2487.

Hughey, B. J., Klinkowstein, R. E., Shefer, R. E., Skipper, P. L., Tannenbaum, S. R., and Wishnok, J. S. (1997). Design of a compact 1MV AMS system for biomedical research. *Nucl. Instrum. Methods Phys. Res. Sect. B* **123:**153–158.

Hurst, S., Loi, C. M., Brodfuehrer, J., and El-Kattan, A. (2007). Impact of physiological, physicochemical and biopharmaceutical factors in absorption and metabolism mechanisms on the drug oral bioavailability of rats and humans. *Expert Opin. Drug Metab. Toxicol.* **3:**469–489.

Jessome, L. L., and Volmer, D. A. (2006). Ion suppression: A major concern in mass spectrometry. *LC/GC* **24:**498–510.

Jones, J., Tomlinson, K., and Moore, C. (2002). The simultaneous determination of codeine, morphine, hydrocodone, hydromorphone, 6-acetylmorphine, and oxycodone in hair and oral fluid. *J. Anal. Toxicol.* **26:**171–175.

King, R., Bonfiglio, R., Fernandez-Metzler, C., Miller-Stein, C., and Olah, T. (2000). Mechanistic investigation of ionization suppression in electrospray ionization. *J. Am. Soc. Mass Spectrom.* **11:**942–950.

Korfmacher, W. A. (2005). Bioanalytical assays in a drug discovery environment. In *Using Mass Spectrometry for Drug Metabolism Studies* (Korfmacher, W. A., Ed.). CRC Press, Boca Raton, FL, pp. 1–34.

Lanckmans, K., Sarre, S., Smolders, I., and Michotte, Y. (2007). Use of a structural analogue versus a stable isotope labeled internal standard for the quantification of angiotensin IV in rat brain dialysates using nano-liquid chromatography/tandem mass spectrometry. *Rapid Commun. Mass Spectrom.* **21:**1187–1195.

Lappin, G., and Garner, R. C. (2005). The use of accelerator mass spectrometry to obtain early human ADME/PK data. *Expert Opin. Drug Metab. Toxicol.* **1:**23–31.

Lappin, G., Kuhnz, W., Jochemsen, R., Kneer, J., Chaudhary, A., Oosterhuis, B., Drijfhout, W. J., Rowland, M., and Garner, R. C. (2006a). Use of microdosing to predict pharmacokinetics at the therapeutic dose: Experience with 5 drugs. *Clin. Pharmacol. Ther.* **80:**203–215.

Lappin, G., Rowland, M., and Garner, R. C. (2006b). The use of isotopes in the determination of absolute bioavailability of drugs in humans. *Expert Opin. Drug Metab. Toxicol.* **2:**419–427.

Larger, P. J., Breda, M., Fraier, D., Hughes, H., and James, C. A. (2005). Ion-suppression effects in liquid chromatography–tandem mass spectrometry due to a formulation agent, a case study in drug discovery bioanalysis. *J. Pharm. Biomed. Anal.* **39:**206–216.

LeBlanc, G. (1999). Current trends and developments in sample preparation. *LC-GC* **17:**S30–S37.

Lees, P., Cunningham, F. M., and Elliott, J. (2004). Principles of pharmacodynamics and their applications in veterinary pharmacology. *J. Vet. Pharmacol. Ther.* **27:**397–414.

Lin, J., Sahakian, D. C., de Morais, S. M., Xu, J. J., Polzer, R. J., and Winter, S. M. (2003). The role of absorption, distribution, metabolism, excretion and toxicity in drug discovery. *Curr. Topics Med. Chem.* **3:**1125–1154.

MacGowan, A. P. (2001). Role of Pharmacokinetics and pharmacodynamics: Does the dose matter? *Clin. Infect. Dis.* **33(Suppl. 3):**S238–239.

McFadden, J. R., Holliman, C. L., and Buchholz, L. M. (2005). LC-MS/MS support of a sub-pharmacologic human dosing study using triple-quadrupole mass spectrometry. In: *Proceedings of the 53rd ASMS Conference on Mass Spectrometry and Allied Topics*, San Antonio, TX.

Meibohm, B., and Derendorf, H. (1997). Basic concepts of pharmacokinetic/pharmacodynamic (PK/PD) modelling. *Int. J. Clin. Pharmacol. Ther.* **35:**401–413.

Meirav, O., Sutton, R. A. L., Fink, D., Middleton, R., Klein, J., Walker, V. R., Halabe, A., Vetterli, D., and Johnson, R. R. (1990). Application of accelerator mass spectrometry in aluminum metabolism studies. *Nucl. Instrum. Methods Phys. Res. Sect. B* **52:**536–539.

Miners, J. O., Smith, P. A., Sorich, M. J., McKinnon, R. A., and Mackenzie, P. I. (2004). Predicting human drug glucuronidation parameters: Application of in vitro and in silico modeling approaches. *Annu. Rev. Pharmacol. Toxicol.* **44:**1–25.

Obach, R. S., Baxter, J. G., Liston, T. E., Silber, B. M., Jones, B. C., MacIntyre, F., Rance, D. J., and Wastall, P. (1997). The prediction of human pharmacokinetic parameters from preclinical and in vitro metabolism data. *J. Pharmacol. Exp. Ther.* **283:**46–58.

Poggesi, I. (2004). Predicting human pharmacokinetics from preclinical data. *Curr. Opin. Drug. Discov. Devel.* **7:**100–111.

Posyniak, A., Zmudzki, J., and Semeniuk, S. (2001). Effects of the matrix and sample preparation on the determination of fluoroquinolone residues in animal tissues. *J. Chromatogr. A* **914:**89–94.

Raunio, H., Taavitsainen, P., Honkakoski, P., Juvonen, R., and Pelkonen, O. (2004). In vitro methods in the prediction of kinetics of drugs: Focus on drug metabolism. *Altern. Lab. Anim.* **32:**425–430.

Richter, B. E. (1999). Current trends and developments in sample preparation. *LC-GC* **17:**S22–S28.

Ritschel, W. A., Vachharajani, N. N., Johnson, R. D., and Hussain, A. S. (1992). The allometric approach for interspecies scaling of pharmacokinetic parameters. *Comp. Biochem. Physiol. C* **103:**249–253.

Rowland, M., and Tozer, T. (1995). *Clinical Pharmacokinetics.* Lippincott Williams & Wilkins, Philadelphia.

Sandhu, P., Vogel, J. S., Rose, M. J., Ubick, E. A., Brunner, J. E., Wallace, M. A., Adelsberger, J. K., Baker, M. P., Henderson, P. T., Pearson, P. G., and Baillie, T. A. (2004). Evaluation of microdosing strategies for studies in preclinical drug development: Demonstration of linear pharmacokinetics in dogs of a nucleoside analog over a 50-fold dose range. *Drug Metab. Dispos.* **32:**1254–1259.

Sarapa, N., Hsyu, P. H., Lappin, G., and Garner, R. C. (2005). The application of accelerator mass spectrometry to absolute bioavailability studies in humans: Simultaneous administration of an intravenous microdose of 14C-nelfinavir mesylate solution and oral nelfinavir to healthy volunteers. *J. Clin. Pharmacol.* **45:**1198–1205.

Schentag, J. J. (1999). Antimicrobial action and pharmacokinetics/pharmacodynamics: The use of AUIC to improve efficacy and avoid resistance. *J. Chemother.* **11:**426–439.

Shah, V. P., Midha, K. K., Findlay, J. W., Hill, H. M., Hulse, J. D., McGilveray, I. J., McKay, G., Miller, K. J., Patnaik, R. N., Powell, M. L., Tonelli, A., Viswanathan, C. T., and Yacobi, A. (2000). Bioanalytical method validation—A revisit with a decade of progress. *Pharm. Res.* **17:**1551–1557.

Shah, V. P., Yacobi, A., Barr, W. H., Benet, L. Z., Breimer, D., Dobrinska, M. R., Endrenyi, L., Fairweather, W., Gillespie, W., Gonzalez, M. A., Hooper, J., Jackson, A., Lesko, L. J., Midha, K. K., Noonan, P. K., Patnaik, R., and Williams, R. L. (1996). Evaluation of orally administered highly variable drugs and drug formulations. *Pharm. Res.* **13:**1590–1594.

Steimer, J. L., Ebelin, M. E., and Van Bree, J. (1993). Pharmacokinetic and pharmacodynamic data and models in clinical trials. *Eur. J. Drug Metab. Pharmacokinet.* **18:**61–76.

Thomas, V. H., Bhattachar, S., Hitchingham, L., Zocharski, P., Naath, M., Surendran, N., Stoner, C. L., and El-Kattan, A. (2006). The road map to oral bioavailability: An industrial perspective. *Expert Opin. Drug Metab. Toxicol.* **2:**591–608.

Tong, X. S., Wang, J., Zheng, S., Pivnichny, J. V., Griffin, P. R., Shen, X., Donnelly, M., Vakerich, K., Nunes, C., and Fenyk-Melody, J. (2002). Effect of signal interference from dosing excipients on pharmacokinetic screening of drug candidates by liquid chromatography/mass spectrometry. *Anal. Chem.* **74:**6305–6313.

Turteltaub, K. W., Felton, J. S., Gledhill, B. L., Vogel, J. S., Southon, J. R., Caffee, M. W., Finkel, R. C., Nelson, D. E., Proctor, I. D., and Davis, J. C. (1990). Accelerator mass spectrometry in biomedical dosimetry: Relationship between low-level exposure and covalent binding of heterocyclic amine carcinogens to DNA. *Proc. Natl. Acad. Sci. USA* **87:**5288–5292.

van De Waterbeemd, H., Smith, D. A., Beaumont, K., and Walker, D. K. (2001). Property-based design: Optimization of drug absorption and pharmacokinetics. *J. Med. Chem.* **44:**1313–1333.

Vogel, J. S., and Love, A. H. (2005). Quantitating isotopic molecular labels with accelerator mass spectrometry. In: *Methods in Enzymology—Biological Mass Spectrometry* (Burlingame, A. L., Ed.). Academic Press, New York, pp. 402–422.

Walter, P. S., Chalk, S., and Kingston, H. M. (1997). Overview of microwave-assisted sample preparation. In: *Microwave Enhanced Chemistry* (Kingston, H. M., and Haswell, E. J., Eds.). American Chemical Society, Washington, DC, pp. 55–222.

Wang, S., Cyronak, M., and Yang, E. (2007). Does a stable isotopically labeled internal standard always correct analyte response? A matrix effect study on a LC/MS/MS method for the determination of carvedilol enantiomers in human plasma. *J. Pharm. Biomed. Anal.* **43:**701–707.

Wang, S., Mei, H., Ng, K., Workowski, K., Astle, T., and Korfmacher, W. (2002). Development of an automated homogenizer-autogizer and its application in brain uptake studies. In: *Proceedings of the 50th ASMS Conference on Mass Spectrometry and Allied Topics*, Orlando, FL.

Weaver, R., and Riley, R. J. (2006). Identification and reduction of ion suppression effects on pharmacokinetic parameters by polyethylene glycol 400. *Rapid Commun. Mass Spectrom.* **20:**2559–2564.

Wells, T. G., Mortensen, M. E., Dietrich, A., Walson, P. D., Blasier, D., and Kearns, G. L. (1994). Comparison of the pharmacokinetics of naproxen tablets and suspension in children. *J. Clin. Pharmacol.* **34:**30–33.

Yu, C., and Cohen, L. H. (2003). Tissue sample preparation—Not the same old grind. *LC/GC* **21:**1038–1048.

Yu, C., Penn, L. D., Hollembaek, J., Li, W., and Cohen, L. H. (2004). Enzymatic tissue digestion as an alternative sample preparation approach for quantitative analysis using liquid chromatography–tandem mass spectrometry. *Anal. Chem.* **76:**1761–1767.

3

Quadrupole, Triple-Quadrupole, and Hybrid Linear Ion Trap Mass Spectrometers for Metabolite Analysis

Elliott B. Jones

Applied Biosystems, Foster City, California

Mass Spectrometry in Drug Metabolism and Pharmacokinetics. Edited by Ragu Ramanathan
Copyright © 2009 John Wiley & Sons, Inc.

3.1   INTRODUCTION

Metabolism studies are an important portion of the drug discovery process. Because of the high cost of bringing a drug to market, drug metabolism studies are being performed at earlier stages of the drug development process. In addition, throughput is a concern, as an increasing amount of information is desired in progressively shorter time frames. Because of its inherent sensitivity and ability to obtain structural information, liquid chromatography–mass spectrometry (LC-MS) has become an invaluable tool in drug metabolism studies (Fernandez-Metzler et al., 1999; Clarke et al., 2001; Hop and Prakash, 2005; Ma et al., 2006). Both triple-quadrupole and ion trap mass spectrometers have proven to be excellent tools for detection and confirmation of metabolite species in both in vivo and in vitro systems. The sensitivity and specificity of triple-quadrupole precursor ion, constant-neutral-loss, and multiple-reaction monitoring scans have been commonly used for metabolite profiling and characterization. However, triple-quadrupole tandem mass spectrometry (MS/MS) scans suffer from a lack of sensitivity, thus making it difficult to obtain structural information from low-level drug-derived components. The main benefit of an ion trap mass spectrometer lies in its ability to obtain sensitive full-scan and MS/MS spectra, which makes this instrument ideally suited for metabolite identification and structural elucidation.

A new generation of linear ion trap mass spectrometers has been developed and exhibits increased performance compared to traditional three-dimensional (3D) ion traps (Hopfgartner et al., 2003; Douglas et al., 2005). A further evolution of the triple-quadrupole family and ion trap class of instruments is the production of the hybrid triple-quadrupole/linear ion trap (QQQ/LIT) platform. Hybrid instruments of this nature allow for operation in space and not just in time when performing MS/MS analysis. This feature allows for increased performance compared to classical ion traps. A powerful combination possible on a hybrid LIT/QQQ instrument is the ability to use highly sensitive and selective precursor ion, constant neutral loss, and multi-MRM as a survey scan for dependent LIT MS/MS. Compared to a simple MS experiment, these comprehensive triple-quadrupole and LIT modes can be more complex to setup.

In this chapter, QQQ/LIT theory, operation, and method setup are discussed. For the reader's convenience, current and up-to-date references are included, but this chapter is not meant to represent a comprehensive review of metabolism analysis by LC–MS.

3.2   QUADRUPOLE AND TRIPLE-QUADRUPOLE THEORY

A quadrupole, used as part of a mass spectrometer, can be found in either a single- or triple-element linear configuration. Single-quadrupole instruments are of limited use for metabolite identification because of relatively poor sensitivity, only moderate resolving power, and the inability to perform MS/MS experiments for structural information. The development of the triple-quadrupole mass spectrometer greatly improved the selectivity and the signal-to-noise ratio (S/N), and has made triple-quadrupole systems the mass spectrometer of choice for many challenging analytical situations in the bioanalytical world (Yost and Enke, 1978, 1979). In this configuration, a set of three quadrupoles (two quadrupole mass filters separated by a quadrupole collision cell) is arranged in series. In some instruments, a radio-frequency (RF)–only focusing quadrupole is used as an ion-focusing element prior to the first mass-filtering quadrupole. This focusing quadrupole helps keep the ions radially concentrated after initial free jet expansion just after the interface. The positive ions then move based on increasingly more negative potential to the first mass-resolving quadrupole. This element can run in either a RF-only mode (radial focusing) or in a RF/direct current (DC) mode, which allows for mass filtering.

The unique ability of the RF/DC field in a quadrupole is that at a specific DC amplitude a stable transmission of a narrow window of mass-to-charge (m/z) through the quadrupole is possible. This transmission is generally set at about unit resolution (0.7 at full width at half maximum, FWHM) for most commercial mass spectrometers. All ions with a higher or lower m/z will become destabilized radially during the transit through the rods. The RF and DC amplitudes are calibrated versus m/z to allow for various experiments. This basic design illustrates a quadrupole's inherent strength and weakness as a mass analyzer. In the case of an experiment in which a continuous stream of ions of a specific m/z is necessary, the quadrupole has a very high duty cycle (efficiency) of transmitting ions to a detector. However, in the case of an experiment in which a mass scan is required, such as a single MS scan or product ion scan, the duty cycle is much lower. Because a quadrupole must sequentially sweep over a series of quantized field steps to produce a scan, many ions are lost during this sequential process. Figure 3.1 represents RF/DC operation of a single-quadrupole element acting in a mass-resolving mode.

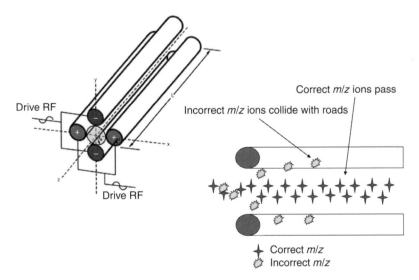

Figure 3.1. *Schematic of RF/DC operation of a single quadrupole.*

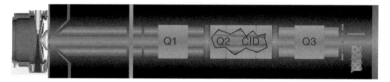

Figure 3.2. Schematic of ion path for hybrid QQQ/LIT (QTRAP) instrument. The ion path has three quadrupoles (Q1, Q2, and Q3) like a standard triple-quadrupole instrument, but Q3 also has ion trap capabilities.

A schematic of a triple-quadrupole ion rail is shown in Fig. 3.2. Ions are generated in the atmospheric pressure ion source and transmitted through the orifice and skimmer into the focusing quadrupole. The focusing quadrupole, which is conventionally called Q0, acts as a radial focusing element by using a RF-only field in the presence of a moderate pressure of gas (Douglas and French, 1992). From Q0, radially focused ions enter the first quadrupole (Q1), which has the ability to operate in a RF transmission mode or in a RF/DC mass-resolving mode. If Q1 is set to transmit a single m/z, these ions are then accelerated into Q2 (high-pressure collision cell). The laboratory energy of the ion, represented as electron volts, and the collisions, with the neutral gas atoms or molecules in the collision cell, induce fragmentation of the precursor ion.

The most critical tuning values on a triple-quadrupole and triple-quadrupole/ hybrid linear ion trap are the declustering potential (DP) and collision energy (CE). A DP of 40 represents a voltage gradient of 40 V from the orifice plate to the ground skimmer. The higher the DP, the more an ion is accelerated through the interface. The ions will collide with other gas-phase molecules present in this region, and at high DP values, these collisions can produce fragmentation in much the same way as the collision cell. A low DP value might avoid excess fragmentation; however, efficient declustering of the analyte ions from solvent and buffer may not be achieved, resulting in less than optimal sensitivity. The CE governs the ion energy of an ion entering the collision cell. Higher CE values impart more energy to the ion, which fragments when the analyte hits the high pressure of collision activated dissociation (CAD) gas in the cell. The higher the value of CE, the greater the amount of secondary fragmentation and lower m/z fragments are produced. The high-energy fragmentation of the collision cell gives a much richer fragment pattern than the low-energy processes typical of a 3D ion trap. A comparison of 3D ion trap and triple-quadrupole MS/MS spectra for dextromethorphan is shown in Fig. 3.3.

Several scan modes are unique to the triple-quadrupole instrument, and most of these modes are superior in duty cycle versus an ion trap, Fourier transform (FT), or time-of-flight (TOF) mass spectrometers. Different elements of the triple-quadrupole perform different operations for each scan mode. These scan modes, each of which will be described in detail, are single-reaction monitoring (SRM) or multiple-reaction monitoring (MRM), precursor ion scanning (PIS), and constant-neutral-loss scanning (NLS). These scan modes and applications for structural elucidation have been described in detail (Yost and Enke, 1978, 1979).

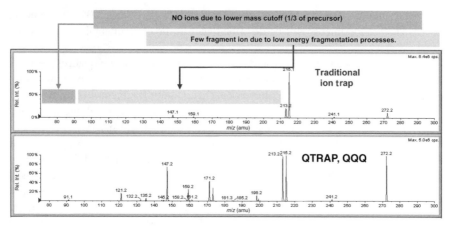

Figure 3.3. *Comparison of 3D ion trap (top) and triple-quadrupole/QTRAP (bottom) MS/MS spectra of dextromethorphan. The triple-quadrupole fragmentation occurs in the Q2 collision cell, which is also used in the hybrid QTRAP instrument, and yields much more fragmentation due to the higher energy collisions.*

In the SRM/MRM mode, Q1 and Q3 operate in the mass-filtering mode. The precursor ion is selected in Q1, the ions travel through the collision cell with a CE optimized for fragmentation, and a specific fragment, usually the most intense, is selected in Q3. The utility of this mode is that an ion can be transmitted to the detector with a very high efficiency/sensitivity and selectivity through the MS/MS process. Because an ion must have both a certain precursor mass as well as fragment, this mode allows for excellent selectivity and hence removal of interfering species. This experiment is the premier method of high-sensitivity quantitation of small molecules in a complex matrix (Chapter 2). A schematic of an MRM scan is shown in Fig. 3.4. The SRM/MRM mode can be used for quantitation of parent and metabolites over a range of relevant time points (Chapter 2) as well as a

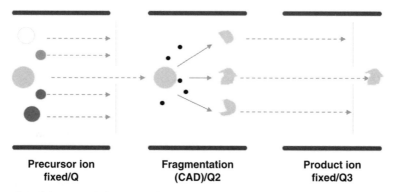

Figure 3.4. *Schematic of MRM scan. Q1 is set to pass precursor ion (s) which are fragmented in Q2. However, the entire product ion mass spectrum of the selected precursor ion is not obtained, rather only a selected product ion is set to pass through Q3 and hit the detector.*

method of qualitative survey for unknown metabolites (Li et al., 2005; Zhang et al., 2005). These experiments will be described later in the experimental design section of this chapter.

Precursor ion scanning (PIS) is another scan unique to triple-quadrupole mass spectrometers (at least on a LC time scale) and is used to find structurally similar compounds by identifying all species that produce a common fragment. In this mode, Q3 is fixed on a selected fragment ion while Q1 is operated in the resolving mode and performs a mass scan through a range using a user-defined step size (m/z increment). For each m/z step, the ions generated are passed to the collision cell using a defined CE and fragmented. If the selected fragment is present, that is, passed through Q3 to the detector, a signal is recorded at the m/z for the precursor ion that passed through Q1. The majority of the background ions will not produce a fragment at the same m/z as selected for Q3. As a result, precursor ion scans are very useful at finding structurally similar compounds, even when present at low levels in biological matrices. Figure 3.5 depicts ion selection for a precursor ion scan experiment.

A useful and complementary scan to precursor ion scan is a constant-neutral-loss scan (NLS). As the name implies, this scan detects the loss of a neutral fragment. It is important to note that not all fragments are charged, and without a charge, little can be done directly with those species in the MS. When a singly charged ion fragments, often both charged and neutral fragments are produced. If this neutral species is common to a number of metabolites, then using a NLS scan to look for these structurally related compounds can be selective and sensitive in the same manner as the precursor ion scan. In a neutral loss scan, Q1 and Q3 both step through a mass range, but at a fixed m/z difference corresponding to the neutral fragment. A signal is registered at the detector only when an ion loses a fragment of the specified neutral mass. In essence, a NLS can be thought of as an indirect PIS, as the neutral is being detected via the mass difference between Q1 and Q3. Figure 3.6 depicts the triple quadrupole operating in this mode. For example, if Q1 is set to

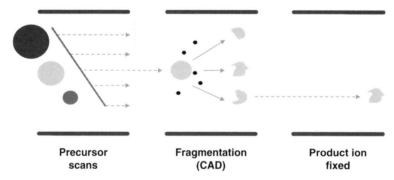

| Precursor
scans | Fragmentation
(CAD) | Product ion
fixed |

Figure 3.5. *Schematic of precursor ion scan (PIS). Q3 is set to pass a fixed product ion and Q1 is set to scan through a desired mass range and all the ions within that desired mass range are fragmented in the collision cell (Q2). Thus, a precursor ion mass spectrum shows all the precursor ions that fragmented to form the product ion set to pass through Q3.*

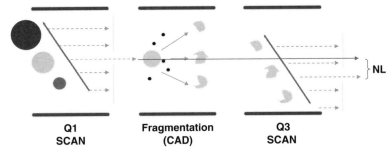

Figure 3.6. *Schematic of constant neutral loss scan (NLS). Q1 and Q3 scan through a defined mass range, but Q3 is set to scan a fixed mass difference below Q1. Only compounds that generate the specified neutral will hit the detector and register a signal.*

pass an ion at an m/z of 400 and a NLS of 100 is set in the experiment, Q3 would then be set to transmit any fragment created at an m/z 300 from the collision cell. If an ion hits the detector, a certain counts per second would then be recorded at the m/z of the Q1 increment. The instrument would then proceed to step both Q1 and Q3 up in mass at the same rate and increment, maintaining the offset of the neutral, that is, Q1 set at m/z 401 and Q3 set at m/z 301 for the previous example. An example of a NLS experiment as it relates to metabolism studies is in the case of glucuronidation. The observed species would be drug compound $+176$. Upon fragmentation, the metabolite loses 176 as a neutral, but not as an m/z 176 ion. Therefore, to detect glucuronides, a neutral loss scan of 176 can be performed.

All the previously described, unique triple-quadrupole scans—SRM/MRM, precursor ion, and neutral loss—filter out a significant amount of background noise, allowing for detection of species present at very low levels.

The collision cell is an integral part of the previously discussed scan types. Due to improvements in technology and understanding of ion motion, collision cells in modern instruments are enclosed, rather than open. This enclosed cell allows the collision cells to operate at higher pressures, thus improving fragmentation efficiency. However, one drawback of high-pressure collision cells is slower transit times for ions as they move through this region. Without modification, greater than 20 ms would be required for fragment ions to be created and exit the cell. With transit times of this magnitude, high-sensitivity MRM, PIS, and NLS data can be difficult or impossible to obtain on a LC time scale. Older QQQ instruments often have very poor resolution and sensitivity in precursor ion and NLS modes. The most effective way to overcome this limitation is to use an axial field in the collision cell to move ions along the axis. The LINAC (linear accelerator) collision cell provides this acceleration, resulting in transit times of 2 ms or less. Reducing the transit time through the collision cell greatly improves the quality of PIS and NLS data. Furthermore, the LINAC cell allows analysis of multiple analytes (multiple MRMs) in a single experiment, as short dwell times can be utilized, and the instrument can cycle through all transitions on an LC time scale.

When using PIS or NLS, it is important to set the appropriate scan speed to acquire the best quality data. A PIS is little more than hundreds of MRMs performed by increasing Q1 mass by regular steps. Scanning from 100 to 200 amu with a step of 0.2 amu yields the following number of steps:

$$\frac{\text{High mass} - \text{low mass}}{\text{Step size}} = \text{Total number of steps during a scan}$$

In this case $200 - 100/0.2 = 500$ steps or 500 MRMs with Q1 setting starting at 100 and increasing to 100.2, 100.4, and so on, up to 200 with a fixed Q3 for each step. The total scan time can be calculated by multiplying the number of steps by the ion transit time. If the fastest time that a collision cell can expel ions is 2 ms (in the case of a LINAC or pulsing cell), then use of a dwell time per step of less than that limit would cause a reduction in sensitivity as well as a reproducible mass shift of about $+0.2$ amu. The reason for the mass shift is that if a mass spectrometer is set to scan faster than 2 ms dwell, by the time the ions reach the detector, Q1 has already moved on to the next step, that is, increased by 0.2 amu. Because this shift is very reproducible and mass accuracy of a few tenths is sufficient for metabolism analysis, this mass shift is not a huge problem. In many cases, software can also correct for this mass shift. However, use of dwell times greater than the transit time is usually optimal for sensitivity and quality data. Because these values can vary, it is best to try different scan times and dwell/step combinations for a desired sensitivity and mass accuracy. This same rule also applies to NLS scanning as well as large sets of MRMs.

3.3 ION TRAP AND LINEAR ION TRAP THEORY

As mentioned in the previous section, triple-quadrupole instruments are very good at finding low levels and structurally related compounds in the presence of biological matrices as well as being the gold standard technique for quantitation. Ion trap mass spectrometers, on the other hand, have the capabilities to obtain high-sensitivity full-scan MS and MS/MS spectra; therefore, they are widely used for qualitative analysis, such as structural elucidation and unknown identification. For complete metabolite identification, it is important to have both the sensitivity and selectivity of triple-quadrupole instruments and the full-scan data quality of ion traps.

The first commercial ion trap mass spectrometers were introduced in the early 1980s. Since the introduction, ion traps have become one of the most popular types of mass analyzers due to sensitivity, MS^n capabilities, relatively inexpensive price, and compact size. The details of ion trap theory are covered in a number of references (Nourse and Cooks, 1990; March, 1997, 1998, 2000b) and only the general details will be given here.

Traditional 3D quadrupole ion traps utilize three electrodes: two end caps and a ring electrode. The end caps are usually at ground potential and a RF voltage is applied to the ring electrode to generate a quadrupole field to store ions. Helium gas is present in the trap at a pressure of ~ 1 mtorr to cool the ions and improve trapping efficiency (Stafford, 2002). The ions are mass selectively ejected by scanning the amplitude of the RF voltage applied to the ring electrode, causing ions of increasing m/z to be unstable and exit the trap for detection (March, 2000a). Because ions of a mass range can be collected in the trap, high-sensitivity full-scan spectra can be acquired with minimal ion loss compared to other mass analyzers.

Because a 3D ion trap consists of a single element, MS/MS is performed in time. The ion trap must (1) collect the ions, (2) isolate the precursor ion of interest by ejecting all nonessential ions, (3) perform resonant excitation on the precursor ion to cause fragmentation, and (4) sequentially eject the (fragment) ions for detection. Steps 2–4 are repeated for MS^n, though in step 4 all ions are ejected except the one that will be further fragmented. Resonant excitation is performed by applying a small voltage (few hundred millivolts) across the end caps. Upon resonant excitation, the kinetic energy of the ions increases, causing collisions with the helium gas, and when the energy is high enough, fragmentation occurs. One notable limitation of MS/MS performed in an ion trap is the low mass cutoff rule that is inherent to the fundamental operating principle of ion trap mass spectrometers. For typical, commercially available ion trap instruments, ions below one-third of the trapped (precursor) m/z will not be trapped; for example, for a precursor of m/z 300, fragments below about 100 will not be stable in the trap and therefore not be detected. Because these lower mass ions cannot be detected, it is often necessary to perform MS^3 experiments to observe low-mass fragment ions. Although changes can be made to alter the low-mass cutoff, implementation of these changes (such as pulsed-Q) has other trade-offs and limitations.

In recent years, linear (or 2D) ion trap mass spectrometers have been developed to improve detection limits over the traditional ion traps (Hager, 2002; Schwartz et al., 2002). In addition, LITs have greater ion storage capacities that increase the number of ions that can be trapped, and hence detected, without space charge effects.

Because metabolite detection and identification are often performed in biological matrices, low-level metabolites can be difficult to detect when using an ion trap mass spectrometer. The main reason for this limitation is a lack of ability to perform highly selective scans, such as triple-quadrupole PIS and NLS. The importance of having both ion trap functionalities and triple-quadrupole scan functions was demonstrated by splitting the LC flow to two different mass spectrometers—an ion trap and a triple quadrupole (Liu et al., 2002). To achieve the advantages of both instruments in a single box, a hybrid QQQ/LIT instrument has been developed (Hager, 2002, 2003, 2004; Hager and Le Blanc, 2003; Hager and Yves Le Blanc, 2003). This hybrid instrument combines the selective and sensitive triple-quadrupole functions, such as PIS, NLS, and MRM, with the full-scan capabilities of an ion trap, which is very useful for structural identification (Hopfgartner et al., 2004; Hopfgartner and Zell, 2005).

A schematic of the QQQ/LIT ion path is shown in Fig. 3.7. Similar to a typical triple-quadrupole mass spectrometer, three quadrupoles are present in a QQQ/LIT mass spectrometer. The difference between a hybrid instrument and a triple quadrupole is that Q3 can be operated in standard triple-quadrupole mode or used as a LIT. When the LIT is active, an entrance potential barrier and an exit potential barrier are created to axially contain the ions. Along with the exit and entrance barriers, a radial

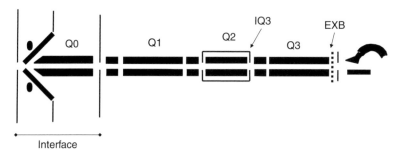

Figure 3.7. *Schematic of ion path for hybrid QQQ/LIT (QTRAP) instrument. The ion path has three quadrupoles (Q1, Q2, and Q3) like a standard triple-quadrupole instrument, but Q3 also has ion trap capabilities.*

RF field, similar to the one used during quadrupole operation, is also used. As its name implies, this field helps to radially contain ions. A diagram depicting Q3 used as an ion trap is shown in Fig. 3.8. All three trapping fields are active from fill to scanning during LIT operation. Only after a scan is complete, the trapping fields are lowered to allow residual ions to be removed before the next scan. To eject the ions for detection, the drive and auxiliary RF voltages are simultaneously ramped with an increase in amplitude, which causes ions of successively increasing m/z to become unstable and be ejected from the end of the trap to the detector. A timing diagram showing activation of the various voltages throughout the trapping and ejection cycle is shown in Fig. 3.9. A hybrid quadrupole LIT is capable of scanning at $250-4000$ amu/s. FWHM resolution is dependent upon scan speed and varies from about 0.3 amu at 250 amu/s to approximately 0.6 amu at 4000 amu/s. Figure 3.10 shows the differences in resolution for reserpine using different scan

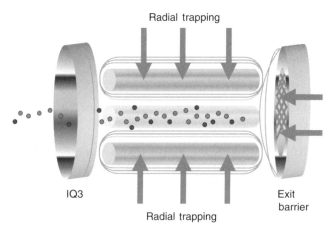

Figure 3.8. *Diagram of Q3 functioning as LIT. Radial trapping is achieved using the standard quadrupole RF and an additional RF. Potentials on IQ3 and the exit barrier grid trap ions in the z axis.*

Ions enter the trap with a wide range of internal energy levels. This cooling time delay is used to allow the ions to reach the the same energy level for improved resolution.

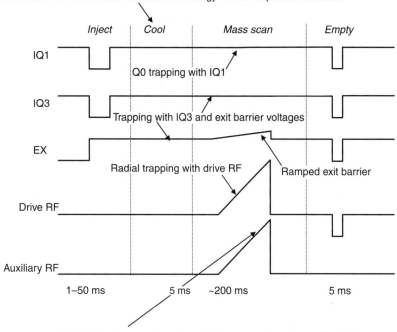

This timing diagram shows ions being scanned out by auxiliary RF.

Figure 3.9. *Timing diagram of QTRAP.*

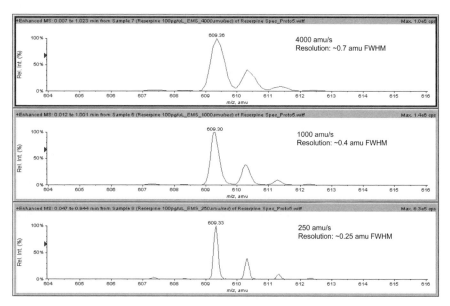

Figure 3.10. *Resolution comparison for reserpine at different LIT scan speeds.*

speeds. Unlike in 3D traps, helium is not present in QQQ/LIT to cool the ions. Instead, residual nitrogen from the collision cell is injected into the ion trap to maintain a pressure of 3×10^{-5}–4.5×10^{-5} torr to cool the ions and also to act as the CAD gas during in-trap fragmentation for MS^3. The instrument can be easily and rapidly switched between triple-quadrupole and LIT scans, in about 5 ms, allowing both LIT and quadrupole scans to be acquired within the same cycle.

In a hybrid QQQ/LIT mass spectrometer, MS/MS is performed in space. Q1 selects the precursor m/z of interest using a quadrupole DC isolation, as in a normal triple-quadrupole or QqTOF product ion scan. This ion is fragmented in Q2 (high-pressure collision cell), and the fragment ions are accumulated in Q3 operating in the LIT mode. As previously mentioned, in a 3D ion trap, each of these steps is performed separately in time. Because precursor ion selection is performed in Q1, higher analyte trapping efficiency can be achieved because the ion trap is filled only with the ions of interest. Also because a LIT can hold many more ions (as much as 70X) compared to a 3D trap, sensitivity can be improved based on increased trap capacity. Another advantage, over traditional ion traps, is that fragmentation in a QQQ/LIT occurs in the collision cell. A collision cell allows for higher energy fragmentation compared to the low-energy process in a standard ion trap. Higher energy fragmentation allows stronger bonds to be broken for better primary fragmentation as well as a secondary fragmentation (fragments of fragments). These secondary fragments are much less common when fragmentation takes place within a trap. A comparison of MS/MS spectra from triple-quadrupole and ion trap mass spectrometers is shown in Fig. 3.3. If higher energies are used in a standard ion trap, ions can be ejected from the trap and lost for detection.

3.4 LINEAR ION TRAP SCAN FUNCTIONS

1. *EMS.* The "enhanced MS" scan is a general MS1 scan that generates a typical mass spectrum covering a defined mass range. In this mode, Q1 and Q2 are operated in the RF-only mode, and there is no collision energy; Q3 acts as an ion trap. The ions pass through the interface, travel through Q1 and Q2, and are then trapped by Q3 and ejected to the detector. An EMS scan is much more sensitive than a Q1 or Q3 single MS scan because the ions can be "concentrated" in the trap.

2. *EPI.* An "enhanced-product-ion" scan is a LIT hybrid product ion scan. In this scan mode, the precursor ion is isolated in Q1, fragmented in Q2, and the fragments are trapped in Q3 with a subsequent trap scan. As previously mentioned, performing MS/MS analyses in this mode has several advantages over performing MS/MS either in the triple-quadrupole mode or using a 3D ion trap. The duty cycle for a triple-quadrupole instrument is too slow to acquire high-quality MS/MS data on an LC time scale, especially if the analyte is present at low concentrations. If a 3D ion trap is used, detecting low-level analytes in the presence of high background/matrix signals is difficult because no isolation occurs prior to trapping ions, resulting in a trap filled mostly with background ions. The duty cycle is also slightly longer versus a QQQ/LIT because isolation and fragmentation occur

in a single element. Finally, traditional ion traps have a low mass cutoff, typically at about one-third the mass of the precursor ion. Figure 3.11 shows the intensity difference between a triple-quadrupole MS/MS spectrum and a LIT spectrum for haloperidol. The intensity of the LIT MS/MS spectrum is about 120 times greater than the intensity for the triple-quadrupole spectrum. Furthermore, the LIT spectrum is acquired in a few hundred milliseconds, versus the few seconds required for the triple-quadrupole spectrum.

3. *ER.* The "enhanced-resolution" scan is used to obtain mass spectra with higher mass-resolving power. The mass resolution is increased by reducing the scan speed and by reducing the space charge effects. Typically, ions are collected in Q3 operating in the LIT mode and scanned out at a rate of 250 amu/s. Under ER scan conditions, resolution on the order of <0.25 amu (FWHM) can be achieved. To help reduce the space charge effects, Q1 is used to isolate ions with a narrow range of mass-to-charge ratio. The isolated ion packet is then transmitted through Q2 with no fragmentation and into the trap. The trap operates at the reduced scan speed with a smaller than normal step to generate a high-resolution spectrum. With external calibration, mass accuracy in the range of 50 ppm can be achieved.

4. *MS^3.* It is possible to acquire high-sensitivity MS^3 data using the QQQ/LIT. MS/MS is performed as described above for EPI scans. However, after fragmentation occurs in Q2 and the fragments are collected in the Q3 ion trap, a quadrupole DC isolation is used to destabilize all the ions except the target ion. Only the target ions will have the stability to remain in the trap during the isolation for further manipulation. The isolated packet of ions is then excited using the same resonant mass specific excitation used in ion trap scans. However, this excitation is performed for a much longer period of time and at a lower energy, causing collision of the excited ions with the nitrogen present in the trap. This excitation/fragmentation process is very similar to the other standard, single-element ion traps, occurs on similar time scales, and results in low-energy fragmentation. It should also be noted that in the MS^3 mode the resulting spectra exhibit low mass cutoff like the 3D ion traps. However, MS^3 is rarely necessary, as the MS/MS fragmentation using the QQQ/LIT is high-energy triple-quadrupole fragmentation, which generates a pretty complete fragmentation pattern without the necessity to perform MS^3.

5. *EMC.* The "enhanced multiply charged" scan removes singly charged ions from the trap, resulting in an enrichment spectrum of multiply charged ions. In analysis of proteins and peptides, the target compound is often multiply charged and the singly charged signals are due primarily to the background/matrix. By removing these extraneous signals, the S/N for the analyte is improved, resulting in better data. The EMC scan has been described in detail elsewhere (Thomson and Chernushevich, 1998; Chernushevich et al., 2003) and only a general description is given here. To perform this experiment, the exit barrier potential is manipulated differently compared to a typical LIT experiment. After the fill step, the potential barrier is lowered enough to allow singly charged ions to slowly leak out; high-charged-state ions are held efficiently due to greater repulsion; that is, the barrier is 2 times greater for a +2 ion versus a +1 ion. After a period of time, the ion

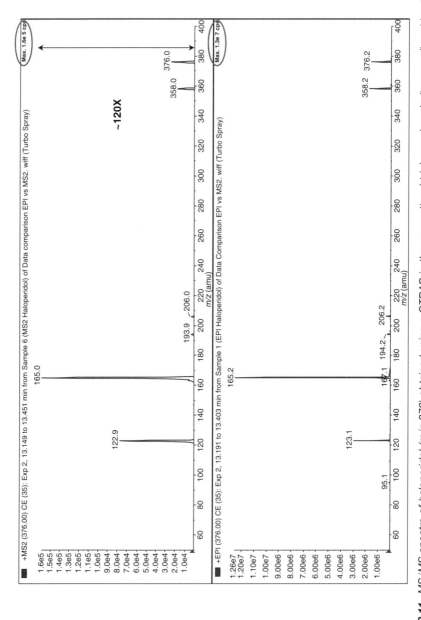

Figure 3.11. MS/MS spectra of haloperidol (m/z 376) obtained using a QTRAP in the conventional triple quadrupole (top panel) and ion trap (bottom panel) modes. The intensity of the ion trap MS/MS spectrum is approximately 120 times greater than the intensity of the triple-quadrupole MS/MS spectrum.

population becomes enriched for higher charged state ions, which are then scanned out to produce a spectrum of mainly multiply charged species. A comparison of mass spectra of a bovine serum albumin (BSA) digest acquired with and without the use of the EMC function is shown in Fig. 3.12.

6. *TDF.* "Time-delayed fragmentation" helps simplify complex MS/MS spectra by reducing the amount of secondary fragment ions observed. In typical MS/MS spectra, the precursor ion is fragmented using acceleration between Q1 and Q2, which is a high-energy process. Because of this high-energy process, it is possible that secondary, and even tertiary, fragmentation can occur. TDF is a three-step process that includes ion activation, ion relaxation, and fragment collection to provide a spectrum consisting of mainly primary fragments. The TDF process has been described in detail (Hager, 2003). In this scan mode, the precursor ion is accelerated between Q2 and Q3 to promote fragmentation. A portion of the nitrogen introduced into Q2 (collision cell) for regular MS/MS experiments is leaked in to Q3 and used as a collision gas. Q3 is set up to operate in the ion trap mode and to trap only ions within a small range of the precursor; that is, smaller fragments are not trapped. The user sets a delay time, usually on the order of 5 ms. During this delay, the trapped precursor has the opportunity to relax and dissociate into fragments. Because much less collision gas is present in comparison to the collision cell, relaxation of high-energy species that can dissociate to form secondary ions is limited. The kinetics of this mechanism of fragmentation allow formation of the secondary ions during the first few milliseconds and the primary ions after that. All ions below the precursor are dumped from the trap during the initial secondary fragmentation stage. After the specified delay time, the primary ions are trapped and then scanned out to produce a spectrum of mainly primary fragments. A comparison of MS/MS spectra of tamoxifen obtained using a typical triple quadrupole MS/MS and TDF scan functions is shown in Fig. 3.13.

Triple-quadrupole fragmentation is highly dependent upon the CE. While most compounds exhibit adequate fragmentation using a CE of 30–35 eV, some analytes require substantially more to generate sufficient fragmentation, while others may be completely obliterated by this energy. When acquiring qualitative data, such as metabolite identification, it can be difficult to determine the appropriate collision energy to use. To help overcome this problem, it is possible to use a collision energy spread (CES) that changes the collision energy during the trap fill and during an EPI scan. This feature allows fragmentation resulting from a CE range to be observed. The CE and CES are specified by the user and the trap fills with ions from fragmentation at energies of CE and CE $\pm$ CES. For example, if the CE is set at 30 with a CES of 15, the MS/MS spectrum will show a summation of ions from fragmentation at 15, 30, and 45. Figure 3.14 shows spectra from three different CEs, and Fig. 3.15 compares the summation of the three CEs and the spectrum using CES. As can be seen from the figure, there is no difference between the summed spectra at three different CEs and the spectrum acquired using CES.

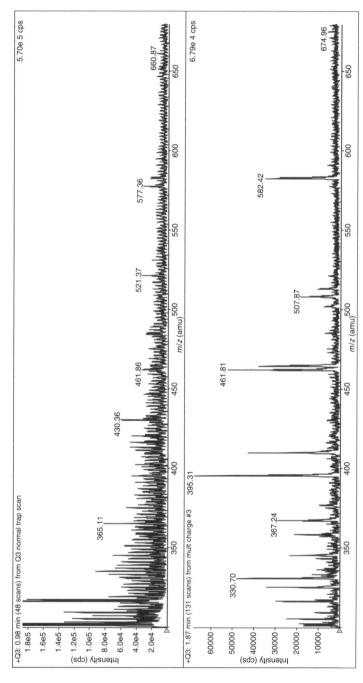

Figure 3.12. Mass spectra of BSA digest obtained using a QTRAP with (bottom panel) and without (top panel) EMC scan function. In the bottom spectrum, all the singly charged background ions are removed to enhance the multiply charged ions from the BSA digest.

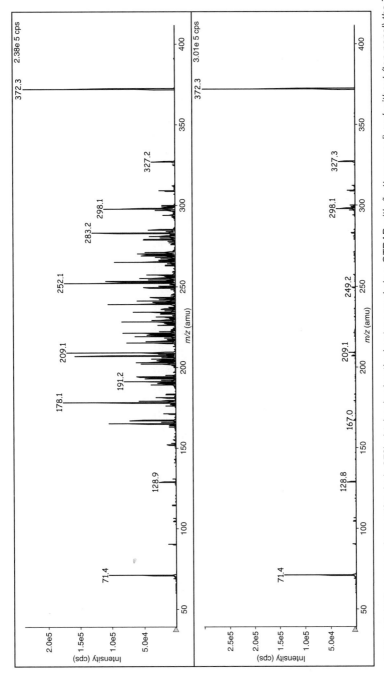

Figure 3.13. MS/MS spectra of tamoxifen (m/z 372) obtained using the ion trap mode in a QTRAP with (bottom panel) and without (top panel) the TDF scan function.

139

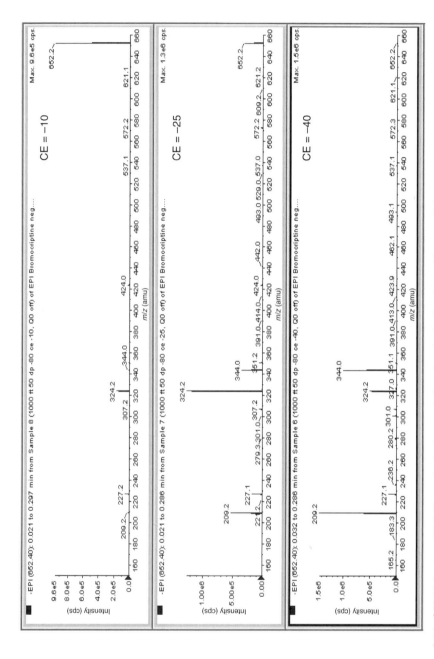

Figure 3.14. QTRAP ion trap MS/MS spectra for bromocriptine. Acquired using three different collision energies: −10 eV (top), −25 eV (middle), and −40 eV (bottom).

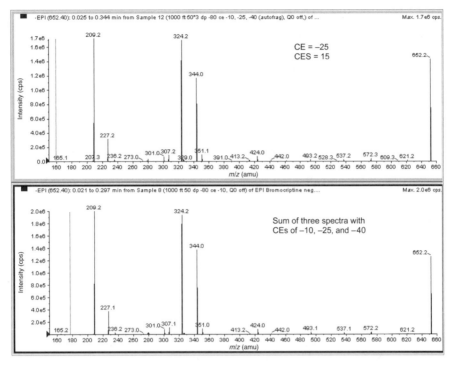

Figure 3.15. QTRAP ion trap MS/MS spectrum of bromocriptine acquired using a collision energy of 25 eV with a CES of 15 (top). The bottom spectrum results from overlaying and summing the three separate MS/MS spectra shown in Figure 3.14.

3.5 EXPERIMENTAL DESIGN

With a good understanding of theory and operation of the instrument, one can now start the application of these modes to different compounds for optimal metabolite detection and confirmation. In thinking about the wide range of scan modes offered on a hybrid LIT/QQQ, such tools tend to have their own strengths and weaknesses. However, use of these modes together always affords a much more complete detection of metabolites than just a single mode within itself. In the following section, a detailed explanation will be provided for optimal experiment setup for each scan type as well as discussion of the accompanying strengths and weaknesses. In terms of metabolite detection, the triple-quadrupole and LIT scan modes can be utilized in the following experiments: EMS (LIT single MS mode), PIS based on parent compound fragmentation, PIS based on phase II conjugate loss (sulfate, phosphate charged fragment loss), NLS based on parent compound fragmentation, NLS based on phase II conjugate neutral fragmentation (glucuronide or glutathione), and theoretical MRM based on a complete set of biotransformations for the parent compound (Xia et al., 2003). An explanation of how these modes can be coupled to automatic product ion experiments (IDA, information-dependent acquisition) will also be covered.

3.5.1 EMS (Single-LIT MS)

A single MS experiment is by far the most common mode used in metabolite detection. A key advantage to an EMS scan is its speed compared to a triple-quadrupole MS experiment. The faster sampling rate allows for more scans across a chromatographic peak, resulting in better detection of closely eluting species. Also, the EMS scan function provides the opportunity to couple ultrapressure liquid chromatographic (UPLC) separation techniques, which inherently deliver sharper chromatographic peaks (<30 s), with mass spectrometric detection. Because the trap can hold all the ions and then sequentially ejects them, sensitivity is better than a quadrupole, in which all other ions that are not at specific m/z increment during the scan are lost. Trap scans also allow for medium to high resolution by using a much lower scan rate, yet without significant sacrifices in duty cycle. General unit resolution (~0.7 m/z at 50% height) is often enough, in the small-molecule arena, for isotope pattern recognition/identification. Because true accurate mass measurement (<5 ppm) is not possible using quadrupole or ion trap mass spectrometers, it is not necessary to perform higher than unit mass resolution scans on an ion trap instrument. Of course, accurate mass data acquired on a FTMS (Chapter 5) or TOF (Chapter 4) instrument can be a valuable tool for molecular formula identification of unknowns.

The key limitation to EMS experiments is background issues. Because electrospray is the most universal LC–MS ionization source, it has the capability to ionize a lot of the background compounds/matrix, which can be observed as extraneous peaks in an EMS scan. Furthermore, it is possible to produce clusters, which further add to the peaks in the spectrum, making it difficult to find low-level metabolites in the presence of matrix. A comparison of a precursor scan and an EMS scan for a metabolite resulting from oxidation of buspirone is shown in Fig. 3.16. Figure 3.16 clearly shows a dioxidation of buspirone in both EMS and precursor scan modes. The disadvantage when using the EMS scan is that the S/N is much poorer than when using the selective precursor scan. Furthermore, because of the selectivity of the precursor scan, the resulting spectrum shows signals that have a high probability of being metabolites, which helps simplify data interpretation. As seen in Fig. 3.16, the EMS spectrum shows several signals, with the majority related to the background/matrix and not irrelevant to the metabolite analysis. In the case of an IDA experiment, the fewer, more relevant signals of the precursor scan generate much more relevant automatic product ion scan data than the complex EMS scan survey. However, the precursor ion scan did not pick up the very strong trioxidation signal that is observed at m/z 434 in the EMS spectrum. This metabolite was missed because the m/z 122 fragment used as the precursor mass had been modified, thereby shifting the mass so this fragment is no longer generated during fragmentation of the precursor ion at m/z 434. Other scan functions, such as NLS and MRM, have comparable strengths and limitations to detecting metabolites while potentially missing others, illustrating the necessity to use a variety of scan types for complete metabolite detection and identification.

In terms of practical experimental design, many factors need to be set correctly for a good EMS–IDA result. First, because most background ions are observed in the

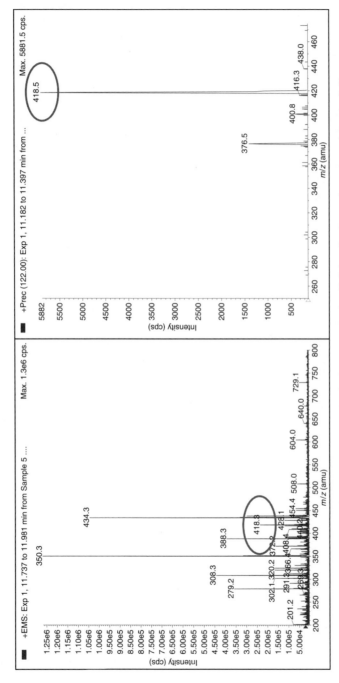

Figure 3.16. Comparison of mass spectra obtained using precursor ion (right) and enhanced mass spectrometry (left) scans.

low-mass region, the lower the MS scan range, the more electrospray ionization/ atmospheric pressure chemical ionization (ESI/APCI) background ions can complicate the analysis, as well as IDA efficiency. Second, the declustering potential needs to be optimized for all of these experiments to ensure good sensitivity. If analyzing for only phase I metabolites, then the optimal DP for the parent compound is usually a good value for general metabolite survey. The metabolic modifications can always change the nature of the molecule; however, the metabolites are generally more similar than not to the parent compound and its DP value is sufficient for detection. If detection of phase II metabolites is important, it is critical to use a low DP value for experiments to prevent the conjugation from being fragmented off the molecule as it travels through the interface. These lower and softer DP values will need to be carried over for experiments in the IDA cycle to allow for phase II-modified metabolite isolation in Q1.

Using appropriate IDA thresholds and settings can also help optimize the EMS–IDA method. One setting that is especially critical for EMS survey experiments is the use of inclusion and exclusion lists. An exclusion list is a list of all notable background or unimportant signals in an LC–MS run that IDA should never submit for further MS/MS analysis. In general, phthalates, salts, and incubation background signals should be added to this list to ensure IDA triggers mainly on relevant signals. Running an EMS experiment on an incubation matrix blank will yield a fairly good list of general background to input to the exclusion list for most experiments. Another critical point is the use of an inclusion list for EMS–IDA experiments. This list would include theoretically predicted metabolite m/z ions (e.g., $+16$ for oxidation, -14 for demethylation) that IDA should trigger on with a higher priority than the general EMS observed signals. A simple script and operator list of general metabolite modifications can generate a theoretical metabolite inclusion list for the operator. This list will increase the IDA hit rate for predicted metabolites by a notable amount; however, it can also reduce the probability that IDA will detect unexpected species. In the case of other QTRAP experiments (PIS, NLS, and multi-MRM), inclusion or exclusion lists are not useful or necessary due to the high degree of selectivity generated by MS–MS, resulting in excellent IDA efficiency.

Another useful tool for better IDA efficiency is a technique called dynamic background subtraction (DBS). This tool helps to improve IDA efficiency as well as reduce the need for inclusion and exclusion lists when using EMS survey scans. With DBS active, the IDA software subtracts the previous two or three survey spectra from the current survey scan. As a result, any constant background ions are removed from the survey scan used by IDA to trigger MS/MS analysis. In essence, IDA is looking at any signal that is increasing but not present in previous scans. As the instrument scans through a peak, the subtraction will remove the major signals in the back half, since the major signal is now less than the previous scans. As a result, there is increased probability that coeluting compounds, or a lower level analyte in the presence of a higher level analyte, will both be submitted for MS/MS analysis. Figure 3.17 is a pictorial representation of how DBS affects IDA selection. Notice the increased opportunity to trigger and perform MS/MS on the low-level analyte because it is on the down slope of the major analyte. The minor analyte signal is on the increase, and the major analyte signal gets subtracted out by DBS. The top panel of Fig. 3.18

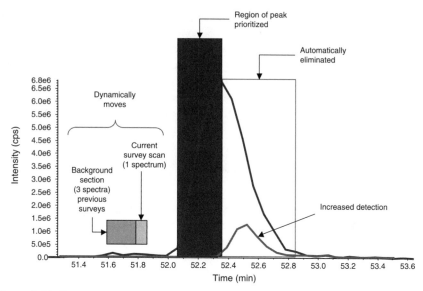

Figure 3.17. Pictorial representation of how DBS works to minimize background peaks and enhance actual signals in EMS–IDA experiments.

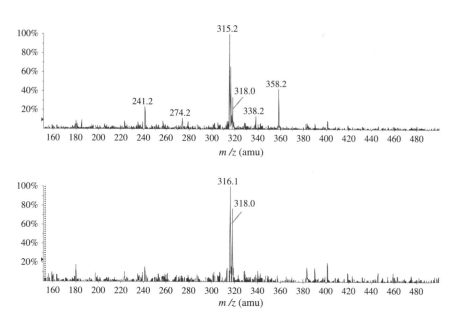

Figure 3.18. EMS survey scan spectra obtained without (top panel) and with (bottom panel) DBS. The spectrum on the top is a full-scan, single-MS ion trap spectrum. The analyte signal, m/z 316, is obscured by background. Therefore, this signal would not be selected by IDA for MS/MS analysis. When the recurring background signals are subtracted out using DBS, the resulting spectrum is shown on the bottom. Notice that the m/z 316 analyte signal now dominates the spectrum, resulting in IDA selection for MS/MS analysis.

shows a typical EMS spectrum, while the bottom panel shows a DBS subtracted spectrum. The analyte signal at m/z 316 is completely obscured by the high background peak at m/z 315 in the regular EMS spectrum. Because the ion at m/z 315 is a background peak present in nearly all spectra, it is subtracted out by DBS, thus allowing IDA to trigger on the relevant m/z 316 ion. This algorithm only affects how IDA selects signals, but the observed spectrum is not altered. In other words, IDA would "see" the spectrum in the bottom panel of Fig. 3.18, but the spectrum in the top panel is saved to the data file and made available to the user.

3.5.2   Precursor Ion Scan: Precursor Fragment and Phase II Conjugates

Precursor ion scanning (or parent ion scanning) has proven to be a very powerful method of finding low-level metabolites, even in the presence of complex matrix background. In this mode, the first mass-resolving quadrupole is set to scan just as in a triple-quadrupole single MS; however, the collision cell and last quadrupole are set up to produce and filter a key fragment of the metabolite and parent. Any time an ion is transmitted from the first quadrupole (Q1) that produces a fragment ion matching the selected mass of the last quadrupole (Q3), a signal is recorded at the corresponding Q1 mass. The spectrum produced is essentially of precursor ions that can generate the structurally unique fragment. This mode allows for a highly relevant and simplified spectrum compared to single MS. Also, because background noise is reduced, it is generally possible to obtain significantly improved S/N versus single MS. Figure 3.19 shows a comparison of the S/N for spectra of a specific metabolite obtained using PIS, NLS, and multi-MRM scans. There is a clear S/N improvement for the precursor scan over single MS. This improvement is achieved mainly due to an overall reduction in noise from the filtering achieved by Q3 during a precursor ion scan. The other scan functions also have improved S/N versus single MS, and analysis using these scans is explained in the following paragraphs. It is important to note that the selective triple-quadrupole scans do not exhibit more intense absolute signal strength compared to single MS; rather the ability to remove a significant amount of background signals results in increased S/N and therefore improved limit of quantitation (LOQs).

A key factor for any of these fragmentation-based modes of detection is the use of appropriate precursor and neutral fragments. For PIS experiments, the chosen fragment should be less likely to be metabolized so that its mass will not change and prevent its detection. A good general rule is to use a small, intense fragment from a more localized portion of the molecule which will tend to have a lower risk of mass change. The fragment should also be moderately unique, as some common fragments are often present in the background. An example of such common background fragment is the ion at m/z 91. The fragment should also have adequate intensity to allow for good sensitivity. Of course, these represent ideal requirements and it is not always possible to have each of these factors fulfilled for all analyses. Good results can be obtained even if a few of these factors are slightly nonoptimized. In the case of a highly probable oxidation or

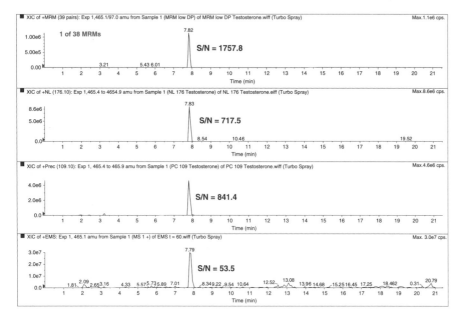

Figure 3.19. *Comparison of S/N for glucuronide metabolite of testosterone detected using different IDA survey scans.*

other biotransformation, it is also possible to use a fragment changed by the modification for a search of metabolites based on that biotransformation. For example, buspirone has a strong fragment at m/z 122. Because oxidation on this fragment is highly probable, a precursor ion scan for $122 + 16$ is a good strategy to detect oxidated metabolites of buspirone. Figure 3.20 shows a fragment/structure analysis for glyburide. In this case, fragment choices are relatively straightforward. The charged fragments of m/z 169, 352, 369, and 395 are all possible precursor fragment choices. The fragment of m/z 169, however, is the best choice because it corresponds to a small portion of the molecule. By choosing a small fragment for the precursor scan, there is the greatest chance to detect the most metabolites by minimizing the possibility that a biotransformation occurs on this fragment. A strategy for detecting metabolites where the modification occurs on the 169 fragment will be discussed in Section 3.5.3.

For precursor ion scans, the appropriate ion path and source parameters should be utilized to ensure maximum fragmentation. These parameters are usually best estimated using the parent compound. It is important to note that tuning estimates based on the parent may not be ideal for the metabolites; however the parent drug collision energies and tuning parameters are the only reasonable method of estimation for the metabolites. The declustering potential of the parent drug can also be used for metabolite analysis, although a low value will be necessary if phase II conjugation analysis is desired. Optimizing parameters for phase II metabolite identification will be discussed in the next section.

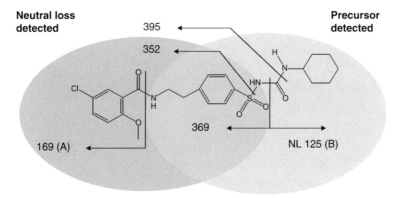

Figure 3.20. *Proposed fragmentation scheme for glyburide.*

3.5.3 Constant Neutral Loss Scan: Precursor Fragment and Phase II Conjugates

Constant neutral loss scans for metabolite analysis fall into two categories: neutral loss based on phase II conjugation and neutral loss based on the fragmentation pattern of the parent compound. A good method for determining a general compound neutral loss is to again examine the fragmentation of the parent compound and find an intense, high-m/z fragment. This strong fragment must have a corresponding neutral loss, which is the difference of the parent mass minus the fragment. The collision energy for the charged fragment can be used for this NLS. It is important not to use a common neutral loss pathway, such as water (H_2O) or carbon dioxide (CO_2). It is also a good idea to interpret the fragmentation of the parent molecule so that, if possible, the PIS and NLS are based on fragments from opposite sides of the molecule. By choosing appropriate fragments, detection of metabolic transformations missed by one of the modes due to a shift in a chosen fragment will be detected in the other. When molecules have multiple sights of charging, it is possible to use a pair of precursor experiments for better coverage as well.

For the case of glyburide (Fig. 3.20), a good strategy is to use a precursor ion scan for the fragment ion of m/z 169. This scan allows for detection of all metabolic changes on the molecule from the amide bond over to the neutral ring. For best complementary coverage of metabolic species, a NLS based on the m/z 125 fragment allows for detection of any modifications on the m/z 369 charge part of the molecule. This neutral loss experiment is especially critical for any changes on the m/z 169 ring, which cannot be directly detected by the precursor ion scan of m/z 169. There are two possible limitations to this approach. First, any drastic change in the molecule might vary the collision energy required for fragmentation and sensitivity might be reduced, possibly to the point of detection not being possible. The second limitation, which is more likely, is that if two transformations take place, changing both fragment masses, this overlapping strategy can still miss such metabolites. However, even though each scan mode has some general limitations, the improved sensitivity and selectivity based on triple-quadrupole survey modes provide a vast improvement in metabolite detection versus a simple single-MS experiment. Also, it is possible when using either NLS or PIS to make a theoretical prediction for metabolic fragment mass change, improving the ability to detect species missed by general parent fragment analysis. A simple example of this theoretical prediction is to increase the NLS of 125 by 16 or the PIS of 169 by 16 to detect any oxidative variations. These specific modifications can be very useful; however, they do need to be more limited in most cases, as they only detect species with a certain base metabolic change.

Constant neutral loss and precursor ion scans can also be used to detect generic phase II metabolic changes. Many possibilities exist in this category, as any added moiety that can fragment in a predictable manner can be scanned for using this method. The most common case is, of course, a neutral loss of m/z 176 corresponding to glucuronide formations. On most QTRAP instruments, a low declustering potential of about 30 V is needed to keep the conjugate intact as it moves through the interface region. Using a low collision energy ($\sim$30 eV) helps to fragment the glucuronide from the molecule, allowing detection of the conjugate charged fragment by a NLS. With any phase II conjugation, there is usually a range of bond formations to the parent drug that are possible. The declustering potential and collision energy for a

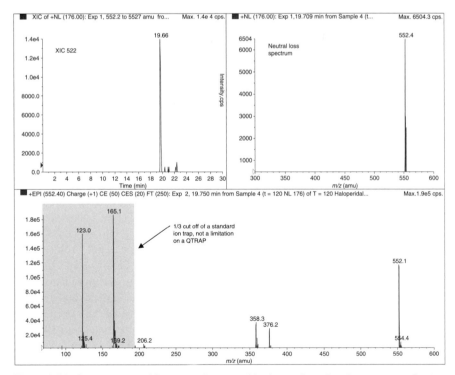

Figure 3.21. Constant-neutral-loss scan (top panels) triggered product ion spectrum (bottom panel) of haloperidol glucuronide.

similar chemical standard should be determined and used for optimal sensitivity for conjugate analysis. In the case of a sulfate or phosphate conjugate, either a precursor ion scan for that species in the negative mode or a NLS in the positive mode can be used. In general, precursor ion scan sensitivity appears to be slightly better than NLS; however, it can depend upon many conditions. On a QTRAP instrument, it is possible to do a negative precursor ion scan followed by positive-mode ion trap product ion scan for detection of sulfonation and phosphorylation phase II species. The positive product ion scan is necessary because most parent drugs ionize better in the positive mode; therefore the fragmentation patterns between the sulfonated and phosphory-lated metabolite can be compared for structural information.

Figure 3.21 shows an example of the power of a QTRAP-type instrument. In the case of this haloperidol glucuronide conjugate ($[M + H]^+$ at m/z 552), detection at a level of about 1% compared to parent was accomplished using the NLS. With the selectivity of the triple-quadrupole for neutral loss of 176, corresponding to glucur-onidation, a QTRAP product ion scan was automatically acquired using IDA. The structural information, with rich high-energy fragmentation along with no low-mass cutoff, provides a complete fragmentation spectrum without the need for slower and less sensitive MS^3 or MS^4 scans as required on a standard ion trap.

3.5.4   Multiple Theoretical MRMs

With the advent of fast ion transit collision cells, large cycles of MRMs (up to 200) can be completed in very short periods of time. A new and exciting method of quali-tative metabolite surveys is the use of a large set of theoretically predicted MRMs (Beaudry et al., 1999). This mode provides unparalleled detection limits for predicted metabolites compared to traditional ion trap MS. In comparison to PIS, NLS, and single-stage MS, MRM scanning allows species present at 3–50 times lower level to be detected due to the inherent superior S/N ratio of MRM scanning. Generally, an MRM experiment is a powerful method of quantitation for drugs and related species in complex biological matrices. With the advent of simple software algorithms used to create a pair of MRMs for each predicted metabolite, a large set of such transitions can be used as a powerful method of detection. For example, for oxidation of a drug with a strong MRM transition from 400 to 100, a theoretical transition from 416 to 116, as well from 416 to 100, will need to be created to ensure detection of all possible forms of the oxidation; that is, taking into account that the modification does and does not occur on the fragment. The strength of this exper-iment is based on the dwell time spent on each MRM compared to a PIS or NLS. If analyzing for common haloperidol metabolites, including glucuronidation plus three oxidations, it would be necessary to scan from about 250 to 600 amu. A scan time of 2 s with a step size of 0.3 amu and a dwell time of 2 ms/step ensure good sensitivity. For a group of 60 theoretical MRMs, which translates into 30 meta-bolic species, a dwell time of over 30 ms per MRM transition, or "step," could be used to maintain the same cycle time of 2 s. Fundamentally, increasing the dwell time increases the S/N. Therefore, the ability to use dwell times per step that are approximately 15 times longer than that used in any other scan types results in the ability of the theoretical MRM experiment to detect metabolites present at lower levels than any other LC–MS/MS-based techniques.

With the strong sensitivity of multiple MRMs as a method of metabolite detection, one can assume that this method is superior to all others. However, like any analytical technique, a number of limitations exist for this type of experiment. First, any metabolites that are not predicted will not be detected. Single MS, precursor, and neutral loss scans are not subject to this limitation. Second, the more modifications that take place on the parent, the more number permutations of MRMs are required. For example, to detect an oxidation and methylation metabolite, as shown in the following table, four theoretical MRMs would be required:

Purpose of MRM Scan	Q1 (m/z)	Q3 (m/z)
Confirms modification of fragment	Parent + 30	Fragment
Confirms oxidation of fragment	Parent + 30	Fragment + 16
Confirms methylation of fragment	Parent + 30	Fragment + 14
Confirms oxidation and methylation of same fragment	Parent + 30	Fragment + 30

Even with these notable limitations, multi-MRM is a powerful and essential experiment for identification of metabolites present at lowlevel. Figure 3.22 shows an example of detection of a variety of low-level buspirone metabolites by using a set of 27 theoretical MRMs. It is especially important with this type of experiment to have sensitive ion trap product ion scans to confirm and validate the nature of these trace-level metabolites.

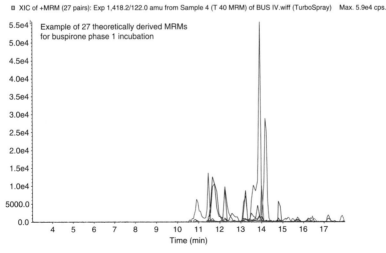

Figure 3.22. Theoretical MRM–IDA analysis of buspirone metabolites.

3.5.5 IDA Experiment Structure

A traditional workflow for metabolite analysis using LC–MS is the following: initial analysis using precursor and neutral loss scanning on a triple-quadrupole LC–MS system; single-stage MS analysis using an ion trap instrument; and confirmation/ structural elucidation using ion trap MS data (Li et al., 2007). The introduction of hybrid QQQ/LIT instruments allows simultaneous acquisition of both triple-quadru- pole survey scans followed by sensitive ion trap MS/MS data acquired in an IDA fashion. Furthermore, because the high-energy triple-quadrupole fragmentation gen- erally eliminates the need for MS^n, analysis time is further reduced. A typical IDA experiment workflow is shown in Fig. 3.23.

There are many options when doing an IDA experiment. First, the user can select either one or two survey scans, which can be single-stage MS, PIS, NLS, or multi-MRM. A wide range of governing criteria can be set for such exper- iments. The primary parameters for IDA selection are mass window and threshold. The threshold allows the user to set an intensity limit, in which any signal below a certain value will be excluded from selection. This value will change depending on the signal strength of the given survey scan mode. For example, in the case of a NLS, signals greater then 2000 counts per second (CPS) might be well worth trig- gering MS^2. However, in the case of single-stage MS, a threshold of 1×10^6 CPS might be more appropriate. Setting the threshold above the general background will help to greatly eliminate extraneous MS^2 data. A second critical criterion is the overall mass range of ion selection. If the mass range is set too low, excessive

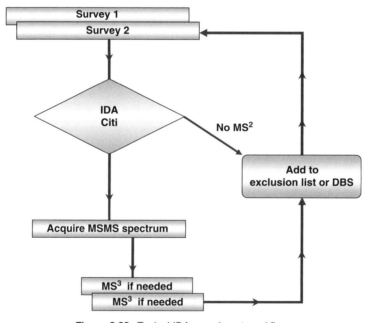

Figure 3.23. *Typical IDA experiment workflow.*

selection of low-mass background ions, which are generally not relevant, will occur, making data analysis more difficult. The software tool DBS helps enhance ion selection when using an MS survey scan. Details of this software are presented in Section 3.5.1. Other features of the IDA software are isotope pattern match, which allows for a high priority of acquisition for any signal that has an isotope pattern matching the parent molecule. This feature is mainly useful for compounds that exhibit a unique isotope pattern due to presence of atoms such as Cl or Br.

Combinations of these surveys can also be used together. For example, a precursor ion scan and a single-stage MS can be set up to occur within the same cycle. One limitation of mixing a single-stage MS with the other modes relates to differences in intensity. Because intensity is one of the fundamental criteria used by IDA to select which peaks will be selected for MS/MS analysis, wide differences in intensity from one survey compared to another can cause IDA to miss significant peaks from the less intense signals. This disparity is the case when simultaneously using MS and precursor scans as survey scans. The raw intensity from MS can be 10–100 times stronger in intensity versus a signal found in a precursor ion scan. This greater intensity will cause the IDA algorithm to miss the weaker signals from the precursor scan and almost exclusively trigger on the single MS spectrum. However, because the precursor scan is extremely selective, the much weaker signals are much more significant candidates for IDA due to the implied structural similarity; therefore, this selection is not advantageous. Fortunately, an advanced software feature has been developed to compensate for this limitation. A user can input a value of 0.1 ms for the fixed filled time of the MS experiment but keep DFT (dynamic fill time) active. DFT will artificially scale the strength of the MS spectra to a much lower value than normally observed. In an ideal case, this scaling should cause the MS to be generally lower in intensity than the average significant precursor signals. This scaling of the MS data intensity results in priority for IDA to trigger on the triple-quadrupole scan if any signal is observed, and the MS signals only when the triple-quadrupole scan is not detecting any notable species. This combination experiment saves time because the best of both experiments can be acquired within the same run. If two triple-quadrupole surveys are combined as survey scans, no scaling adjustment is usually necessary. However, because triple-quadrupole-based scans often have long scan cycle times, two such surveys together might cause insufficient sampling across a standard chromatographic peak. Furthermore, long scan cycle times can limit the optimum utility of UPLC-based separations because UPLC peak widths are shorter than that observed with conventional pressure liquid chromatography separations.

3.5.6 Automation of Incubation Screening

A QTRAP instrument provides a wide range of options and greater power for metabolite identification. One aspect, which can often be a struggle, is the ability to automate much of this analysis for increased throughput, especially as metabolite identification is moved to earlier stages of the drug discovery process. Recent work by Rourick et al. (2002) has generated a new software tool for automation in this area. The general idea behind this work is to couple early stage incubation stability assessment with online comprehensive metabolite ID. Standard automated quantitation software used for stability and PK screening was expanded to gauge how much metabolic turnover the parent drug achieved over a given time frame. The user could then set up the software to automatically switch to a qualitative analysis for samples exhibiting appropriate metabolism, as determined by parent MRM area reduction. Because the software has already established fragmentation and ion path tuning voltages for the quantitation, this information can be further used to automatically create precursor, neutral loss, multi-MRM, and MS methods within the inclusion list. The flow chart shown in Fig. 3.24 lists all the steps involved in automated metabolite analysis of incubation samples.

The first step in this automated process is injection from a plate of tuning standards to determine DP and major fragments with corresponding CEs. The software then builds a quick MRM quantitation experiment and determines the percent area loss for the parent drug at a specified time point. If the percent loss is exceeded, the system will switch a valve and begin acquiring on a long column for adequate metabolite separation. Based on the early tuning and a list of theoretical metabolite transformations, PIS–, NLS–, MS–, and multi-MRM–IDA experiments are created. A strong low-mass fragment is used for the precursor ion and MRM methods, and a strong but unique high-mass fragment is used for the neutral loss. Using the same list of theoretical metabolic transformations, an IDA inclusion list is generated for the MS^{1}–IDA experiment as well. This suite of IDA methods is then submitted and acquired for the $T = 0$ control and a later time point for which information about metabolites is desired.

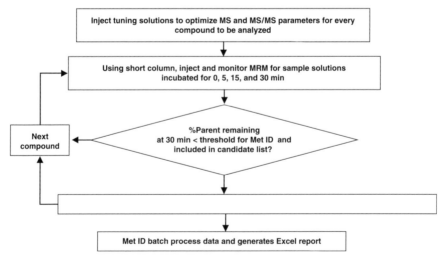

Figure 3.24. *Automated metabolite identification process.*

3.6 SUMMARY

The advent of a hybrid QQQ/LIT instrument combines the triple-quadrupole selectivity and sensitivity with ion trap full-scan sensitivity in one platform. Furthermore, the ability to use a combination of these scans within the same cycle allows simultaneous detection and structural information of even trace-level species. The mechanism of fragmentation by use of a collision cell also helps to overcome the low-mass cutoff found on standard 3D ion traps, resulting in richer and more descriptive fragmentation. Each of the scan modes possible on a QTRAP has significant benefit for metabolite identification. Precursor ion and neutral loss scans allow for low-level, selective detection of most metabolites. Phase II PIS and NLS allow for specific and selective detection of general conjugate formation. In this case, high sensitivity is critical for distinction between general background conjugation and true drug conjugate formation. The new and novel tool of theoretical multi-MRM metabolite survey scans permits the lowest possible detection for predicted metabolites. Often, this mode detects the highest number of metabolites due to its unprecedented sensitivity as a qualitative mode of metabolite detection. As with a traditional 3D ion trap, the simple single-stage MS IDA experiment is a quite valuable and simple method for high-level metabolite detection for general screening. Except for the lack of accurate mass measurements achieved by a high-resolution instrument, all modes of mass-based metabolite analysis are provided at the utmost performance within this hybrid platform.

ACKNOWLEDGMENTS

The author would like to thank Gary Impey and Yves Le Blanc of AB/SCIEX for data and experimental design.

REFERENCES

Beaudry, F., Le Blanc, J. C. Y., Coutu, M., Ramier, I., Moreau, J.-P., and Brown, N. K. (1999). Metabolite profiling study of propranolol in rat using LC/MS/MS analysis. *Biomed. Chromatogr.* **13**:363–369.

Chernushevich, I. V., Fell, L. M., Bloomfield, N., Metalnikov, P. S., and Loboda, A. V. (2003). Charge state separation for protein applications using a quadrupole time-of-flight mass spectrometer. *Rapid Commun. Mass Spectrom.* **17**:1416–1424.

Clarke, N. J., Rindgen, D., Korfmacher, W. A., and Cox, K. A. (2001). Systematic LC/MS metabolite identification in drug discovery. *Anal. Chem.* **73**:430A–439A.

Douglas, D. J., Frank, A. J., and Mao, D. (2005). Linear ion traps in mass spectrometry. *Mass Spectrom. Rev.* **24**:1–29.

Douglas, D. J., and French, J. B. (1992). Collisional focusing effects in radio frequency quadrupoles. *J. Am. Soc. Mass Spectrom.* **3**:398–408.

Fernandez-Metzler, C. L., Owens, K. G., Baillie, T. A., and King, R. C. (1999). Rapid liquid chromatography with tandem mass spectrometry-based screening procedures for studies on the biotransformation of drug candidates. *Drug Metab. Dispos.* **27**:32–40.

Hager, J. W. (2002). A new linear ion trap mass spectrometer. *Rapid Commun. Mass Spectrom.* **16**:512–526.

Hager, J. W. (2003). Product ion spectral simplification using time-delayed fragment ion capture with tandem linear ion traps. *Rapid Commun. Mass Spectrom.* **17**:1389–1398.

Hager, J. W. (2004). Recent trends in mass spectrometer development. *Anal. Bioanal. Chem.* **378**:845–850.

Hager, J. W., and Le Blanc, J. C. (2003). High-performance liquid chromatography-tandem mass spectrometry with a new quadrupole/linear ion trap instrument. *J. Chromatogr. A* **1020**:3–9.

Hager, J. W., and Yves Le Blanc, J. C. (2003). Product ion scanning using a Q-q-Q linear ion trap (QTR.AP) mass spectrometer. *Rapid Commun. Mass Spectrom.* **17**:1056–1064.

Hop, C. E. C. A., and Prakash, C. (2005). Metabolite identification by LC-MS: Applications in drug discovery and development. In *Identification and Quantification of Drugs, Metabolites and Metabolizing Enzymes by LC-MS* (Chowdhury, S. K., Ed.). Elsevier, Amsterdam, The Netherlands, pp. 123–158.

Hopfgartner, G., Husser, C., and Zell, M. (2003). Rapid screening and characterization of drug metabolites using a new quadrupole-linear ion trap mass spectrometer. *J Mass. Spectrom.* **38**:138–150.

Hopfgartner, G., Varesio, E., Tschaeppaet, V., Grivet, C., Bourgogne, E., and Leuthold, L. A. (2004). Triple quadrupole linear ion trap mass spectrometer for the analysis of small molecules and macromolecules. *J. Mass Spectrom.* **39**:845–855.

Hopfgartner, G., and Zell, M. (2005). Q Trap MS: A new tool for metabolite identification. In *Using Mass Spectrometry for Drug Metabolism Studies* (Korfmacher, W. A., Ed.). CRC Press, Boca Raton, FL, pp. 277–304.

Li, A. C., Alton, D., Bryant, M. S., and Shou, W. Z. (2005). Simultaneously quantifying parent drugs and screening for metabolites in plasma pharmacokinetic samples using selected reaction monitoring information-dependent acquisition on a QTrap instrument. *Rapid Commun. Mass Spectrom.* **19**:1943–1950.

Li, A. C., Gohdes, M. A., and Shou, W. Z. (2007). "N-in-one" strategy for metabolite identification using a liquid chromatography/hybrid triple quadrupole linear ion trap instrument using multiple dependent product ion scans triggered with full mass scan. *Rapid Commun. Mass Spectrom.* **21**:1421–1430.

Liu, D. Q., Xia, Y. Q., and Bakhtiar, R. (2002). Use of a liquid chromatography/ion trap mass spectrometry/triple quadrupole mass spectrometry system for metabolite identification. *Rapid Commun. Mass Spectrom.* **16**:1330–1336.

Ma, S., Chowdhury, S. K., and Alton, K. B. (2006). Application of mass spectrometry for metabolite identification. *Curr. Drug Metab.* **7**:503–523.

March, R. E. (1997). An introduction to quadrupole ion trap mass spectrometery. *J. Mass Spectrom.* **32**:351–369.

March, R. E. (1998). Quadrupole ion trap mass spectrometry: Theory, simulation, recent development and applications. *Rapid Commun. Mass Spectom.* **12**:1543–1554.

March, R. E. (2000a). Quadrupole ion trap mass spectrometer. In *Encyclopedia of Analytical Chemistry* (Meyers, R. A., Ed.). Wiley, Chichester, pp. 11848–11872.

March, R. E. (2000b). Quadrupole ion trap mass spectrometry. A view at the turn of the century. *Int. J. Mass Spectrom.* **200:**285–312.

Nourse, B. D., and Cooks, R. G. (1990). Aspects of recent developments in ion-trap mass spectrometry. *Anal. Chim. Acta* **228:**1–21.

Rourick, R. A., Jenkins, K. M., Walsh, J. P., Xu, R., Cai, Z., and Kassel, D. B. (2002). Integration of custom LC/MS automated data processing strategies for the rapid assessment of metabolic stability and metabolite identification in drug discovery. In *Proceedings of the 50th ASMS Conference on Mass Spectrometry and Allied Topics*, Orlando, FL.

Schwartz, J. C., Senko, M. W., and Syka, J. E. P. (2002). A two-dimensional quadrupole ion trap mass spectrometer. *J. Am. Soc. Mass Spectrom.* **13:**659–669.

Stafford, G. (2002). Ion trap mass spectrometry: A personal perspective. *J. Am. Soc. Mass Spectrom.* **13:**589–596.

Thomson, B. A., and Chernushevich, I. V. (1998). A new scan mode for the identification of multiply charged ions. *Rapid Commun. Mass Spectrom.* **12:**1323–1229.

Xia, Y. Q., Miller, J. D., Bakhtiar, R., Franklin, R. B., and Liu, D. Q. (2003). Use of a quadrupole linear ion trap mass spectrometer in metabolite identification and bioanalysis. *Rapid Commun. Mass Spectrom.* **17:**1137–1145.

Yost, R. A., and Enke, C. G. (1978). Selected ion fragmentation with a tandem quadrupole mass spectrometer. *J. Am. Chem. Soc.* **100:**2274–2275.

Yost, R. A., and Enke, C. G. (1979). Triple quadrupole mass spectrometry for direct mixture analysis and structure elucidation. *Anal. Chem.* **51:**1251A–1264A.

Zhang, M. Y., Pace, N., Kerns, E. H., Kleintop, T., Kagan, N., and Sakuma, T. (2005). Hybrid triple quadrupole–linear ion trap mass spectrometry in fragmentation mechanism studies: Application to structure elucidation of buspirone and one of its metabolites. *J. Mass. Spectrom.* **40:**1017–1029.

4

Applications of Quadrupole Time-of-Flight Mass Spectrometry in Reactive Metabolite Screening

Jose M. Castro-Perez

Waters Corporation, Milford, Massachusetts

Mass Spectrometry in Drug Metabolism and Pharmacokinetics. Edited by Ragu Ramanathan
Copyright © 2009 John Wiley & Sons, Inc.

159

4.1 INTRODUCTION

Reactive metabolite screening in drug discovery is an integral part of the early screening process as it helps to detect possible new chemical entities which may undergo bioactivation. Such behavior can pose a threat to drug programs and is widely accepted to be a mechanism for xenobiotic-induced toxicity. Having the capability to detect biologically reactive electrophiles during early discovery phase allows pharmaceutical companies to re-optimize the compound in question and minimize the risk of toxicity at the development stage. This chapter consists of a detailed description of the type of instruments used for the screening of reactive metabolites. The term Q-TOF is used to describe a type of hybrid mass spectrometry (MS) system in which a quadrupole mass filter (Q) is used in conjunction with a time-of-flight mass analyzer (TOF). In the Q-TOF, the quadrupole is used either in narrow-bandpass mode or wide-bandpass mode to transmit the ions into the TOF analyzer for mass measurement. The narrow-bandpass operation is typical when using MS/MS acquisition modes in which the mass of a compound is known and it will then be selected for MS/MS. Other ions, which are previously not selected, will not be allowed to go through to the collision cell and subsequently into the TOF area. This is achieved by adjusting the RF and DC potentials applied to the rods. In the wide-bandpass mode, all ions will be allowed into the collision cell and subsequently into the TOF analyzer. For this to happen, the quadrupole is operated in the RF-only mode.

4.2 HIGH-RESOLUTION CHROMATOGRAPHY

A very important factor in the detection of GSH conjugates is the chromatographic separation method used. Today, there is a need to produce results faster without compromising quality. Over the past 30 years, high-performance liquid chromatography (HPLC) has proven to be the most widely used technique in laboratories worldwide. During all this time, the HPLC hardware has remained stagnant in terms of development. However, advances have been made on column chemistry, column size, and particle size and some of the detectors that couple to HPLC such as ultraviolet and fluorescence detectors. In terms of column particle size, major developments have been made in going from 5 to 1.7 μm.

The van Deemter equation is an empirical formula that describes the relationship between linear velocity (flow rate) and plate height (or column efficiency) (van Deemter et al., 1956). The particle size is one of the variables used in the van Deemter equation. As illustrated in Fig. 4.1, as the particle size decreases to less than 2.5 μm, the efficiency is increased. Furthermore, when using smaller size particles, the efficiency is not affected with increasing flow rates or linear velocities.

By the use of smaller size particles, speed and peak capacity can be extended to new horizons. This separation technology is termed ultraperformance liquid chromatography, or UPLC™ (Churchwell et al., 2005; Johnson and Plumb, 2005; Kamel and Prakash, 2006; Mensch et al., 2007; Pedraglio et al., 2007; Goodwin et al., 2007; Plumb et al., 2007). By using UPLC technology, higher LC flow rates

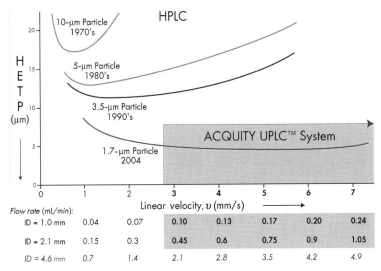

Figure 4.1. Van Deemter curve. A 1.7-μm partricle column has a large optimal chromatographic functional area even at higher flow rates.

can be employed to achieve faster separations with excellent peak capacities for very complicated matrices such as plasma, bile, urine, and feces. UPLC is further illustrated in Fig. 4.2, which compares the separation of in vitro metabolites of dextromethorphan using HPLC and UPLC.

The design and development of sub-2-μm particles present a significant challenge, and researchers have been very active in this area to capitalize on the advantages of smaller size particles (Wu et al., 2001; Jerkovitch et al., 2003; Wyndham et al., 2003). To provide enhanced mechanical stability for UPLC operation, the column hardware is required to have a smoother interior surface and be able to retain the smaller size particles and resist clogging. Packed-bed uniformity is also critical, especially when shorter columns are required to maintain resolution while accomplishing the goal of faster separations. Achieving high-peak-capacity separations with smaller particles require operating at a much greater pressure range than required by conventional HPLC instrumentation. The calculated pressure drop at the optimum flow rate for maximum efficiency across a 15-cm-long column packed with 1.7-μm particles is in the region of 15,000 psi. Therefore, a pump, capable of delivering solvent smoothly and reproducibly at very high pressures, which can also compensate for solvent compressibility and operate in both the gradient and isocratic separation modes, is required. Sample introduction is also critical. Conventional injection valves, either automated or manual, are not designed to work at extreme pressures. To protect the column from experiencing extreme pressure fluctuations, the injection process must be relatively pulse free. The swept volume of the device also needs to be minimal to reduce potential band spreading. A fast injection cycle time is needed to fully capitalize on the speed afforded by UPLC, which in turn requires a high sample capacity. To benefit from increased sensitivity, low volume injections with minimal carryover are also required.

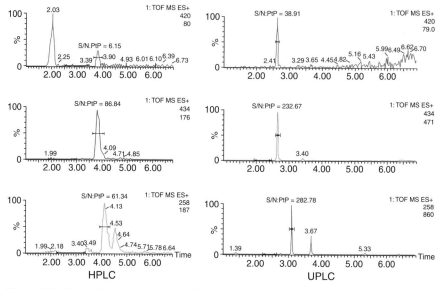

Figure 4.2. Separation of in vitro metabolites of dextromethorphan with HPLC vs. UPLC (bottom panel: desmethyl-dextromethorphan (m/z 258); middle panel: desmethyl-dextromethorphan-glucuronide (m/z 434); top panel: di-desmethyl-dextromethorphan-glucuronide (m/z 420); where S/N:PtP = peak-to-peak signal-to-noise ratio).

The detectors used with UPLC systems have to be able to handle very fast scanning methods because peak half-height widths of around 1 s are typically obtained with columns packed with 1.7-μm particles. In order to accurately and reproducibly integrate an analyte peak, the detector sampling rate must be high enough to capture enough data points across the peak. Conceptually, the sensitivity increase for UPLC

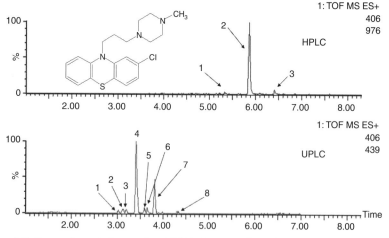

Figure 4.3. Comparison between the HPLC (top panel) and the UPLC (bottom panel) approaches for doubly hydroxylated metabolites of prochlorperazine.

detection should be 2–3 times higher than HPLC separations, depending on the detection technique. MS detection is significantly enhanced by UPLC; increased peak concentrations, with reduced chromatographic dispersion at lower flow rates (no flow splitting), improve ionization efficiencies.

Due to the demands of running fast chromatography, scientists are used to making compromises, and one of the most common scenarios involves sacrificing resolution for speed. UPLC overcomes such problems, as illustrated in Fig. 4.3, which shows a separation of eight double hydroxylated metabolites of prochlorperazine diuretics in under 5 min. The same separation, on a 2.1×100-mm 5-μm C_{18} HPLC column, yields a very different separation with most metabolites coeluting under the same analyte peak. For analyses involving in vivo metabolism, where a large number of small but important metabolites may be present, speed is of a secondary importance, and peak capacity and resolution are the priority.

4.3 TIME-OF-FLIGHT MASS SPECTROMETER

Time-of-flight mass spectrometers are very accurate "clocks." Ions from the first mass analyzer (quadrupole) enter the pusher region (Fig. 4.4) where they are subjected to a high voltage (typically $\sim$9.1 kV) before being ejected orthogonally into the flight tube. The flight tube, which is the ion path, is of a defined length. In traversing the flight tube, the heavier ions will take longer than the lighter ions. Hence, ions are separated by mass. The time between the pusher signal and an ion's arrival at the detector (i.e., its time of flight) relates to the m/z value of interest.

The basic principle of time of flight is highlighted by the following formula

$$v^2 = \frac{2zeU}{m_i} \tag{4.1}$$

where z is the number of charges on an ion, $+1$, $+2$, etc; U is the accelerating voltage, 9100 V; e is the electronic charge, 1.602×10^{-19} C; and m_i is expressed in kilograms per ion (mass $\times 1.66022 \times 10^{-27}$ kg). Therefore an ion's velocity (v) is inversely proportional to the square root of the m/z value and the ions of smaller m/z value arrive first and those of larger value arrive last at the detector. Thus, by measuring the "flight time" of the ions along the TOF tube, the m/z values can be deduced and a mass spectrum can be measured. The relationship between an ion's velocity (v) and the time (t) necessary to travel the flight tube (L) can be expressed using the formula

$$t = \frac{L}{v}$$

For example, the time required for singly charged ions of mass 500 and 1000 to travel a flight tube of length 2 m are 3.4 and 4.8 microseconds, respectively. One of the main advantages of TOF MS systems is the fact that thousands of spectra, with exact mass, can be acquired in a very short time.

The other advantage of using the configuration shown in Fig. 4.4 is the use of a reflectron. The reflectron allows ions of different kinetic energies to be tightly

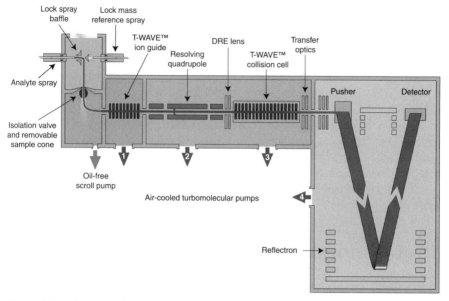

Figure 4.4. *Schematic of an orthogonal Q-TOF mass spectrometer. In this example, an ion beam is produced by electrospray ionization (ESI). The solution introduced to the mass spectrometer-ESI source may be an effluent from a liquid chromatography column or simply an infusion solution of an analyte.*

bunched signals at specific time intervals. In effect, this means that the ions of different m/z value are inadequately separated—the resolution of the analyzer is impaired. As shown in Fig. 4.4, a set of rings which are held at increasing potential make up the reflectron. Faster ions travel further into the reflectron compared with slower ones before being reversed and sent back out. Thus, the slower ions get a chance to "catch up" with the faster ones so that, when both slower ones and faster ones reach the detector, they arrive at about the same time. The result is an apparent reduction in the velocity spread in the ions of any one particular m/z value and leads to a major improvement in the resolution of the analyzer.

As shown in Fig. 4.4, for almost all drug metabolism applications, ions are produced with an electrospray ionization (ESI) source. Ions formed by ESI have little or no internal energy and exhibit no tendency to fragment in the ion source, thus allowing one to obtain molecular mass information from the stable molecular ions. However, a molecular mass gives relatively little structural information. By subjecting the molecular ions to fragmentation, product ions are produced which are characteristic of the structure of the original (precursor) ion and therefore of the original molecule. The ability to obtain an unequivocal relative molecular mass, molecular formula, and structural information makes the hybrid mass spectrometer a powerful tool for investigating single substances or mixtures of unknown substances.

Over the years, the Q-TOF instrumentation has become an important mass spectrometer of choice for scientists working in the drug metabolism arena. This is due to the fact that Q-TOF offers very high sensitivity in the full-scan MS and

MS/MS modes of operation. Additionally, when looking at unknown samples such as the case of metabolite identification, having exact mass measurements and high resolution helps to decipher simple and complicated biotransformations.

4.4 EXACT MASS MEASUREMENTS AND HIGH-RESOLUTION MASS SPECTROMETRY

In many areas of science (chemistry, biochemistry, etc.), and particularly in drug metabolism studies, there is a need to determine the molecular formula of metabolites of interest. This is where the use of MS for accurate mass measurement has transformed the way in which scientists work. Today, it is very easy to obtain an exact mass measurement of a known or unknown substance.

4.4.1 Atomic and Molecular Mass: Fragment Ion Mass

The actual mass of an atom is very small indeed. For example, a hydrogen atom weighs about 10^{-24} g. Instead of using such an absolute scale, a relative integer mass scale is easier to handle. On this relative scale, all atomic masses have values near to integers. For instance, the absolute masses for hydrogen (1×10^{-24} g) and deuterium (2×10^{-24} g) are relatively 1 and 2, respectively. However, sometimes an integer value is all which may be required. There are some cases, especially in drug metabolism, when trying to solve challenging problems, where exact mass measurement is required. Therefore, by definition, on this accurate mass scale, carbon (^{12}C) is given the value of exactly 12, that is, 12.00000. The accurate masses of some elemental combinations are compared with their respective integer values in Table 4.1.

Decomposition of ions gives fragment ions which can also be mass measured with exact mass. Figure 4.5 illustrates a simple example in which the molecular ion of verapamil fragments to give a diagnostic product ion at m/z 303. The protonated molecular mass at 455.2910 corresponds to the molecular formula (or elemental composition) of $C_{27}H_{38}N_2O_4$ while the fragment ion mass at 303.2073 corresponds to the elemental composition, $C_{18}H_{27}N_2O_4$.

4.4.2 Importance of Exact Mass Measurement

Every element found in nature is very unique because it has a very specific mass. Elements are therefore combined randomly to give rise to compounds. In turn, these particular compounds are also very unique because the exact mass measurement

TABLE 4.1. Relative Integer Masses and Exact Masses for Some Elemental Combinations

Elements	Exact Mass	Integer Mass
CO	27.9949	28
N_2	28.0061	28
C_2H_4	28.0313	28

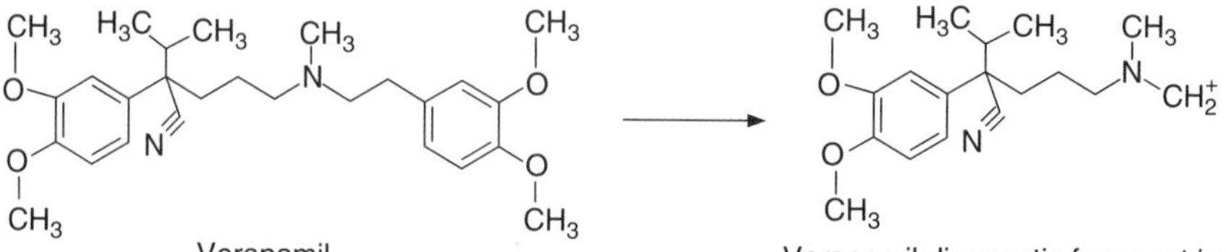

Figure 4.5. *Example of exact mass measurement for fragment ion of verapamil.*

will give us an indication with respect to the type and number of elements that make up a particular compound as well as the number of rings and double bonds [r + d or simply double-bond equivalence (DBE)] present in that particular compound (DBE = $1 + c - 0.5h + 0.5n$, where c is number of carbons, h is the number of hydrogens, n is the number of nitrogens, and other monovalent elements such as F, Cl, Br, and I are treated as hydrogens). The accuracy of an exact mass measurement is usually quoted as the mass difference or "error" between the value measured experimentally and the calculated mass of the compound.

The mass accuracy is usually described as either the absolute mass error in milli-daltons (or thousandths of a mass unit) or as a part-per-million error which is the ratio of the absolute error and the calculated mass multiplied by one million. For instance, if an ion has a calculated or theoretical mass of 400.000 and the measured mass is 400.002, then the error is 2 mDa. The ppm error is calculated by dividing the 2-mDa error by the calculated mass of 400.000 and multiplying by one million to give 5 ppm. Note that if the same absolute error of 2 mDa was found for a compound of theoretical mass of 800, the ppm error would be 2.5 ppm. For the same absolute error, the ppm error varies with mass.

Time-of-flight technology allows exact mass measurement to be performed. A mass can be measured to four decimal places, compared to one decimal place using quadru-pole or ion trap instrumentation. It is possible to have combinations of atoms which have the same nominal mass (or integer). But a differentiation can be made using the monoisotopic values. For example, the exact mass of a hydrogen is 1.0078. Two organic molecules may have the same nominal mass but have different elemental com-position. If such compounds can be mass measured with sufficient accuracy, then it is possible to determine the elemental composition. The specific data acquired using TOF therefore give a high level of confidence when confirming target compounds. In the case of unknowns it is possible to predict an elemental composition with confidence.

Exact mass is currently used in a range of application areas such as metabolism studies, impurity profiling, environmental analysis, synthetic chemistry (open access), industrial applications, clinical screening for drugs of abuse, and natural products, where rapid confirmation and low-level detection with high level of confidence are required.

4.4.3   Resolution of Mass Spectrometers and TOF

Mass resolution (R) is a mass spectrometric instrumental performance parameter defined as the separation of two masses (M_1, M_2) by a mass spectrometer. The

100%
50%
0.05 Da
m/z
499.9
500.0
500.1
Mass = 500
Peak width (at 50%) = 0.05 Da
Resolution (FWHM) = 500/0.05 = 10000

Figure 4.6. *Definition for FWHM used in TOF mass spectrometers.*

most common definition of R is given by Equation (4.2), in which $\Delta M = M_1 - M_2$ and $M = M_1$ or M_2:

$$R = \frac{M}{\Delta M} \tag{4.2}$$

Higher mass resolution results in narrower peaks and greater precision in determining mass-to-charge ratios. Therefore, mass resolution is directly proportional to mass accuracy. As described in Fig. 4.6, in all mass spectrometers, except in magnetic sectors, the resolution is measured as the full width at half maximum (FWHM).

Overall, exact mass is a very valuable piece of information for compound identification. The information generated by exact mass measurements will help with the

+2.0157 amu
+1.9793 amu
Diazepam

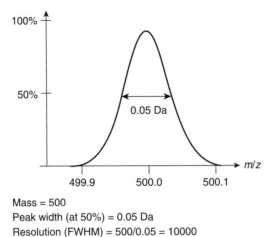

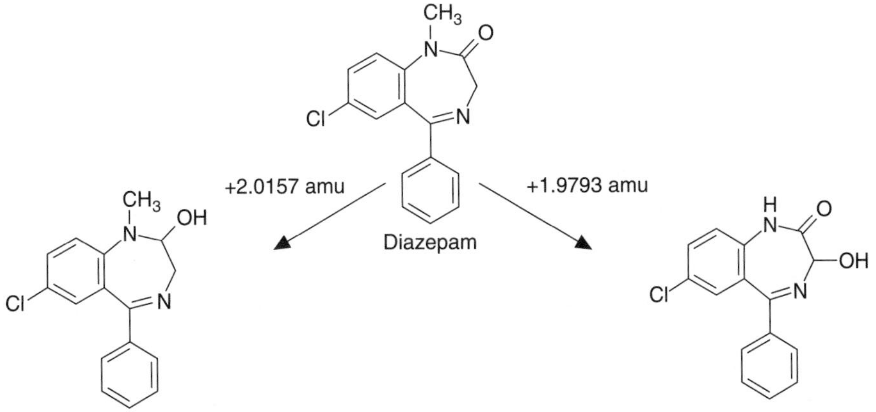

Figure 4.7. *Metabolites of diazepam with same nominal mass but different elemental compositions.*

determination of unknown structures in the MS/MS mode. Exact mass will make a huge difference to the analyst when reduction of false positives is required or verification of a particular compound is needed as in the case of the metabolites of diazepam. As shown in Fig. 4.7, one metabolite is formed via an N-dealkylation followed by a hydroxylation to give a mass difference of 1.9793 Da from that of the diazepam while the other metabolite is formed by reduction of the oxygen to give a mass difference of 2.0157 Da from that of the diazepam. The mass difference between these two metabolites is 0.0364 Da, and a mass spectrometer with a resolving power of 8000 ($286/0.0364 = 7857$) is required to separate the metabolites.

To underscore the utility of accurate mass analysis for distinguishing isomeric metabolites, the metabolism of rabeprazole (used as an inhibitor of gastric acid secretion, marketed under the tradename Aciphex®) is considered. Following incubation of rabeprazole with microsomes, two metabolites with nominal m/z of 344 are formed (Miura et al., 2006; Setoyama et al., 2006). These metabolites, shown below, are the aldehyde (m/z 344.1069) and sulphide (m/z 344.1433) metabolites of rabeprazole:

Rabeprazole ($C_{18}H_{21}N_3O_3S$)
[M+H]+ at m/z 360.1382

Rabeprazole-sulphide ($C_{18}H_{21}N_3O_2S$)
[M+H]$^+$ at m/z 344.1433

Rabeprazole-aldehyde ($C_{17}H_{17}N_3O_3S$)
[M+H]$^+$ at m/z 344.1069

As shown in Fig. 4.8, the mass difference between the two metabolites is 36.4 mDa and a mass spectrometer with a resolving power of at least 9500 ($M/\Delta M = 344/0.0364 = 9450$) is required to separate between these two metabolites. MS/MS fragmentation combined with accurate mass measurement is the preferred method for structural elucidation of metabolites, especially when it can help to correlate the elemental composition determined in the MS mode. In the example shown in Fig. 4.8, exact mass measurements of the precursor ions are used as lock mass to measure the exact mass of each fragment ion. Exact mass measurements of the fragment ion at m/z 226 help to narrow down the sites of modifications and allows one to distinguish between rabeprazole–sulphide and rabeprazole–aldehyde.

Increasing the resolution also allows the separation and differentiation of endogenous compounds in biological matrices from a drug of interest or its metabolites. A comparison of mass spectra of sulfacetamide obtained using a quadrupole mass spectrometer (unit mass resolution) with that obtained using a time-of-flight mass spectrometer ($R = 10,000$) is shown in Fig. 4.9. While at unit mass resolution the

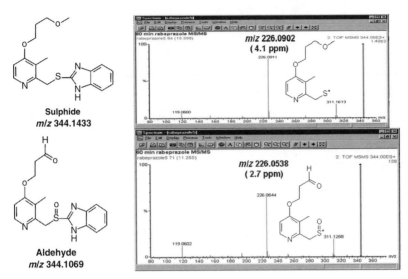

Sulphide
m/z 344.1433

Aldehyde
m/z 344.1069

Figure 4.8. *Example of exact mass use for rabeprazole in MS and MS/MS mode. (Data courtesy of Lars Weidolf, AstraZeneca, Sweden.)*

sulfacetamide and the contaminant peaks are not separated, at a mass resolving power of 10,000 the contaminant and the sulfacetamide peaks are well separated.

In Fig. 4.10, a total ion chromatogram (TIC) from a biological sample is compared with extracted ion chromatograms (XICs) for the analyte of interest (*m/z* 211.97)

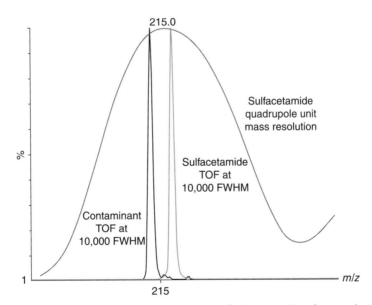

Figure 4.9. *Comparison of low- and high-resolution MS for separation of contaminants from analyte of interest.*

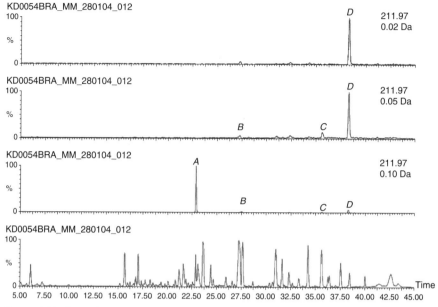

Figure 4.10. *Example of extracted ion chromatograms with different resolution extraction windows. The bottom trace represents the TIC. If we focus on compound D, it is apparent that as the extracted ion chromatogram window is tightened from 0.1 to 0.02 Da, A, B, and C (endogenous peaks) gradually disappear. Therefore, the high resolution of the TOF eliminates interferences and provides a chromatogram with better signal-to-noise ratio.*

obtained using increasing mass resolution. Increasing the mass resolution from 2000 (211.97/0.1) to 10,000 (211.97/0.02) clearly allows the separation of the peak of interest (*D*) from all the nominal mass-to-charge ratio endogenous peaks (*A*, *B*, and *C*). Therefore high mass resolution can tremendously simplify post-acquisition data mining and help to focus on the peaks of interest. The power of high resolution can be further leveraged for data mining by processing the data using ACD/IntelliXtract (Antler, 2008; McGibbon, 2008; McGibbon et al., 2008) or similar liquid chromatography (LC)–MS software tools as elegantly described in a recent review (Balogh, 2006). Section 4.4.5 further describes the utility of software tools for smart LC–MS data mining.

4.4.4 Automation for Exact Mass Measurements

All TOF mass spectrometers measure accurate mass by reference to standard substances, otherwise known as "lock mass." This is an internal reference which is introduced via the ion source by means of a second sprayer system (Fig. 4.11) or coinfused with the sample using a single sprayer (Williams et al., 2006). However, the latter method is not preferred due to dilution of analyte with reference compound and issues of suppression of ionization (Herniman et al., 2004). The reference compound

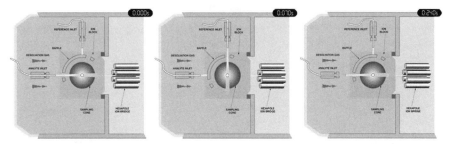

Figure 4.11. Diagram showing the use of Lockspray to introduce a lock mass solution into an ESI source of a Q-TOF mass spectrometer.

may be of any kind; a well-known example is leucine enkephalin, which may be used for both positive- and negative-mode ionization. As shown in Fig. 4.11, when using the dual-sprayer ionization source, the rotor will be moving back and forward at predetermined time intervals so that both reference and analyte data are acquired by the mass spectrometer. For exact mass measurements, the lock mass reference substance and the analyte under investigation have to be present in the mass spectrometer at the same time for near-simultaneous mass measurements. Ions from the lock mass reference standard have known mass and ions from the analyte or unknown (sample) are mass measured by interpolation between successive masses due to the lock mass reference standard.

4.4.5 Data-Processing Algorithms for Metabolite Detection and Identification

The identification of drug metabolites in drug discovery and development is a difficult and time-consuming process. Traditionally, the manual task of sifting through paper copies of multiple, complex data sets to confirm the presence of predicted biotransformations is very labor intensive. LC–MS has become a standard analytical tool in drug metabolism laboratories over the last decade. However, unlike UV detection, the typical LC–MS total ion chromatogram (TIC) often shows little obvious evidence of analytes in the background signal of a complex biological sample matrix (Fig. 4.10). Each spectrum in the chromatographic time frame must be individually checked for evidence of new components.

Although time consuming, the process of checking each spectrum allows the confirmation of expected metabolites based on prior knowledge of the experienced metabolism scientist. However, unexpected components are also common and are not so easily identified. For several years, the pharmaceutical industry has been very successful in applying hybrid quadrupole orthogonal time-of-flight mass spectrometers to solve drug metabolism problems (Hop, 2004; Leclercq et al., 2005). In turn, this has allowed scientists to obtain MS- and MS/MS-mode exact mass data to identify metabolites with great confidence. The bottleneck is no longer in producing analytical data; it has shifted to the processing and interpretation of

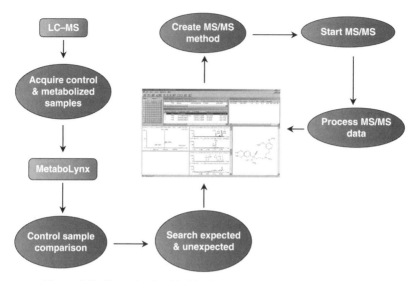

Figure 4.12. *Steps involved in MetaboLynx for metabolite identification.*

these data sets to extract useful information for decision making. The acquisition of exact mass data is the key to maximizing the capability of the software to identify metabolites with more confidence.

Automated software algorithms such as Waters MetaboLynx™ Application-Manager detects putative biotransformations for expected and unexpected putative metabolites (Nassar and Adams, 2003; Mortishire-Smith et al., 2005). The Application-Manager automatically runs samples scheduled for analysis by LC–MS and processes the resulting data (Fig. 4.12). Results are reported via a data browser that enables the chromatographic and mass spectrometric evidence that supports each automated metabolic assignment.

This automated algorithm operates by comparing and contrasting each sample with a control sample, although unexpected metabolite searching may still be performed in the absence of a suitable control. Samples, from in vitro incubations or in vivo dosing experiments, can be quickly analyzed by LC–MS, followed by a multidimensional data search which correlates retention time, m/z value, intensity, and components from alternative detection technologies (e.g., diode array UV or radiochemical monitoring). Comparison of analyte data with the control sample allows filtering of matrix-related peaks, which would otherwise produce an unmanageable list of false metabolite peaks.

Isotopic cluster analysis can be built into the automated processing method and is used to target potential metabolites with the desired isotope ratios (Kind and Fiehn, 2006). For example, chlorine or bromine or radiolabeled drugs/metabolites can be pinpointed, at low levels, within a complex matrix background to dramatically enhance specificity and increase confidence in metabolite identification (Jindal and Lutz, 1986; Shirley et al., 1997).

4.4.6 Exact Mass Data Processing

Exact mass measurements enable the elemental composition of detected peaks to be confirmed for "known" drugs and corresponding metabolites using both MS and MS/MS spectra. For unknowns the number of plausible elemental compositions may be restricted to a small number (or uniquely identified) with the aid of additional chemical information, for example, the molecular formula of the parent drug and knowledge of possible metabolic pathways.

Certain software algorithms such as the one mentioned previously uses the high-quality exact mass data from the high-resolution TOF systems to report calculated elemental compositions within the results browser. Good scientific practice mandates the use of an internal reference or lock mass to obtain valid exact mass measurements (sub-5-ppm). Exact mass measurement gives greater confidence in the confirmation of expected metabolites and allows the prediction of the elemental composition of unknowns. The Q-TOF mass spectrometer can be set up to automatically acquire exact mass LC–MS and exact mass MS–MS data to within the recommended mass accuracy guidelines of 5 ppm.

Under metabolite identification data-processing algorithm control, both LC–MS and LC–MS/MS protocols can be performed sequentially—without operator intervention. The primary LC–MS data are acquired and correlated with a control sample; detected components are then scheduled for LC–MS/MS. Software applications managers are capable of automatically generating and executing an LC–MS/MS sample list for confirmation of the putative metabolites. LC–MS/MS spectra are presented electronically in a results browser for comparison with the parent drug spectrum to highlight common product ions and neutral losses. Alternatively, this algorithm can process data from a data-directed analysis (DDA) experiment or neutral loss as in the case of glutathione (GSH) assays—collecting LC–MS and MS/MS information within a single run—thereby eliminating waste of precious samples on repeat analyses.

4.4.7 Removing False Positives

One of the major obstacles in the use of automated data-processing algorithms is being able to remove unwanted endogenous peaks which are not drug related (Hakala et al., 2006). Typically, depending on the different biological matrices used, there may be hundreds of endogenous ions which may be detected. Clearly, exact mass will help to remove most of these "unwanted peaks." A recent development in this arena has been the use of exact mass data filters or mass defect filter (MDF).

Exact mass filter exclusion based on the decimal places of a parent drug, is a post processing filter which allows complete removal of unexpected entities (ions) which do not agree with the criteria preset by the user. Such a filter is fully adjustable once the samples have been processed. This process can dramatically reduce the number of ions in the analyte sample by filtering out the vast majority of matrix-related ions. This will also allow use of very low threshold values to detect low-level metabolites without having to go through the very tedious and long task of manual exclusion of false positives. Typically, extracted ion chromatogram windows of 0.1 mDa allow the user to get better integration parameters and differentiate drug-related peaks from endogenous peaks (Fig. 4.10). A narrower XIC window will also make the entire control comparison process a lot more accurate for both control and analyte samples.

4.4.8 Exact Mass Filter Window

An exact mass filter window is an extremely accurate and specific filter which is based on the exact mass and mass deficiencies of the parent drug (Zhang et al., 2003; Zhu et al., 2007). Since the elemental composition (specific number of C, H, N, O, and so on) of the parent drug is known, the mass deficiency specific to the parent drug is also known. For example, clozapine (Fig. 4.13) contains 18 carbons, 19 hydrogens, 4 nitrogens, and a chlorine ($C_{18}H_{19}N_4Cl$), which equates to a monoisotopic protonated mass of 327.1376 Da. If an alkyl group is removed due to biotransformation (N-dealkylation, a common metabolic route), the mass is shifted by -14.0157 Da to a monoisotopic mass of 313.1219 Da. The mass difference between clozapine and its N-dealkylated metabolite is 0.0157 Da ($0.1376 - 0.1219 = 0.0157$ Da). Therefore, by limiting the exact mass filter to around 20 mDa, only N-dealkylated metabolite is detected and all the other entries/ions which fall outside this window are excluded.

With this in mind, the following can be proposed:

- All metabolites have masses within 0.25 Da of the parent ion's decimals.
- Most metabolites are in general within 0.1 Da of the parent ion's decimals, except when there are major metabolic cleavages leading to much smaller fragments. As an example, the biggest phase II biotransformation, glutathione conjugation, will lead to a mass defect difference of 0.07 Da compared to that of the parent drug.
- Most metabolites fall within a 0.180-Da window of the parent compound, even after cleavages yielding to smaller fragments (metabolites) (Fig. 4.14).

Therefore, using these hypothetical assumptions it is possible to remove "unwanted false positives" from the data processing step. However, if a compound contains

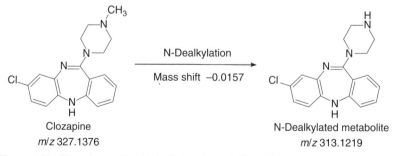

Figure 4.13. *Clozapine and its N-dealkylated metabolite with corresponding mass shift.*

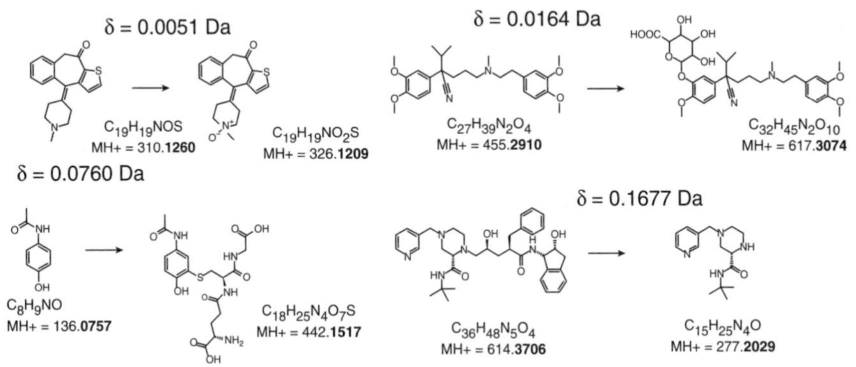

δ = 0.0051 Da

C$_{19}$H$_{19}$NOS
MH+ = 310.**1260**

→

C$_{19}$H$_{19}$NO$_2$S
MH+ = 326.**1209**

δ = 0.0164 Da

C$_{27}$H$_{39}$N$_2$O$_4$
MH+ = 455.**2910**

→

C$_{32}$H$_{45}$N$_2$O$_{10}$
MH+ = 617.**3074**

δ = 0.0760 Da

C$_8$H$_9$NO
MH+ = 136.**0757**

→

C$_{18}$H$_{25}$N$_4$O$_7$S
MH+ = 442.**1517**

δ = 0.1677 Da

C$_{36}$H$_{48}$N$_5$O$_4$
MH+ = 614.**3706**

→

C$_{15}$H$_{25}$N$_4$O
MH+ = 277.**2029**

Figure 4.14. *Variety of metabolites for ketotifen, verapamil, indinavir, and para-acetoaminophenol including cleavages which take place giving a maximum mass deficiency of 167.7 mDa away from the parent compound. Mass deficiency shifts are very specific for each metabolite and parent drug.*

elements other than C, N, H, and O then the cleavage metabolites formed through loss of the site containing such elements (halogens, sulfur, etc.), may not exhibit a linear mass defect (Mortishire-Smith et al., 2007). Therefore, exact mass filtering or mass defect filtering set around the m/z of the parent $+/-$ defined tolerances may result in false negative results or miss some of the key metabolites as shown in Fig. 4.15. To overcome some of the limitations associated with MDF, intelligent filtration approach was designed (Mortishire-Smith et al., 2007). The intelligent MDF relies on the prediction of plausible metabolic cleavages upfront with the aim of generating multiple mass defect filters for each new chemical entity to be

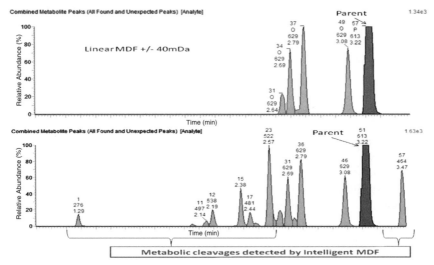

Figure 4.15. *Comparison of linear MDF (top panel) vs. intelligent MDF (bottom panel) for in-vitro formed human metabolites of indinavir.*

interrogated (Fig. 4.15). The intelligent MDF is designed to help remove the matrix ions and avoid filtering out metabolites. To account for the most possible "known and unknown" metabolites, combinations involving phases I and II metabolic possibilities are included along with the parent.

4.5   REACTIVE DRUG METABOLISM AND Q-TOF TECHNOLOGY

Toxicity is a major cause for the withdrawal of drugs from the market and it is a major concern for pharmaceutical researchers. Reactive metabolism is certainly a very "hot topic" within the whole approach to drug metabolism. The downstream consequences of not identifying reactive metabolites may be financially catastrophic. There is an increasing drive to have early prediction of the metabolic fate and interactions of candidate drug molecules. Factors such as metabolic stability, toxic metabolite production, and P450 inhibition and induction are all routinely monitored to prevent compounds with "poor" pharmacokinetic properties from progressing forward onto clinical trials.

Over recent years, a number of drugs have been withdrawn due to adverse reactions (Lazarou et al., 1998; Lasser et al., 2002). Adverse drug reactions (ADRs) are a major clinical problem (Naisbitt et al., 2001; Williams et al., 2002; Park et al., 2005). ADRs account for significant morbidity and mortality. Idiosyncratic drug reactions are very life threatening, and as a consequence several drugs have been removed from the market (Guzey and Spigset, 2004). It is believed that the majority of these reactions may be caused by immune-mediation, which is caused by immunogenic conjugates formed by the reaction of a reactive metabolite from a drug with cellular proteins (Uetrecht, 2003; Nassar and Lopez-Anaya, 2004). Not all reactive metabolites will be detoxified, and the ones which manage to escape this process may undergo covalent binding to macromolecules in certain target cells (Kalgutkar et al., 2002, 2005, 2007). Even though there is no direct link between toxicity and formation of reactive metabolites, there is plenty of evidence indicating that idiosyncratic drug reactions are due to reactive metabolites (Park et al., 2000, 2001; Baillie and Kassahun, 2001; Naisbitt et al., 2001; Evans et al., 2004). It is rather challenging and difficult to predict which reactive metabolites will be toxic. Moreover, it is rather difficult to quantify the amount of reactive metabolite formed. However, there seems to be a correlation between the amount of a particular reactive metabolite formed and the actual risk of idiosyncratic reactions. For example, patients producing higher levels of reactive metabolites may be more susceptible to developing idiosyncratic drug reactions. High levels of these reactive metabolites in certain patients may be due to the presence of abnormally high levels of the enzyme which activates the drug to form reactive metabolite(s). The rate at which biotransformation and detoxification are formed depends on genetic variations.

Research over the past three decades in the field of biological reactive intermediates has provided a wealth of information on the functional groups that may be converted by either phase I or phase II enzymes to electrophilic metabolites

(Baillie and Kassahun, 2001). Despite the rather extensive literature on the mechanisms by which drugs and other foreign compounds undergo metabolic activation, studies show that numerous functional groups which have not been recognized as precursors to reactive intermediates also can undergo bioactivation. Therefore, the detection and identification of biologically reactive metabolites have become an essential requirement in the discovery and development of drug candidates within the pharmaceutical and biotechnology industry.

In most cases, reactive metabolites are electrophiles, which are susceptible to covalent binding with nucleophilic portions of proteins, DNA, and cellular GSH. The generation of GSH conjugates is a clearance pathway for the body to remove reactive metabolites. Such reactions in vivo may lead to enzyme deactivation, genotoxicity, hapten formation triggering immune responses, compromised cellular function due to GSH depletion and increased oxidative stress, apoptosis, and cumulative damage to long-lived proteins and ultimately cause cell death (Naisbitt et al., 2001; Park et al., 2001, 2005). Several analytical strategies exist to detect reactive metabolite formation. Most prevalently used strategies are GSH depletion assays, time-dependent P450 inhibition assays, covalent binding of radiolabeled material, and GSH and cyano trapping of reactive species with LC–MS/MS analysis (Chen et al., 2001; Samuel et al., 2003a; Yan and Caldwell, 2004; Ma and Subramanian, 2006).

The most commonly used reactive metabolism assays involve GSH-trapping experiments. GSH is a tripeptide (γ-L-glutamyl-L-cisteinylglycine) present in mammalian systems. GSH plays an important role in the detoxification of electrophilic foreign compounds and chemically reactive intermediates, which may arise during the biotransformation of xenobiotics/drugs. There are two types of GSH conjugations which are fairly common; displacement reactions and nucleophilic additions (Michael reactions). In displacement reactions the GSH displaces an electron withdrawing group, such as halogens, nitriles, and caboxylic acids (Fig. 4.16). Michael addition (Fig. 4.17) includes the nucleophilic addition to an α, β unsaturated carbonyl compound. Also, conjugated aldehydes, esters, nitriles, amides, and nitro compounds can all act as the electrophilic acceptor components. It is worth pointing out the fact that in certain specific cases steric hindrance of the double bond may prevent Michael addition. For this reason, GSH conjugation is an important biotransformation to consider in drug metabolism and biochemical toxicology. There are several methods available (Guzey and Spigset, 2004; Nassar and Lopez-Anaya, 2004; Madsen et al., 2007; Yan et al., 2007; Zheng et al., 2007) to identify reactive metabolites by the use of chemical trapping agents, such as reduced GSH or cyanide trapping, to form stable adducts that are amenable to characterization by LC-tandem MS (Evans et al., 2004; Soglia et al., 2006; Yan et al., 2007). One of the most commonly used methods is glutathione (GSH) trapping with hepatic sub cellular fractions in the presence of cytochrome P450 cofactor NADPH or microsomal sub cellular fractions in the presence of cytochrome P450 fortified with GSH at concentrations ranging from 0.5–5 mM. Using this trapping methodology, the electrophilic metabolites are trapped via the cysteine sulfhydryl group to form the GSH conjugate.

Other less common but highly effective trapping techniques include cyanide trapping via iminium ion formation and methoxylamine or semicarbazide for reactive

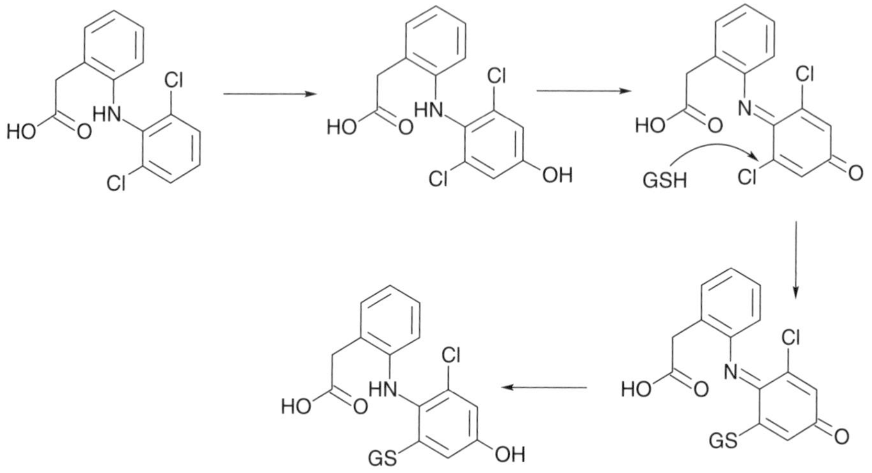

Figure 4.16. *Conjugation of GSH with diclofenac via displacement reaction.*

intermediate aldehydes. The cyanide anion (CN^-) is a "hard" nucleophile which acts as a very powerful trapping agent for certain types of reactive electrophiles such as compounds containing alicyclic amines which may have the tendency to form iminium ions that may exhibit reactivity towards macromolecules. Potassium cyanide is typically used as the reagent for the trapping assay at a concentration of 1 mM which is then added to the incubation mixture for the amount of time desired. This incubation may give rise to cyanide adducts which produces, under CID conditions, a diagnostic neutral loss of 27 (Gorrod et al., 1994). Methoxylamine or semicarbazide trapping is also another effective trapping chemical for reactive aldehydes which have the propensity to react with the lysine residues on proteins. The main aim here is the capability of these reagents to form a Shiff base with the aldehydes and trap the reactive intermediates.

The detection, identification, and quantitative analysis of GSH conjugates have advanced with time as new analytical technologies and techniques become available (Baillie and Davis, 1993). GSH conjugates are chemically and thermally unstable, and they are very polar. Therefore, analysis using gas chromatography (GC)–MS requires

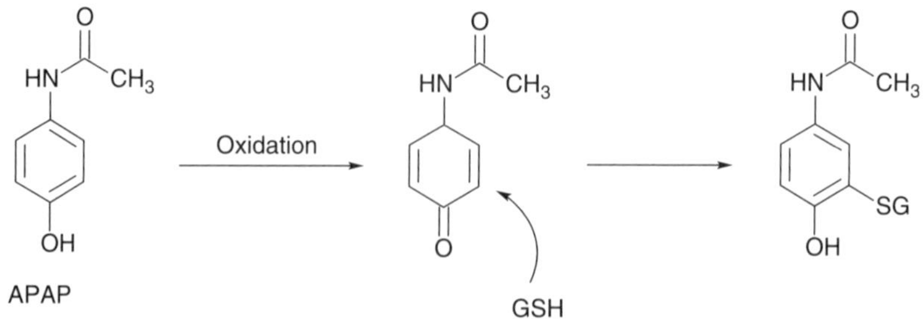

Figure 4.17. *Conjugation of GSH with acetoaminophen (APAP) via Michael addition.*

extensive derivatization of GSH conjugates (Wolf et al., 1980) and does not produce a molecular ion in direct chemical ionization MS (Nelson et al., 1981). Characterization by fast atom bombardment (FAB) or plasma desorption (PD) MS gives reliable molecular weight information but limited diagnostic fragment ions (Haroldsen et al., 1988; Deterding et al., 1989; Bean et al., 1990; Ramanathan et al., 1998).

The advent of atmospheric pressure ionization (API) in the early 1990s allowed LC to be directly coupled to MS and by the mid 1990s API-based MS became a common tool in most drug metabolism laboratories. The enhanced selectivity and sensitivity made LC–MS the instrument of choice for both quantitative and qualitative analysis (Feng, 2004; Korfmacher, 2005; Zhou et al., 2005). GSH conjugates, under collision-induced dissociation (CID) conditions, fragment to give a characteristic diagnostic loss corresponding to the pyroglutamic acid (129 Da) moiety (Fig. 4.18). Therefore, constant-neutral-loss scanning for 129 has become a screening tool for GSH conjugates (Chen et al., 2001; Samuel et al., 2003b; Yan and Caldwell, 2004; Mutlib et al., 2005; Zheng et al., 2007). Having said this, the loss of the pyroglutamic acid moiety is not always the diagnostic neutral loss of interest (Baillie et al., 1993). This particular neutral loss very much depends on the position where the GSH adduct is attached in the molecule. Other common diagnostic neutral losses are 307 and 147 for thioethers (aliphatic and benzylic) and thioesters, respectively. In addition, another common screening technique for GSH is to monitor the γ-glutamyl-dehydroanalyl-glycine ion (m/z 272) using the precursor ion scan mode in electrospray negative ion mode (Fig. 4.19).

Until most recently, GSH assay involving neutral loss of 129 has mostly been carried out using triple-quadrupole mass spectrometers operated under nominal mass conditions. This approach suffers from low sensitivity because both

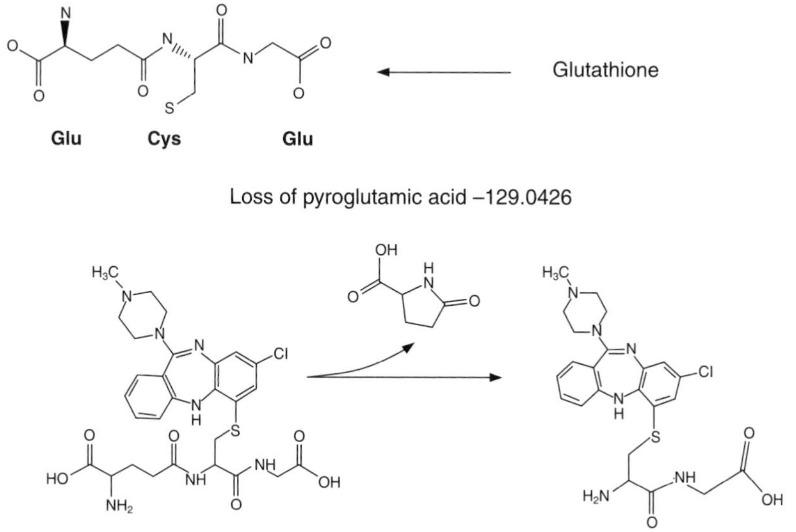

Figure 4.18. Example of neutral loss experiment showing loss of pyroglutamic acid moiety.

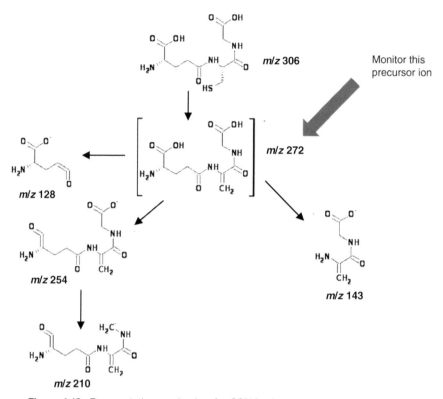

Figure 4.19. *Fragmentation mechanism for GSH in electrospray negative ion mode.*

quadrupoles (Q1 and Q3) are operated in the full-scan mode (Fig. 4.20). Another disadvantage of using a triple-quadrupole-based GSH screen is the poor selectivity because the neutral loss of 129 Da is not exclusive for GSH adducts and endogenous components from biological matrices can give a false-positive signal.

From Fig. 4.21, it can be observed that only one component gave rise to the true loss of m/z 129. In turn this will mean that more time will have to be spent reviewing meticulously the data acquired to remove false positives. Furthermore, due to the fact that not always the loss of m/z 129 will be present due to the nature of the

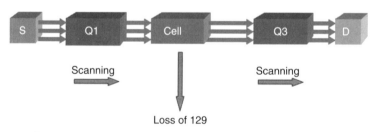

Figure 4.20. *Schematic of neutral loss experiments with triple-quadrupole mass spectrometer.*

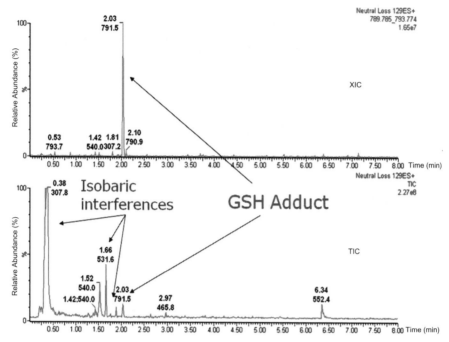

Figure 4.21. Nominal mass isobaric interferences for the GSH adduct of nefazodone when tandem quadrupole mass spectrometry is used.

conjugation, this would translate in the requirement of multiple neutral loss experiments to obtain maximum coverage for all types of new chemical entities. The low duty cycle of scanning instruments may hinder detection of low level GSH adducts. The other significant but important factor is that most of the GSH adducts may manifest as the doubly charged species $[M + 2H]^{2+}$ which does not fragment during CID to provide the appropriate neutral loss of interest but giving rise to singly charged species. Most often a subsequent product ion MS/MS or comparison with control sample is necessary for further verification. Therefore, exact mass measurement can play a pivotal role in determining whether the putative GSH is real or not. Exact mass measurements can be used to readily exclude false positives and remove unwanted interferences to make the GSH or other trapping assays truly high throughput.

Conventionally, metabolite identification and more specifically reactive metabolite screening typically uses an array of chromatographic and mass spectrometric methods, and may require multiple injections of the same sample. This is to ensure that enough information has been collected to detect all metabolites and to have sufficient fragmentation information available to elucidate structures.

With this in mind, an efficient and rapid approach would be to use the combination of UPLC and Q-TOF technology. The idea behind this approach is to have better ways to filter unwanted "false positives" with minimal repeat of injections.

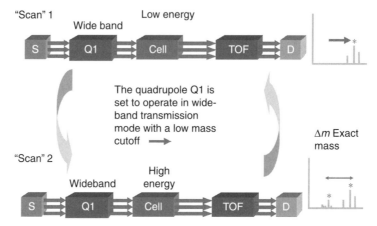

Figure 4.22. *Schematic of exact mass neutral loss experiments with Q-TOF mass spectrometer.*

A method previously described by Wrona et al. (2005) and Bateman et al. (2007) enables such collection of both parent and fragment information from a single injection. This alternating low/high energy data acquisition (MS^E) uses two scan functions interleaved such that the first (low collision energy, or low CE) function contains data from the intact metabolites and the second (high collision energy, or high CE) function contains data from the fragment ions. As shown in Fig. 4.22, the first quadrupole (Q1) is operated in the wide-bandpass mode to transmit all the ions into the collision cell. Argon is always present in the collision cell and alternate high- and low-energy spectra are generated by switching the collision energy between low and high collision energies. The typical low collision energy used is 5 eV and the high collision energy ramp used range from 15–35 eV. Mass separation is constantly achieved in the TOF analyzer and all ions are detected in the TOF part of the Q-TOF. The low energy spectra contain predominantly unfragmented precursor ions, while the high-energy spectra contain all of the fragmented ions of all precursor ions present at a particular time.

One key advantage of obtaining all of the information in full scan mode is that the less obvious possible doubly charged species for GSH adducts will be detected together with the singly charged species. The comparison between the low energy and high energy trace allows for very accurate neutral losses and precursor ions for specific key diagnostic ions. Confirmation in the low energy data with exact mass and retention time alignment between both low and high energy traces allows for extra confidence in the results obtained. For instance, the exact mass of the pyroglutamic acid fragment loss from a GSH conjugate is 129.0426 Da. The mass spectrometer is set to perform consecutive low and high energy scans and to look for ions separated by neutral loss of 129.0426 Da.

Typically, the mass window is set for $+/-10$ mDa. The purpose of using a narrow mass window is to provide high selectivity for GSH conjugate detection and to exclude false positives from nominal mass endogenous components.

The advantage of this approach is that, since all the data is collected in one run, post-acquisition processing of multiple fragment ions is thus possible. With this approach, the entire data set is then mined post-acquisition for specific metabolite masses, precursor and product ions, and neutral losses as all the necessary data is collected simultaneously. Selectivity for biotransformation of the parent drug is achieved through exact mass measurement. A variety of data processing algorithms can be used to extract metabolite information from these data previously described (Tiller et al., 2008). From a single injection it is possible to obtain neutral loss and precursor ion information with exact mass containing diagnostic losses for reactive metabolites for both neutral and precursor ions acquisitions. In turn, these diagnostic neutral losses and precursor ions may also be used for in-vitro reactive metabolism screening in conjunction with the low energy data to confirm the presence of a reactive electrophile intermediates.

This strategy is clearly defined by the use of a model compound such as nefazodone. Nefazodone is an antidepressant which was approved in the USA in late 1994. In spite of its therapeutic effects there has been a number of cases (55 cases of liver failure (20 fatal) and another 39 cases of less severe liver failure) reported showing hepatobiliary dysfunction and cholestasis (Kalkutkar et al., 2005). In this example the described MS^E approach was utilized for this particular compound from an in-vitro rat microsomal incubation fortified with GSH. This compound undergoes the loss of the pyroglutamic acid as it can be observed in Fig. 4.23. A total of 5 GSH adducts were detected in positive ion mode. These corresponded to; m/z 757 (+O—Cl+GSH), two m/z 791 (+O+GSH), and two m/z 807 (+O_2+GSH). The 5 GSH adducts were confirmed by reviewing the data in the low energy scan using exact mass.

Even though it was possible to detect all GSH adducts for this drug using this approach, this strategy is not restricted to just searching one particular diagnostic loss as all the data is contained within the low and high energy acquisitions we could potentially search for an unlimited number of neutral losses utilizing exact mass to remove "false positives." It is worth mentioning that in some cases the diagnostic neutral losses or precursor ions for GSH are not always generated. Even if this is the case intact full scan exact mass MS data is always available with this approach. This may not be the case with other techniques such as neutral loss scanning with a tandem quadrupole, thus resulting in not detecting a potential GSH adduct. MS^E can be used to confirm the presence of a GSH adduct which does not follow the neutral loss "rules" and further verified in the high energy.

Electrospray negative ion MS^E as previously mentioned may also be carried out by extracting diagnostic the precursor ion of m/z 272 (Fig. 4.24) from the high energy mode and further confirmation in the low energy mode. In this example, the fragment data derived from the precursor ion of m/z 272 (the γ-glutamyl-dehydroanalyl-glycine) is shown but again is not confined to search just one precursor ions but all precursor ions of interest as "all the data" are present. This will include other diagnostic ions such as m/z 254, 210, and 143. Therefore, enhancing the chances of detecting low level GSH adducts.

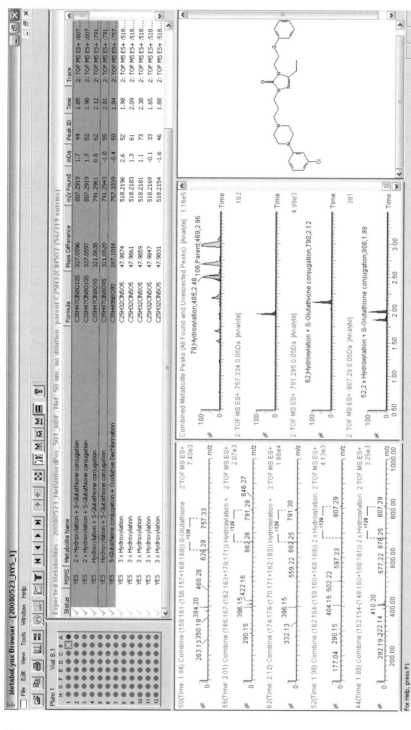

Figure 4.23. Low and high energy acquisitions denoting GSH adducts of nefazodone which correspond to the loss of the pyroglutamic acid and further confirmed in the low energy trace with exact mass.

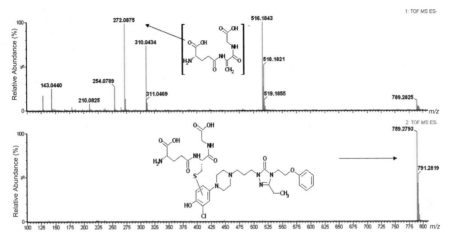

Figure 4.24. LC–MS (bottom panel) and MS/MS (top panel) spectra of GSH adduct of nefazodone obtained using negative ion mode ESI–MS.

4.6 SUMMARY

Until most recently, tandem-quadrupole or linear quadrupole ion trap MS-based GSH assays were considered ideal for high-throughput drug discovery and development. Under a neutral loss (129-Da) mode of operation, a tandem-quadrupole or linear quadrupole ion trap mass spectrometer can be used to detect GSH conjugates without any prior knowledge about the drug or its metabolite which has undergone GSH conjugation. To improve on the sensitivity of the tandem-quadrupole mass spectrometer GSH assays, quadrupole ion trap (Agrawal et al., 2006) and linear ion trap-methods have been evaluated (Zheng et al., 2007). To further improve the sensitivity and specificity, stable-isotope-labeled GSH has been employed in a linear ion MS-based GSH assay (Mutlib et al., 2005).

The advantages of the Q-TOF mass spectrometer over tandem quadrupole, quadrupole ion trap, and linear ion trap are sensitivity, scan speed, and mass accuracy. Especially, the scan speed (duty cycle) achievable with the Q-TOF mass spectrometer allows the utility of UPLC for optimal speed and higher throughput. The combined UPLC and exact mass neutral loss scan GSH assay, described in this chapter, gives the potential to automate the process of high-throughput in vitro and in vivo metabolite identification and GSH conjugate detection and identification. This strategy provides several advantages over other LC–MS/MS methodology based on quadrupoles, quadrupole ion traps, and linear ion traps in terms of the information generated from a single run.

As future direction, and the fact that mass spectrometry has evolved very rapidly in the last decade, other mass spectrometry based technologies such as ion mobility (Thalassinos et al., 2004; Clemmer et al., 2005) may play an important role in the screening of reactive metabolites. This separation stage is orthogonal to the LC and mass spectrometric separations and occurs on an intermediate timescale between the LC and MS. Ion mobility separates ionic species as they drift through a gas under the influence of an electric field. The rate of the drift depends on the factors such as the mass of the ion, its particular charge state and the interaction cross-section of the ion with the gas. Consequently, it is possible to separate ions with the same nominal m/z value if they have different charge states or different interaction cross-sections. This results in a data set with an extra dimension such as drift time information for each ion of interest which in turn will allow researchers to have more valuable tools to search for "difficult to detect" significant metabolites and further remove endogenous isobaric components.

REFERENCES

Agrawal, S., Winnik, B., Buckley, B., Mi, L., Chung, F.-L., and Cook, T. J. (2006). Simultaneous determination of sulforaphane and its major metabolites from biological matrices with liquid chromatography-tandem mass spectroscopy. *J. Chromatogr. B Anal. Technol. Biomed. Life Sci.* **840**:99–107.

Antler, M. (2008). Fragment prediction and assignment for experimental mass spectra. In *Proceedings of the 56th ASMS Conference on Mass Spectrometry and Allied Topics*, Denver, CO.

Baillie, T. A., and Davis, M. R. (1993). Mass spectrometry in the analysis of glutathione conjugates. *Biol. Mass. Spectrom.* **22**:319–325.

Baillie, T. A., and Kassahun, K. (2001). Biological reactive intermediates in drug discovery and development: A perspective from the pharmaceutical industry. *Adv. Exp. Med. Biol.* **500**:45–51.

Balogh, M. P. (2006). Spectral Interpretation, Part II: Tools of the trade. *LCGC* **24**:762–769.

Bateman, K. P., Castro-Perez, J., Wrona, M., Shockcor, J. P., Yu, K., Oballa, R., and Nicoll-Griffith, D. A. (2007). MS^E with mass defect filtering for *in vitro* and *in vivo* metabolite identification. *Rapid Commun. Mass Spectrom.* **21**:1485–1496.

Bean, M. F., Pallante-Morell, S. L., Dulik, D. M., and Fenselau, C. (1990). Protocol for liquid chromatography/mass spectrometry of glutathione conjugates using postcolumn solvent modification. *Anal. Chem.* **62**:121–124.

Chen, W. G., Zhang, C., Avery, M. J., and Fouda, H. G. (2001). Reactive metabolite screen for reducing candidate attrition in drug discovery. *Adv. Exp. Med. Biol.* **500**:521–524.

Churchwell, M. I., Twaddle, N. C., Meeker, L. R., and Doerge, D. R. (2005). Improving LC-MS sensitivity through increases in chromatographic performance: Comparisons of UPLC-ES/MS/MS to HPLC-ES/MS/MS. *J. Chromatogr. B Anal. Technol. Biomed. Life Sci.* **825**:134–143.

Clemmer, D. E., Valentine, S. J., Liu, X., Plasencia, M. D., Hilderbrand, A. E., Kurulugama, R. T., Koeniger, S. L. (2005). Developing liquid chromatography ion mobility mass spectrometry techniques. *Expert Rev. Proteomics.* **2**(4):553–565.

Deterding, L. J., Srinivas, P., Mahmood, N. A., Burka, L. T., and Tomer, K. B. (1989). Fast atom bombardment and tandem mass spectrometry for structure determination of cysteine, *N*-acetylcysteine, and glutathione adducts of xenobiotics. *Anal. Biochem.* **183**:94–107.

Evans, D. C., Watt, A. P., Nicoll-Griffith, D. A., and Baillie, T. A. (2004). Drug-protein adducts: An industry perspective on minimizing the potential for drug bioactivation in drug discovery and development. *Chem. Res. Toxicol.* **17**(1):3–16.

Feng, W. Y. (2004). Mass spectrometry in drug discovery: A current review. *Curr. Drug. Discov. Technol.* **1**:295–312.

Goodwin, L., White, S. A., and Spooner, N. (2007). Evaluation of ultra-performance liquid chromatography in the bioanalysis of small molecule drug candidates in plasma. *J. Chromatogr. Sci.* **45**:298–304.

Guzey, C., and Spigset, O. (2004). Genotyping as a tool to predict adverse drug reactions. *Curr. Top. Med. Chem.* **4**:1411–1421.

Hakala, K. S., Kostiainen, R., and Ketola, R. A. (2006). Feasibility of different mass spectrometric techniques and programs for automated metabolite profiling of tramadol in human urine. *Rapid Commun. Mass Spectrom.* **20**:2081–2090.

Haroldsen, P. E., Reilly, M. H., Hughes, H., Gaskell, S. J., and Porter, C. J. (1988). Characterization of glutathione conjugates by fast atom bombardment/tandem mass spectrometry. *Biomed. Environ. Mass. Spectrom.* **15**:615–621.

Herniman, J. M., Bristow, T. W. T., O'Connor, G., Jarvis, J., and Langley, G. J. (2004). Improved precision and accuracy for high-performance liquid chromatography/Fourier transform ion cyclotron resonance mass spectrometric exact mass measurement of small molecules from the simultaneous and controlled introduction of internal calibrants via a second electrospray nebuliser. *Rapid Commun. Mass Spectrom.* **18**:3035–3040.

Hop, C. E. C. A. (2004). Applications of quadrupole-time-of-flight mass spectrometry to facilitate metabolite identification. *Am. Pharm. Rev.* **7**:76–79.

Jerkovitch, A. D., Mellors, J. S., and Jorgenson, J. W. (2003). Ultra performance liquid chromatography (UPLC): An introduction. *LCGC* **21**:8–14.

Jindal, S. P., and Lutz, T. (1986). Ion cluster techniques in drug metabolism: Use of a mixture of labeled and unlabeled cocaine to facilitate metabolite identification. *J. Anal. Toxicol.* **10**:150–155.

Johnson, K. A., and Plumb, R. (2005). Investigating the human metabolism of acetaminophen using UPLC and exact mass oa-TOF MS. *J. Pharm. Biomed. Anal.* **39**:805–810.

Kalgutkar, A. S., Dalvie, D. K., O'Donnell, J. P., Taylor, T. J., and Sahakian, D. C. (2002). On the diversity of oxidative bioactivation reactions on nitrogen-containing xenobiotics. *Curr. Drug. Metab.* **3**:379–424.

Kalgutkar, A. S., et al. (2005). Bioactivation of the nontricyclic antidepressant nefazodone to a reactive quinone-imine species in human liver microsomes and recombinant cytochrome P450 3A4. *Drug Metab. Dispos.* **33**:243–253.

Kalgutkar, A. S., Obach, R. S., and Maurer, T. S. (2007). Mechanism-based inactivation of cytochrome p450 enzymes: Chemical mechanisms, structure-activity relationships and relationship to clinical drug-drug interactions and idiosyncratic adverse drug reactions. *Curr. Drug. Metab.* **8**:407–447.

Kamel, A., and Prakash, C. (2006). High performance liquid chromatography/atmospheric pressure ionization/tandem mass spectrometry (HPLC/API/MS/MS) in drug metabolism and toxicology. *Curr. Drug. Metab.* **7**:837–852.

Kind, T., and Fiehn, O. (2006). Metabolomic database annotations via query of elemental compositions: Mass accuracy is insufficient even at less than 1 ppm. *BMC Bioinformat.* **7**:234.

Korfmacher, W. A. (2005). Foundation review: Principles and applications of LC-MS in new drug discovery. *Drug Discov. Today* **10:**1357–1367.

Lasser, K. E., Allen, P. D., Woolhandler, S. J., Himmelstein, D. U., Wolfe, S. M., and Bor, D. H. (2002). Timing of new black box warnings and withdrawals for prescription medications. *JAMA* **287:**2215–2220.

Lazarou, J., Pomeranz, B. H., and Corey, P. N. (1998). Incidence of adverse drug reactions in hospitalized patients: A meta-analysis of prospective studies. *JAMA* **279:**1200–1205.

Leclercq, L., Delatour, C., Hoes, I., Brunelle, F., Labrique, X., and Castro-Perez, J. (2005). Use of a five-channel multiplexed electrospray quadrupole time-of-flight hybrid mass spectrometer for metabolite identification. *Rapid Commun. Mass Spectrom.* **19:**1611–1618.

Ma, S., and Subramanian, R. (2006). Detecting and characterizing reactive metabolites by liquid chromatography/tandem mass spectrometry. *J. Mass Spectrom.* **41:**1121–1139.

Madsen, K. G., Olsen, J., Skonberg, C., Hansen, S. H., and Jurva, U. (2007). Development and evaluation of an electrochemical method for studying reactive phase-I metabolites: Correlation to in vitro drug metabolism. *Chem. Res. Toxicol.* **20:**821–831.

McGibbon, G. A. (2008). Information extraction from full-scan LC/MS data using isotope patterns. In *Proceedings of the 56th ASMS Conference on Mass Spectrometry and Allied Topics*, Denver, CO.

McGibbon, G. A., Bayliss, M. A., Antler, M., Lashin, V. (2008). Automated software analysis of isotope cluster mass differences for components in LC-MS datasets. In *Proceedings of the 56th ASMS Conference on Mass Spectrometry and Allied Topics*, Denver, CO.

Mensch, J., Noppe, M., Adriaensen, J., Melis, A., Mackie, C., Augustijns, P., and Brewster, M. E. (2007). Novel generic UPLC/MS/MS method for high throughput analysis applied to permeability assessment in early drug discovery. *J. Chromatogr. B.* **847:**182–187.

Miura, M., Tada, H., Satoh, S., Habuchi, T., and Suzuki, T. (2006). Determination of rabeprazole enantiomers and their metabolites by high-performance liquid chromatography with solid-phase extraction. *J. Pharm. Biomed. Anal.* **41:**565–570.

Mortishire-Smith, R. J., O'Connor, D., Castro-Perez, J. M., and Kirby, J. (2005). Accelerated throughput metabolic route screening in early drug discovery using high-resolution liquid chromatography/quadrupole time-of-flight mass spectrometry and automated data analysis. *Rapid Commun. Mass Spectrom.* **19:**2659–2670.

Mortishire-Smith, R. J., Hill, A., and Castro-Perez, J. M. (2007). Generic dealkylation: A tool for increasing the hit-rate of metabolite identification, and customizing mass defect filters. In *Proceedings of the 55th ASMS Conference on Mass Spectrometry and Allied Topics*, Indianapolis, IN.

Mutlib, A., Lam, W., Atherton, J., Chen, H., Galatsis, P., and Stolle, W. (2005). Application of stable isotope labeled glutathione and rapid scanning mass spectrometers in detecting and characterizing reactive metabolites. *Rapid Commun. Mass Spectrom.* **19:**3482–3492.

Naisbitt, D. J., Williams, D. P., Pirmohamed, M., Kitteringham, N. R., and Park, B. K. (2001). Reactive metabolites and their role in drug reactions. *Curr. Opin. Allergy Clin. Immunol.* **1:**317–325.

Nassar, A. E., and Adams, P. E. (2003). Metabolite characterization in drug discovery utilizing robotic liquid-handling, quadruple time-of-flight mass spectrometry and in-silico prediction. *Curr. Drug. Metab.* **4:**259–271.

Nassar, A. E., and Lopez-Anaya, A. (2004). Strategies for dealing with reactive intermediates in drug discovery and development. *Curr. Opin. Drug. Discov. Devel.* **7:**126–136.

Nelson, S. D., Vaishnav, Y., Kambara, H., and Baillie, T. A. (1981). Comparative electron impact, chemical ionization and field desorption mass spectra of some thioether metabolites of acetaminophen. *Biomed. Mass Spectrom.* **8:**244–251.

Park, B. K., Kitteringham, N. R., Powell, H., and Pirmohamed, M. (2000). Advances in molecular toxicology—Towards understanding idiosyncratic drug toxicity. *Toxicology* **153:**39–60.

Park, B. K., Naisbitt, D. J., Gordon, S. F., Kitteringham, N. R., and Pirmohamed, M. (2001). Metabolic activation in drug allergies. *Toxicology* **158:**11–23.

Park, K., Williams, D. P., Naisbitt, D. J., Kitteringham, N. R., and Pirmohamed, M. (2005). Investigation of toxic metabolites during drug development. *Toxicol. Appl. Pharmacol.* **207:**425–434.

Pedraglio, S., Rozio, M. G., Misiano, P., Reali, V., Dondio, G., and Bigogno, C. (2007). New perspectives in bioanalytical techniques for preclinical characterization of a drug candidate: UPLC-MS/MS in in vitro metabolism and pharmacokinetic studies. *J. Pharm. Biomed. Anal.* **44:**665–673.

Plumb, R., Mazzeo, J. R., Grumbach, E. S., Rainville, P., Jones, M., Wheat, T., Neue, U. D., Smith, B., and Johnson, K. A. (2007). The application of small porous particles, high temperatures, and high pressures to generate very high resolution LC and LC/MS separations. *J. Sep. Sci.* **30:**1158–1166.

Ramanathan, R., Cao, K., Cavalieri, E., and Gross, M. L. (1998). Mass spectrometric methods for distinguishing structural isomers of glutathione conjugates of estrone and estradiol. *J. Am. Soc. Mass Spectrom.* **9:**612–619.

Samuel, K., Yin, W., Stearns, R. A., Tang, Y. S., Chaudhary, A. G., Jewell, J. P., Lanza, T., Jr., Lin, L. S., Hagmann, W. K., Evans, D. C., and Kumar, S. (2003a). Addressing the metabolic activation potential of new leads in drug discovery: A case study using ion trap mass spectrometry and tritium labeling techniques. *J. Mass Spectrom.* **38:**211–221.

Samuel, K., Yin, W., Stearns, R. A., Tang, Y. S., Chaudhary, A. G., Jewell, J. P., Lanza, T., Jr., Lin, L. S., Hagmann, W. K., Evans, D. C., and Kumar, S. (2003b). Addressing the metabolic activation potential of new leads in drug discovery: A case study using ion trap mass spectrometry and tritium labeling techniques. *J. Mass Spectrom.* **38:**211–221.

Setoyama, T., Drijfhout, W. J., van de Merbel, N. C., Humphries, T. J., and Hasegawa, J. (2006). Mass balance study of [14C] rabeprazole following oral administration in healthy subjects. *Int. J. Clin. Pharmacol. Ther.* **44:**557–565.

Shirley, M. A., Wheelan, P., Howell, S. R., and Murphy, R. C. (1997). Oxidative metabolism of a rexinoid and rapid phase II metabolite identification by mass spectrometry. *Drug Metab. Dispos.* **25:**1144–1149.

Soglia, J. R., Contillo, L. G., Kalgutkar, A. S., Zhao, S., Hop, C. E. C. A., Boyd, J. G., and Cole, M. J. (2006). A semiquantitative method for the determination of reactive metabolite conjugate levels in vitro utilizing liquid chromatography-tandem mass spectrometry and novel quaternary ammonium glutathione analogs. *Chem. Res. Toxicol.* **19:**480–490.

Thalassinos, K., et al. (2004). Ion mobility mass spectrometry of proteins in a modified commercial mass spectrometer. *Int. J. Mass Spectrom.* **236**(1–3):55–63.

Tiller, P. R., Yu, S., Castro-Perez, J., Fillgrove, K. L., and Baillie, T. A. (2008). High-throughput, accurate mass liquid chromatography/tandem mass spectrometry on a quadrupole time-of-flight system as a 'first-line' approach for metabolite identification studies. *Rapid Commun. Mass Spectrom.* **22:**1053–1061.

Uetrecht, J. (2003). Screening for the potential of a drug candidate to cause idiosyncratic drug reactions. *Drug Discov. Today* **8:**832–837.

van Deemter, J. J., Zuiderweg, F. J. and Klinkenbergm, A. (1956). A. Longitudinal diffusion and resistance to mass transfer as causes of nonideality in chromatography. *Chem. Eng. Sci.* **5:**271–289.

Williams, D. P., Kitteringham, N. R., Naisbitt, D. J., Pirmohamed, M., Smith, D. A., and Park, B. K. (2002). Are chemically reactive metabolites responsible for adverse reactions to drugs? *Curr. Drug. Metab.* **3:**351–366.

Williams, J. P., Lock, R., Patel, V. J., and Scrivens, J. H. (2006). Polarity switching accurate mass measurement of pharmaceutical samples using desorption electrospray ionization and a dual ion source interfaced to an orthogonal acceleration time-of-flight mass spectrometer. *Anal. Chem.* **78:**7440–7445.

Wolf, D. E., VandenHeuvel, J. A., Tyler, T. R., Walker, R. W., Koniuszy, F. R., Gruber, V., Arison, B. H., Rosegay, A., Jacob, T. A., and Wolf, F. J. (1980). Identification of a glutathione conjugate of cambendazole formed in the presence of liver microsomes. *Drug Metab. Dispos.* **8:**131–136.

Wrona, M., Timo, M., Bateman, K. P., Mortishire-Smith, R. J., and O'Connor, D. (2005). All-in-one analysis for metabolite identification using liquid chromatography/hybrid quadrupole time-of-flight mass spectrometry with collision energy switching. *Rapid Commun. Mass Spectrom.* **19:**2597–2602.

Wu, N., Lippert, J. A., and Lee, M. L. (2001). Practical aspects of ultrahigh pressure capillary liquid chromatography. *J. Chromatogr. A* **911:**1–12.

Wyndham, K. D., O'Gara, J. E., Walter, T. H., Glose, K. H., Lawrence, N. L., Alden, B. A., Izzo, G. S., Hudalla, C. J., and Iraneta, P. C. (2003). Characterization and evaluation of C18 HPLC stationary phases based on ethyl-bridged hybrid organic/inorganic particles. *Anal. Chem.* **75:**6781–6788.

Yan, Z., and Caldwell, G. W. (2004). Stable-isotope trapping and high-throughput screenings of reactive metabolites using the isotope MS signature. *Anal. Chem.* **76:**6835–6847.

Yan, Z., Maher, N., Torres, R., and Huebert, N. (2007). Use of a trapping agent for simultaneous capturing and high-throughput screening of both "soft" and "hard" reactive metabolites. *Anal. Chem.* **79:**4206–4214.

Zhang, H., Zhang D., and Ray K. (2003). A software filter to remove interference ions from drug metabolites in accurate mass liquid chromatography/mass spectrometric analyses. *J. Mass Spectrom.* **38:**1110–1112.

Zheng, J., Ma, L., Xin, B., Olah, T., Humphreys, W., and Zhu, M. (2007). Screening and Identification of GSH-trapped reactive metabolites using hybrid triple quadruple linear ion trap mass spectrometry. *Chem. Res. Toxicol.* **20:**757–766.

Zhou, S., Song, Q., Tang, Y., and Naidong, W. (2005). Critical review of development, validation, and transfer for high throughput bioanalytical LC-MS/MS methods. *Curr. Pharm. Anal.* **1:**3–14.

Zhu, M., Ma, L., Zhang, H., and Humphreys, W. G. (2007). Detection and structural characterization of glutathione-trapped reactive metabolites using liquid chromatography-high-resolution mass spectrometry and mass defect filtering. *Anal. Chem.* **79:**8333–8341.

5

Changing Role of FTMS in Drug Metabolism

Petia A. Shipkova and Jonathan L. Josephs

Bristol-Myers Squibb Pharmaceutical Research Institute, Pennington, New Jersey

Mark Sanders

ThermoFisher Scientific, Somerset, New Jersey

Mass Spectrometry in Drug Metabolism and Pharmacokinetics. Edited by Ragu Ramanathan
Copyright © 2009 John Wiley & Sons, Inc.

5.1 INTRODUCTION

A thorough understanding of a drug candidate's absorption, distribution, metabolism, excretion, and toxicology (ADMET) properties is critical for successful drug development, and there is a desire to obtain this information early in the drug discovery process in order to mitigate the cost of developing candidates that would eventually drop out due to ADMET liabilities. This provides significant analytical challenges, especially in drug discovery, where the number of potential candidates can be large. The ability to rapidly characterize compounds in terms of structure, amount, and distribution becomes essential, and mass spectrometry (MS) has played an ever-increasing role in this endeavor due to its sensitivity, selectivity, and ability to provide rapid structural information

To date, Fourier transform ion cyclotron resonance mass spectrometry (FTICR–MS) has played only a minor role in the field of drug metabolism. The obvious question is why? Commercial FTICR instruments have been available for over 20 years and have been successfully applied to various analytical fields, including proteomics (Bogdanov and Smith, 2005; Pihakari, 2007), in-vitro-based drug screening (Siegel, 2005), accurate mass measurements for drug discovery (Zhang et al., 2005c), structure elucidation of natural products (McDonald et al., 2003), combinatorial chemistry (Zhang et al., 2005c), toxicology, and forensic sciences (Ojanperae et al., 2005). Notably missing from this list are the ADMET applications. While the expectation is that the combination of excellent sensitivity and highest mass accuracy and mass resolution currently available (Marshall, 2000) would make Fourier transform mass spectrometry (FTMS) a must for the pharmaceutical ADMET laboratory; until now FTMS has had little impact. In this chapter, the authors discuss how the new generation hybrid FT mass spectrometers are changing the drug discovery and development paradigm and applications of the technique for metabolite detection and structure elucidation.

5.2 ACCURATE MASS MEASUREMENTS AND METABOLITE CHARACTERIZATION

FTICR–MS is an established technology that provides the highest mass accuracy and mass resolution currently available (Marshall, 2000; Bristow and Webb, 2003; Bristow, 2006) and yet FTICR–MS has not seen widespread incorporation into pharmaceutical ADMET laboratories and workflows. Does this mean that high-resolution accurate mass data is not needed for drug metabolism work? One could certainly make that argument for the majority of metabolite identification and characterization work. The fact that the parent compound structure is known, the majority of biotransformations are well understood, and there are a limited number of them significantly limits the chemical space that needs to be considered during the structure elucidation of metabolites. This has been borne out by the fact that for many years the metabolism scientist has been successful using low-resolution, nominal mass instruments such as the quadrupole and ion trap mass spectrometers. Nominal mass measurements with triple-quadrupole MS/MS (i.e., product ion, precursor ion, or neutral loss scanning, discussed in Chapter 3) or ion trap MS^n-scanning techniques have proven to be sufficient for most metabolite identification needs (Ramanathan et al., 2007a,b). After all, accurate mass determination of an $M + 16$ metabolite is likely to do little more than to confirm the addition of oxygen. However, there are occasions when unexpected metabolites are formed, and deducing their structure is often contingent upon the unequivocal assignment of the fragmentation pathway of the parent compound. While accurate mass is not always required for such assignment, having high resolution and accurate mass data to unambiguously assign empirical formulas to ions in the fragmentation spectra (MS/MS and/or MS^n) can save significant amounts of time, in terms of both the actual assignments and reducing the possibility of an incorrect assignment (Bristow and Webb, 2003; Bristow, 2006). An incorrectly assigned fragment ion could lead to much confusion and wasted time in a metabolite identification study. Typically, accurate mass measurements, for both MS and MS/MS experiments, provide increased productivity on a daily basis; faster manual interpretation of data often reduces the need for additional MS/MS or MS^n experiments to confirm a structure, incorrect assignments are minimized, and the more precise data facilitate the use of automated data interpretation. As a result, most mass spectrometrists and metabolism scientists, even when working within a limited chemical space, would prefer to have high-resolution accurate mass data when performing any kind of structure elucidation work (Bristow and Webb, 2003; Bristow, 2006; Leslie and Volmer, 2007). Additionally, we are finding accurate mass measurements to be an extremely simple and effective tool for locating the presence of drug metabolites. By exploiting the mass defect properties of molecules, structurally related compounds (e.g., metabolites) can be selectively identified in the presence of complex biological mixtures (Chapter 6). This is proving to be a valuable tool for metabolite identification work (Zhu et al., 2006; Bateman et al., 2007). One area in particular is the identification of phase II conjugates, where mass defect filtering can be more selective than the traditional neutral loss and/or precursor scan approaches on the triple-quadrupole mass spectrometer.

5.3 TRADITIONAL FTMS AND METABOLITE CHARACTERIZATION

The overwhelming majority of FTMS applications and publications to date are from the proteomics field, where this technology has brought numerous advantages and improvements (Bogdanov and Smith, 2005; Zimmer et al., 2006). For small molecules, most FTMS applications utilize the unique advantage of externally calibrated robust accurate mass measurements as discussed in detail in recent comprehensive reviews (Heeren et al., 2004; Marshall, 2004; Zhang et al., 2005c). In the pharmaceutical industry, accurate mass measurements are typically performed on pure/isolated drug candidates (Burton et al., 1999; Ziqiang Guan, 2001; Wang et al., 2003) and related degradants (Winger and Kemp, 2001) and/or metabolites (Wang et al., 2003; Saghatelian et al., 2004) and in the analysis of combinatorial chemistry libraries (Nawrocki et al., 1996; Poulsen et al., 2000; Ramjit et al., 2000; Schmid et al., 2000) without (or with minimal) liquid chromatography (LC) separation. The utility of accurate mass measurements for less ambiguous metabolite structure elucidation is not a new concept and has been successfully applied over the last seven or eight years for online LC–MS applications using LC–time-of-flight (TOF) and LC–Q-TOF platforms (Hopfgartner et al., 1999; Zhang et al., 2000; Clarke et al., 2001; Bristow and Webb, 2003; Liu and Hop, 2005). However, these applications have not been readily extended to the traditional FTMS instruments for a number of reasons, as discussed below.

Traditional FTMS instrumentation has been big, expensive, difficult to use, and costly to maintain and the sample throughput was relatively low. To utilize FTMS instruments effectively, experienced operators with a good understanding of the specifics and fundamentals of FT mass spectrometers are required. These requirements placed the cost of FTMS very high and often too high for structural characterization of drug metabolites, especially considering that the less expensive and much easier to use ion traps and triple quadrupoles could provide answers to a majority of drug metabolism problems. The ability to perform multiple LC–MS and LC–MS/MS analyses from a single sample injection (Fitch et al., 2007) combined with the automation (Fandino et al., 2006) and data-dependent MS/MS capabilities of quadrupole ion trap and triple-quadrupole mass spectrometers (Triolo et al., 2005) made the necessary data collection easier and provided the answers in a more timely fashion, compared to the more sophisticated and more precise FT mass spectrometers. While the economics of the situation were probably enough to keep the traditional FTMS instrumentation out of the metabolite characterization arena, there were also technical hurdles to overcome, most notably that these FTMS instruments were fundamentally incompatible with the 1-mL/min LC that is traditionally employed in ADMET laboratories. The incompatibility was in part due to the electrospray source design on the early FTMS instruments. There was a lack of a commercially available high-flow, robust, and user-friendly high-performance liquid chromatography (HPLC) interfaces, but beyond that, the FTMS mass measurements were compromised when coupled to LC. The traditional FTMS instrumentation utilized the ICR cell for all ion manipulations, including isolation, fragmentation, and mass measurement, resulting in relatively slow scan speeds, causing problems for detection of the rapidly changing analyte concentrations in typical LC–MS applications. This was especially pronounced when performing MS/MS experiments, practically excluding the possibility for data-dependent experiments. The ICR cell is sensitive to "space charge" effects, where ions trapped in the cell interact with each other and cause perturbations in the measured mass. To overcome space charging, the instrument is calibrated with a certain target ion population. During mass measurement, as long as the ICR cell ion population is maintained within the target ion population used for calibration, good mass accuracy can be maintained. However, when scanning across an HPLC peak, where there is a marked change in analyte concentration, the ion populations could vary significantly from the calibrated target, resulting in mass shifts. The ion population can also vary drastically due to endogenous ions from biological samples such as plasma, urine, and bile.

Another apparently minor, yet significant, drawback of the older FTMS systems was the free induction decay (FID)–based data system. FTMS instruments do not measure mass-to-charge ratio (m/z) directly as other mass spectrometers do; these instruments use a resonance method similar to nuclear magnetic resonance (NMR) spectrometers. As described in detail (Marshall et al., 1998), magnetic field (x and y directions) and electric potentials (z direction) are used to trap the ions in the ICR cell. Once trapped, all ions are simultaneously excited by a radio-frequency pulse. What is measured is the image current of the ions as they pass by a detector and the FID of the signal as the excited ions relax back to the original state. The transient ion image current signal is a composite of the cyclotron frequencies of all the ions in the cell. This resulted in a large LC–FID data set which was cumbersome to manipulate. In order to obtain a mass spectrum, a fast Fourier transform (FFT) was required, where each frequency component was processed to yield mass/intensity data.

For these reasons, the traditional FTMS technology has not been amenable to high-throughput automation without compromising the traditional high flow rates, speed, sensitivity, and chromatographic separation that ADMET scientists are accustomed to achieving with conventional HPLC, TOF, triple-quadrupole, and ion trap technologies.

5.4 NEXT-GENERATION FTMS AND METABOLITE CHARACTERIZATION

The breakthrough for the small-molecule FTMS field came with the introduction of the hybrid FTMS instruments, including the quadrupole FTMS from Bruker Daltonics (Billerica, MA), IonSpec (Lake Forest, CA, now Varian), and the Thermo Electron (San Jose, CA) linear ion trap FTMS family, including the LTQ-FT and the LTQ-Orbitrap. To the best of our knowledge, there are no literature reports describing drug metabolism applications using the hybrid quadrupole FTMS systems from Bruker or IonSpec (now Varian). Since the authors have little knowledge of the Bruker and IonSpec FTMS systems, the discussion in this chapter is limited to the hybrid linear ion trap mass spectrometers, the LTQ-FT and the LTQ-Orbitrap.

The LTQ-FT mass spectrometer was introduced in late 2003 and, as expected, the main application discussed in the literature is for the analysis of proteins and peptides (Johnson et al., 2004; Syka et al., 2004). A recent book chapter (van der Greef et al., 2004) and a review article (Brown et al., 2005) discussed the application of the LTQ-FT to metabolomics. FTMS applications to drug metabolism are still very new and drug discovery research laboratories which have recently purchased the instrument are still in the process of developing and validating methods and approaches. A recent publication describes the depth and flexibility of the experimental setup utilizing accurate mass data-dependent exclusion MS^n measurements with a LTQ-FT (Tozuka et al., 2005). We have reported several integrated approaches for determination of metabolic stability, characterization of metabolites and metabolic soft spots (Josephs et al., 2008; Shipkova et al., 2004; Sanders et al., 2006), the use of mass defect filters for detection of metabolites (Ma et al., 2005; Zhang et al., 2005a), and its use for metabonomics studies (Warrack et al., 2004, 2005; Zhang et al., 2005b).

The LTQ-Orbitrap (Makarov, 2000; Hardman and Makarov, 2003; Makarov et al., 2006a,b; Peterman et al., 2006), introduced commercially in June 2005, is another hybrid linear ion trap FT mass spectrometer, where the superconducting magnet and the ICR cell are replaced by an electrostatic trap (C-trap). LTQ-Orbitrap is specifically targeted for small-molecule analysis and is anticipated to have a significant impact in the field of drug metabolism (Peterman et al., 2006; Chen et al., 2007; Lim et al., 2007). In addition to the drug metabolism arena, the recent modification of the LTQ-Orbitrap mass spectrometer with electron transfer dissociation (ETD) (McAlister et al., 2007) has allowed scientists from the metabonomics and proteomics arena to capitalize on the LTQ-Orbitrap technology (Macek et al., 2006; Scigelova and Makarov, 2006; Williamson et al., 2006; Yates et al., 2006; McAlister et al., 2007).

These next-generation mass spectrometers provide the much-needed bridge between the excellent performance of the high-mass-resolving-power, high-mass-accuracy FT mass spectrometers and the well-established, tested, and validated features of quadrupoles and ion traps, including their compatibility with standard HPLC flow rates, ease of operation, high-throughput and automation compatibility, and data-dependent MS^n capabilities. By coupling another mass analyzer (linear ion trap) to the FTMS, many of the ion manipulations (e.g., MS/MS), can be performed in the linear ion trap, while the FTMS is used only for mass measurements. The coupling of the linear ion trap with FTMS made faster scan times possible and made the hybrid instrument more compatible with LC–MS applications. Since all MS/MS experiments are performed in the linear ion traps, mass measurements of the MS/MS product ions in the ICR cell are not compromised by introduction of a collision gas to the high-vacuum region of the ICR cell (typically 2×10^{-10} morr). In addition, the linear ion trap hybrid instrument makes use of the automatic gain control, within the ion trap, to regulate the packet of ions sent to the ICR cell. As a result, the ions are delivered to the ICR cell for mass measurement in a controlled and reproducible fashion, minimizing the possibility of space charging and resulting in improved accuracy and precision. This allows for excellent mass accuracies on a routine basis, even when dealing with the rapidly changing concentrations observed across rapid HPLC gradient or even the much narrower ultra performance liquid chromatography (UPLC) peaks. It has been noted in strategies for comprehensive metabolite identifications that there is great value in having MS^n and accurate mass capability for rapidly confirming expected metabolites or elucidating unusual or unexpected metabolites (Clarke et al., 2001). The combination of traditional high-flow chromatography and robust accurate mass determination for both parent and fragment ions with external calibration generates a very powerful analytical tool.

One of the most attractive features of the LTQ-FT is the outstanding mass accuracy achieved with external calibration in both MS and MS/MS modes. The generally accepted mass accuracy standard for confirmation of elemental composition and publication purposes is 5 ppm (Bristow and Webb, 2003). A recent report suggests that mass accuracies within 5 ppm is actually insufficient for unambiguous formula determination for unknowns in the 200-Da range (Senko et al., 2004). For example, if one were to consider an unknown with a molecular weight (MW) of about 500 Da, with no constraints, there are almost 100 possible elemental formulas within a 5-ppm window (Bristow and Webb, 2003), which is clearly insufficient for identification of complete unknowns, that is, endogenous metabolites representing potential biomarkers, natural products, and so on. Using measured mass alone, the required mass accuracy for obtaining only one or two formula hits for MW = 500 Da is in the mid-ppb range (Senko et al., 2004). Until the introduction of the LTQ-FT, such accuracies have not been demonstrated for high-flow LC–MS with external calibration. However, it should be noted that for drug metabolite identification there is a substantial prior knowledge, such as the elemental composition of the parent, and likely biotransformation processes and pathways. This information, combined with appropriate valence rules, can significantly reduce the number of possibilities, although there will likely be several plausible formulas for masses above m/z 350.

Shown in Table 5.1 are 10 mass measurements of warfarin over a 45-h period in both positive- and negative-ion modes after external calibration (Sanders et al., 2006). These data were obtained during a 4 min LC gradient at 1 mL/min flow with positive/negative switching. Using external calibration, all mass measurements were <0.5 ppm and the range of measurements during the 45-h period was only 0.06 mmu for positive ion and 0.13 mmu for negative ion, demonstrating the stability and robustness of the mass measurements. Such precise measurements greatly simplify the determination of elemental compositions of metabolites and, when coupled with accurate mass data for MS^n fragment ions, significantly enhance the structural elucidation process. Figure 5.1 displays a typical MS/MS spectrum of verapamil, where all fragment ions, regardless of relative intensity, are detected with ≤1 ppm

TABLE 5.1. Ten Accurate Mass Measurements of Warfarin over 45 hours by LC–MS with Pos/Neg Switching; Observed Mass Errors <0.5 ppm Using External Calibration

Time (hr)	Obs. $[M + H]^+$ (309.11214)	Δ (ppb)	Obs. $[M - H]^-$ (307.09758)	Δ (ppb)
0:00	309.11203	356	307.09752	195
0:04	309.11206	259	307.09753	163
4:34	309.11203	356	307.09750	260
4:38	309.11203	356	307.09760	65
25:00	309.11203	356	307.09756	65
25:04	309.11200	453	307.09760	65
29:00	309.11206	259	307.09750	260
29:04	309.11200	453	307.09753	163
45:00	309.11200	453	307.09747	358
45:04	309.11203	356	307.09741	553
	Range: 0.06 mmu	Avg. 366 ppb	Range: 0.13 mmu	Avg. 215 ppb

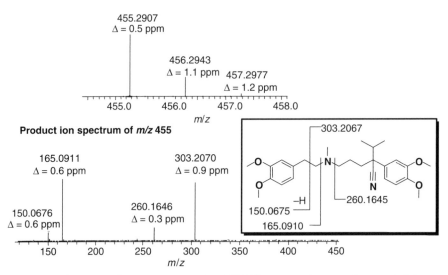

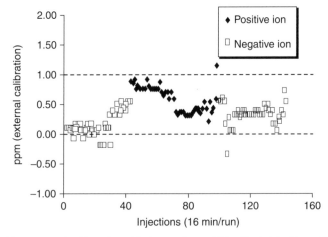

Figure 5.1. *Full-scan MS and data-dependent MS/MS spectra of verapamil generated from a microsomal incubation sample.*

mass accuracy. In contrast to TOF mass spectrometers, the hybrid FTMS instruments provide excellent mass accuracy even for ions with low relative intensity, as shown in Fig. 5.1 for the verapamil isotope peaks, where both the $M+1$ (^{13}C peak, 30% of parent intensity) and $M+2$ (double ^{13}C, 4% of parent intensity) are detected with accuracies ≤ 1.2 ppm. The mass stability of the LTQ-FT is further demonstrated in Fig. 5.2, where mass measurements gathered five weeks post–external calibration are well within 2 ppm (<1 ppm for most cases) for both positive and negative ions (Sanders et al., 2006).

Figure 5.2. *Mass errors for hippuric acid observed over 42 h; 10-μL injection of 50:50 urine/ water every 16 min; external calibration >5 weeks old.*

The commonly used ion trap features, that is, the ability to set a variety of MS^n data-dependent scan events and segments from either a parent mass list or simply for ions observed above a preset threshold, are fully preserved in the hybrid LTQ–FTMS instruments and further enhanced with accurate mass measurements. All ion manipulations, including isolation and fragmentation, are quickly performed in the linear ion trap while only one relatively "long" FT experiment, the actual mass measurement, is performed in the ICR cell. Also, with the hybrid linear ion trap instruments, the linear ion trap is a fully functional mass spectrometer with its own detectors, and as such can be operated independently of the FT detector. This provides for even more efficient data-dependent scanning, and it has been demonstrated that during the time it takes for the FT to gather one high-resolution MS scan, the linear ion trap can acquire up to four nominal mass MS/MS or MS^n scans in parallel (Horning et al., 2004; Peterman et al., 2005). For example, the instrument can be programmed to acquire an accurate mass spectrum at 100,000 resolution in the FT and, while that measurement is being made, over a period of about 1 s, simultaneously ion trap MS/MS data could be acquired on the four largest peaks in the mass spectrum. In addition, these new-generation FTMS instruments take advantage of new developments in data acquisition and digital signal processing that allow for real-time FFT and therefore smaller and easier to handle LC–MS data files. This provides both efficient data review and the ability to perform real-time data-dependent experiments using the accurate mass data as the trigger.

5.4.1 LTQ–FTICR

Table 5.2 shows a representative set of accurate mass measurements of 50 pharmaceutical compounds which were acquired using an eight-scan event experiment on the LTQ–FTICR (Josephs et al., 2004b). Mass spectra from all eight scan events were acquired within 7 s, the width (at the base) of a typical chromatographic peak, as shown in Fig. 5.3 for warfarin (MW 308). The first scan event is a full-scan positive-ion MS acquired in the ion trap. If the expected protonated molecule is detected above a preset threshold, a second scan event, an ion trap (IT) data-dependent MS/MS scan, is triggered. This is followed by a third scan event, where the protonated molecule is accurately measured in the FT, and a fourth scan event, a full-scan MS/MS in the FT.

After all four positive scans are completed (typically within 3 s), the polarity is switched and the fifth scan event records a negative-ion full-scan MS. If the expected protonated molecule is not detected in the positive mode, the second, third and fourth scan events are skipped and the fifth (negative-ion-mode) scan event is triggered. Similar to the positive-ion mode, if the expected [M–H]$^-$ ion is detected in the full-scan MS, IT data-dependent MS/MS (sixth), FT accurate MS (seventh), and FT MS/MS (eighth) scan events are acquired. Clearly demonstrated here is the ability of the LTQ-FT to handle multiple experiments on a chromatographic time scale. One might question the need for such an elaborate data-dependent scheme when apparently all that is needed is an accurate mass determination followed by a data-dependent accurate mass MS/MS spectrum. Apart from the fact that using the

TABLE 5.2. Accurate Mass Measurements Gathered in an 8-scan Event Data Dependant LC–MS Experiment

Name	Theor. $[M + H]^+$	Obs. $[M + H]^+$	Δ (ppm)	Δ (mmu)
Caffeine	195.08765	195.08765	0.00	0.00
Pindolol	249.15975	249.15973	−0.08	−0.02
Tolbutamide	271.11109	271.11102	−0.26	−0.07
Fenoterol	304.15433	304.15427	−0.20	−0.06
Piroxicam	332.06995	332.06989	−0.18	−0.06
Ampicillin	350.11690	350.11685	−0.14	−0.05
Enalapril	377.20710	377.20703	−0.19	−0.07
Reserpine	609.28066	609.28040	−0.42	−0.26
Vinblastine	811.42766	811.42780	0.17	0.14
	Theor. $[M - H]^-$	Obs. $[M - H]^-$	Δ (ppm)	Δ (mmu)
Phenytoin	251.08260	251.08255	−0.20	−0.05
Tolbutamide	269.09654	269.09650	−0.15	−0.04
Diclofenac	294.00941	294.00943	0.07	0.02
Fenoterol	302.13978	302.13977	−0.03	−0.01
Piroxicam	330.05540	330.05542	0.06	0.02
Ampicillin	348.10235	348.10242	0.20	0.07
Enalapril	375.19255	375.19265	0.27	0.10
Reserpine	607.26610	607.26630	0.33	0.20
Vinblastine	809.41310	809.41284	−0.32	−0.26

Data shown is a representative set from 50 pharmaceutical compounds. All determinations with external calibration and errors <0.4 ppm.

IT for the survey scan is more efficient and reliable for triggering the data-dependent experiments on sharp HPLC peaks and an accurate mass selected ion monitoring (SIM) scan provides better accuracy than a full MS scan in the FT (Senko et al., 2004), the main reason for the eight-scan event was the fact that the IT and FT MS/MS spectra were different. As can be seen in Fig. 5.3, the relative ion intensities were not consistent and in some cases low mass ions were completely missing in the FT–MS/MS spectra. One of the unique characteristics of the ion traps is the ability to generate highly reproducible MS/MS spectra from day to day and from instrument to instrument (Josephs and Sanders, 2004) and an extensive (>100,000 spectra) in-house MS/MS library has been built. IT–MS/MS spectra are required for library searching and the FT–MS/MS spectra to provide the molecular formulas for the ions (as shown in Fig. 5.3). This approach also ensured that diagnostic low-mass MS/MS ions would be observed, even if only in the nominal-mass IT scan.

5.4.2   The Time-of-Flight Effect

The reason for the disparity between the MS/MS spectra from the two detectors originates from a "time-of-flight effect," which is also observed in a full-scan MS mode and is a consequence of the spatial separation between the two mass detectors. The result is a discrimination against the lower mass region when scanning a relatively wide mass range (e.g., 85–850 Da) and is a feature on most hybrid FTMS

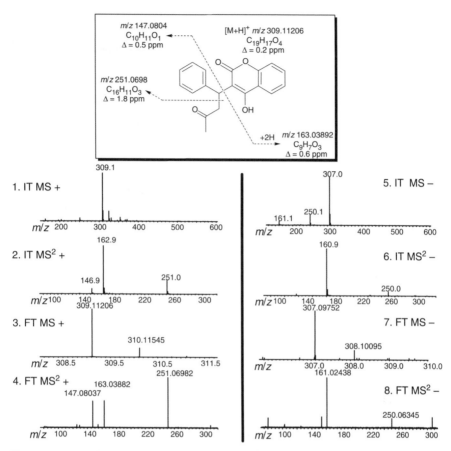

Figure 5.3. Warfarin mass spectra from eight-scan event data-dependent experiment gathered on a chromatographic scale (7 s wide at peak base). Fragments detected with accuracy <2 ppm.

instrumentation. The time-of-flight effect takes place as ions are transferred from the ion trap to the FTICR cell. Ions are guided to the FTICR cell by multipole lenses and have to travel approximately 1 m. It is during this flight that the ions begin to separate according to mass, in a process that forms the basis for time-of-flight mass spectrometers. This results in an elongated ion cloud reaching the FTICR cell instead of a tight ion packet. Ions are trapped in the ICR cell by manipulating a gating voltage at the front of the cell. The timing of the gating voltage is critical in order to capture the ions of interest. An elongated ion packet causes problems because leaving the FTICR cell gate open too long results in the faster moving low-mass ions being reflected from the rear of the cell back out of the open front gate and, therefore, not being captured. In contrast, closing the cell quickly results in the loss of the slower high-mass ions (see Fig. 5.4). Capturing the appropriate ions is a compromise and the result is often less than optimal ion transmission in the low-mass region when scanning over a relatively wide mass range (e.g., m/z 8–850). For metabolite ID

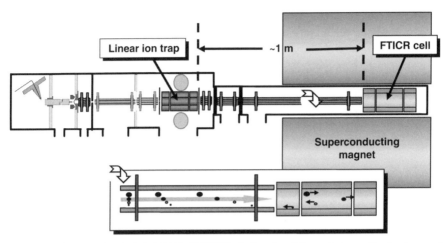

Figure 5.4. *Schematic of a hybrid LTQ–FTICR illustrating the distance between the ion trap and the ICR cell. Shown is an expansion of the ion transfer optics and how ions separate on their way to the ICR cell and how the low-mass ions could enter the cell, be reflected at the rear of the cell, and exit the cell while waiting for the higher mass ions to enter the cell. (Courtesy of ThermoFisher Scientific.)*

work, where the components of interest are typically in the range of 250–650 amu, MS/MS is often required for structure elucidation. Therefore, good low-mass ion transmission without compromising the mass accuracy is important.

To overcome the time-of-flight effect, a "wide-scan" function (Sanders et al., 2005), in which various gate-timing and ion transmission characteristics are optimized without compromising the mass accuracy (Figs 5.5b,c), was implemented. Operating in the wide-scan mode provides significant gains in ion intensity (fourfold or higher) for ions below 200 amu with only a two fold drop in intensity for ions in the range of 300–800 amu. A comparison of MS/MS spectra of buspirone obtained using a LTQ scan (Fig. 5.5a), a "regular" FT scan (Fig. 5.5b), and a wide FT scan (Fig. 5.5c) is shown (Sanders et al., 2005). One trade-off for the wide-scan mode is the longer scan time, which can be overcome by reducing the FTICR mass resolution with very little impact on the mass accuracies. For typical metabolite ID studies mass resolution in the range of 25,000 to 50,000 is sufficient. Although the wide scan provides a significant improvement, the ion transmission in the low-mass region is still compromised in comparison to an ion trap MS/MS spectrum (Fig. 5.5a vs. 5.5c).

5.4.3 LTQ-Orbitrap

Due to its radically different design, the latest hybrid linear ion trap FTMS instrument, the LTQ-Orbitrap (Fig. 5.6), does not suffer from the time-of-flight effect. In this instrument, the superconducting magnet and the ICR cell are replaced by an electrostatic trap (C-trap) and so distances traveled by the ions from one MS device to the other are much smaller; in addition a radically different ion transfer mechanism virtually eliminates any possibility for a time-of-flight effect (Makarov,

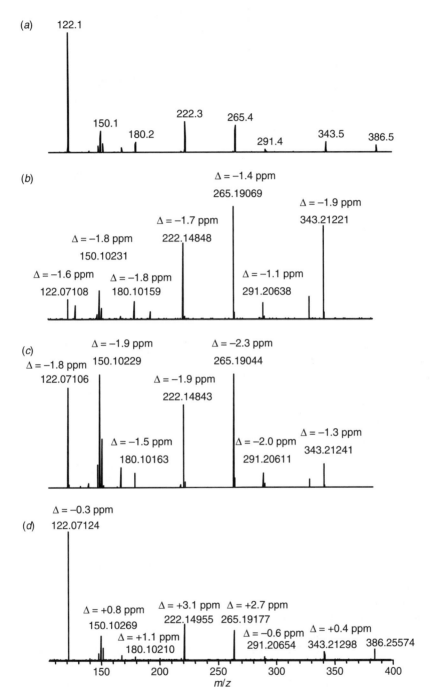

Figure 5.5. Buspirone MS/MS (m/z 386) spectra collected using a (a) linear ion trap, (b) FTICR operated in the regular-scan mode, (c) FTICR operated in the "wide"-scan mode, and (d) Orbitrap.

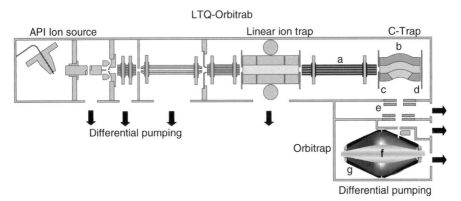

Figure 5.6. *Schematic of a hybrid LTQ-Orbitrap mass spectrometer: (a) transfer octapole; (b) curved rf-only quadrupole (C-trap); (c) gate electrode; (d) trap electrode; (e) ion optics; (f) inner orbitrap electrode; (g) outer orbitrap electrodes (Makarov et al., 2006a).*

2000; Hardman and Makarov, 2003), making the LTQ-Orbitrap especially suitable for small-molecule analyses. For most compounds, the relative intensities of ions in Orbitrap mass spectra are very similar to those spectra from LTQ (IT scan) (Fig. 5.5) (Sanders et al., 2005). As discussed above, this feature becomes very important for building MS/MS libraries and subsequent library searches for related structures (i.e., metabolites, structural analogues, degradants, etc.). Also highlighted in Fig. 5.5 are the excellent mass accuracies observed in the MS/MS mode, even for some very minor fragments.

The LTQ-Orbitrap has resolution and mass accuracy performance close to that of the LTQ–FTICR. As shown in Table 5.3 (column 4), LTQ-Orbitrap accurate mass measurements, using external calibration, for a set of 30 pharmaceutical compounds resulted in less than 2.3 ppm error. The data were acquired with a 4-min, 1-mL/min-flow-rate, positive-mode LC–ESI–MS method where all measurements were performed within 5 h from mass calibration. Mass accuracies below 2–3 ppm, and often below 1 ppm, can be routinely achieved in both the positive- and negative-ion mode (Table 5.3, columns 4 and 5). The long-term mass stability of the LTQ-Orbitrap is not as consistent as observed for the LTQ–FTICR–MS, and the Orbitrap requires more frequent mass calibration; however, mass calibration is a routine procedure that can be accomplished within 5–10 min. Figure 5.7 displays a 70-h (external calibration) mass accuracy plot for three negative ions collected with a LTQ-Orbitrap where the observed accuracy is 2.5 ppm or better with little mass drift for each ion. Overall, for routine accurate mass measurements on the Orbitrap, once-a-week calibration (for the desired polarity) is required; however, considering the ease of the process, more frequent external calibration is not a burden.

The lock mass (LM) option can be used as a substitute for a more frequent calibration. Post column infusion of a known component can be utilized so that the lock mass ion is present in every scan. Lock mass mode accurate mass measurements are also possible for MS/MS experiments, where the lock mass ion is not present in

TABLE 5.3. Accurate Mass Measurements for a set of 30 Pharmaceutical Compounds Collected in Positive Ion Mode (Experimental Error in ppm Shown in Column 4), Negative Ion Mode (Experimental Error in ppm Shown in Column 5), and with Pos/Neg Switching (Data Shown in Columns 6 and 7, Respectively) Using a Lock Mass (LM) for Each Polarity, Reserpine, m/z = 609 for Positive Mode and Hippuric Acid, m/z 178 for Negative Mode

Compound	Composition	MW	Δ (ppm) pos	Δ (ppm) neg	Δ (ppm) pos/ neg + LM	Δ (ppm) pos/ neg + LM
Propronolol	$C_{16}H_{21}NO_2$	259.15723	−1.540		−0.076	
Cortisone	$C_{21}H_{28}O_5$	360.19367	0.909		0.399	
Norfloxacin	$C_{16}H_{18}FN_3O_3$	319.13322	−0.269		0.026	
Pyrilamine	$C_{17}H_{23}N_3O$	285.18411	−0.971		0.081	
Acetophenetidine	$C_{10}H_{13}NO_2$	179.09463	−2.222		0.332	
Diltiazem	$C_{22}H_{26}N_2O_4S$	414.16133	−0.760		0.154	
Paclitaxel	$C_{47}H_{51}NO_{14}$	853.33096	−0.935		−0.119	
Quinidine	$C_{20}H_{24}N_2O_2$	324.18378	−0.292		0.325	
Bumetanide	$C_{17}H_{20}N_2O_5S$	364.10929	0.577	0.724	0.07	0.403
Enalapril	$C_{20}H_{28}N_2O_5$	376.19982	−0.352	0.107	0.141	0.706
Hydrocortisone	$C_{21}H_{30}O_5$	362.20932	0.379		−1.081	
Metoprolol	$C_{15}H_{25}NO_3$	267.18344	−1.666		−0.02	
Ranitidine	$C_{13}H_{22}N_4O_3S$	314.14126	0.034	0.014	0.141	−0.108
Desipramine	$C_{18}H_{21}N_2Cl$	300.13933	−0.204		0.037	
Etoposide	$C_{29}H_{32}O_{13}$	588.18429	0.080		−1.448	
Nadolol	$C_{17}H_{27}NO_4$	309.19401	−0.855		−0.081	
Pindolol	$C_{14}H_{20}N_2O_2$	248.15248	−1.220		−0.059	
Ketoconazole	$C_{26}H_{28}Cl_2N_4O_4$	530.14876	−0.491		−0.052	
Piroxicam	$C_{15}H_{13}N_3O_4S$	331.06268	0.476	1.288	−0.315	−2.145
Sulfasalazine	$C_{18}H_{14}N_4O_5S$	398.06849	−1.045	0.697	0.405	−1.532
L-Aspartyl	$C_{14}H_{18}N_2O_5$	294.12157	−1.147		−0.129	
Dextromethorphan	$C_{18}H_{25}NO$	271.19361	−2.303		0.143	
Fenoterol	$C_{17}H_{21}NO_4$	303.14706	−0.74	0.138	−0.641	−1.032
Progesterone	$C_{21}H_{30}O_2$	314.22458	−0.115		0.231	
Sulpiride	$C_{15}H_{23}N_3O_4S$	341.14093	−0.131	0.078	0.6	−1.475
Timolol	$C_{13}H_{24}N_4O_3S$	316.15691	−0.565		0.381	
Tolbutamide	$C_{12}H_{18}N_2O_3S$	270.10381	−1.574	1.572		−1.039
Verapamil	$C_{27}H_{38}N_2O_4$	454.28316	−0.994		0.407	
Indapamide	$C_{16}H_{16}ClN_3O_3S$	365.06009	−0.229	0.966	1.033	−1.881
Pimozide	$C_{28}H_{29}F_2N_3O$	461.22787	−1.032		−0.549	
		Av. Dev.	0.627	0.477	0.325	0.823

the fragmentation spectrum of the analyte of interest but is stored in the ion transfer optics of the instrument and is later "added" to the fragmentation spectra. The lock mass approach becomes especially valuable when positive/negative switching is desired. On all the ion traps (both 3D and 2D) positive/negative switching experiments have become routine, especially in the fast-scanning linear (2D) ion trap, where there are adequate scan events even for quantitation purposes. In the LTQ–FTICR–MS positive/negative switching is possible without any "sacrifice" of mass accuracy; the only downside to polarity switching is that relatively fewer scans are collected. This is a result of both polarity switching time and splitting

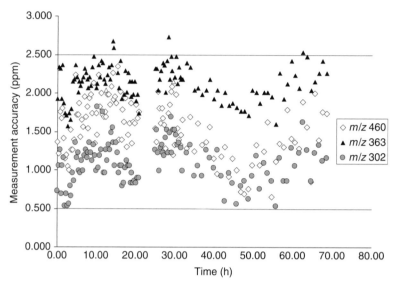

Figure 5.7. *Accurate mass measurements of three ions (pimozide, MH⁺ 460; fenoterol, MH⁺ 302; and bumetanide, MH⁺ 363) over 70 h by LC–MS in negative-ion mode using external calibration.*

data collection between the two modes. With the LTQ-Orbitrap, a settling time of 2–3 h is typically required after a polarity switch before robust external calibration accurate mass measurements can be made. Following a polarity switch, without the 2–3 h settling time, a 20–30-ppm mass shift can be observed. Therefore, for positive/negative switching experiments, the software-supported lock mass option becomes the only alternative and the acquired data have excellent accuracies, as shown in Table 5.3 (columns 6 and 7) for the same set of 30 pharmaceutical compounds. For the reasons discussed above, positive/negative switching comes with overall sacrifice in the number of scans collected across the chromatographic peak and polarity switching could compromise the detection of minor or difficult-to-ionize metabolites.

The sensitivity of the LTQ-Orbitrap is demonstrated using a mixture of synthetic standards of buspirone metabolites spiked into a rat plasma extract. The samples were analyzed with a standard 4.6-mm HPLC column. LC–MS/MS chromatograms for the five-component mixture obtained using the Orbitrap and the LTQ mass spectrometers are compared in Figs 5.8a and b. Also shown are MS/MS (m/z 402) spectra of the oxa-buspirone metabolite ($R_t = 7.4$ min) at 10 pg on column in the Orbitrap (Fig. 5.8c) and LTQ (Fig. 5.8e). As discussed above, the MS/MS fragmentation spectra, obtained in the Orbitrap (Figs 5.8c and d), are very similar to those from the LTQ (Fig. 5.8e), and even at such low concentrations excellent mass accuracies are maintained.

Synthetic standards of five hydroxylated buspirone metabolites were spiked into rat plasma and analyzed on a 4.6 × 150-mm YMC-ODS AQ S3 column at a flow rate of 1 mL/min: LC–MS chromatograms at 10 pg on column for (a) Orbitrap and

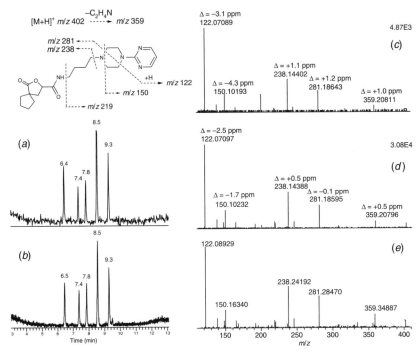

Figure 5.8. Synthetic standards of five hydroxylated buspirone metabolites spiked into rat plasma and analyzed on a 4.6 × 150-mm YMC-ODS AQ S3 column at a flow rate of 1 mL/min; LC–MS chromatograms obtained following injection of 10 pg of each metabolite on column: (a) Orbitrap and (b) linear ion trap. MS/MS (m/z 402) spectra of oxa-buspirone metabolite (R_t = 7.4 min): (c) Orbitrap at 10 pg on column, (d) Orbitrap at 100 pg on column, and (e) linear ion trap at 10 pg on column.

(b) linear ion trap; MS/MS (m/z 402) spectra of oxa-buspirone metabolite (R_t = 7.4 min) on (c) Orbitrap at 10 pg on column, (d) Orbitrap at 100 pg on column, and (e) linear ion trap at 10 pg on column.

Although both hybrid LTQ–FTMS instruments find applications in the drug metabolism area, the LTQ-Orbitrap, with its simplified maintenance, lower cost, and in most cases better sensitivity at low masses, is generally more applicable to small-molecule analysis. The LTQ–FTICR, with the advantage of higher resolution and mass accuracy, covers a wider range of applications, from small molecules to peptides and proteins, where most advantages are found in the analysis of large biomolecules.

5.5 INTEGRATED STRATEGIES FOR METABOLIC STABILITY DETERMINATION AND METABOLIC SOFT-SPOT CHARACTERIZATION

Accurate mass measurements play a major role in distinguishing nominally isobaric ions and structure assignment based on observed fragmentation mechanisms

8 (ppm) for the major fragment peaks observed						
RT (min)	29.0	26.8	26.5	25.1	21.1	20.8
MH+	Verapamil	M1	M2	M3	M4	M5
455.2904	0.6					
441.2748		1.5	1.5	1.5		
398.2200		2.0				
303.2067	0.9		0.9			
291.2067			0.3		0.1	
289.1911				0.6		
277.1911						1.4
260.1650	-1.5	-1.1	-1.5		0.7	0.3
248.1519						1.6
247.1441			0.8			
246.1489					0.0	
234.1056						0.0
165.0910	0.6	0.6	0.4	0.6		
151.0754			0.0			

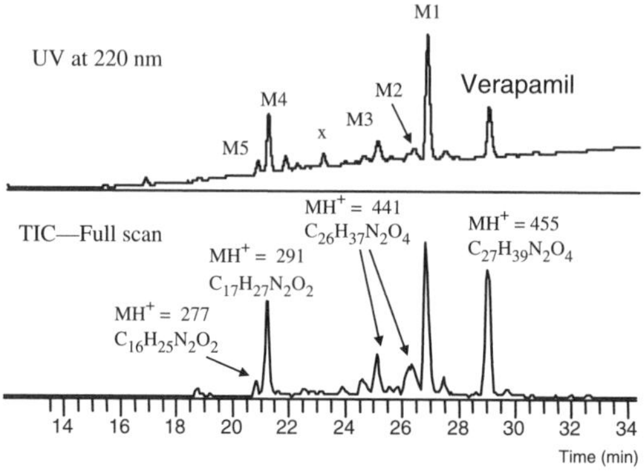

Figure 5.9. *LC–MS (TIC) and UV (220 nm) chromatograms for microsomal incubation of vera-pamil containing parent molecule and detected metabolites. Data-dependent accurate mass measurements for verapamil, metabolite parents and fragment ions are shown in the inset.*

(Clarke et al., 2001; Liu and Hop, 2005). With data-dependent MS^n scans (typically, $n = 2, 3$) and mass accuracy of 2 ppm or better, gathering the data is very efficient and metabolite identifications become significantly less complex. Figure 5.9 represents a typical LC–MS chromatogram of a verapamil microsomal incubation sample obtained using a data-dependent LC–MS/MS method on a LTQ-FT mass spectrometer. The first scan is a "survey" ion trap scan that triggers two data-dependent events when ions of interest or simply ions above a preset ion current threshold are detected. The first event sends the ions to the ICR cell to obtain accurate mass measurements on the parent ions, while the second event performs MS/MS on the specified ions of interest and gathers accurate mass measurements for the product ions. If desired, a third event, performing MS^3 could also be triggered. The inset in Fig. 5.9 shows the data-dependent accurate MS/MS measurements obtained for verapamil and its five major metabolites, where all observed measurements are within 2 ppm.

Accurate mass measurements of MS/MS and MS^n fragment ions, generated in a chromatographic run, can provide valuable structural information and make the technique a part of routine structural determination. For example, loss of a hydroxyl radical (17 Da), which is sometimes generated from *N*-oxide metabolites, can be easily distinguished from the loss of an ammonia molecule (17 Da) and therefore eliminates the possibility of incorrect structure assignments (Chapter 9). Often during MS/MS fragmentation of drug metabolites several fragments with the same nominal mass are produced and the correct identification could be essential for determination of metabolic soft spots or the existence of reactive/undesired metabolites. An example is shown in Fig. 5.10, where two different metabolites of buspirone give a fragment ion of 168 Da nominal mass, and based on accurate mass measurements, these fragments can be assigned to two very different ions, $C_9H_{14}NO_2$ and $C_7H_{10}N_3O_2$, representing oxidation at opposite parts of the molecule.

As stated previously, the ability to rapidly and comprehensively characterize a potential drug candidate's metabolic profile in terms of structure and quantity

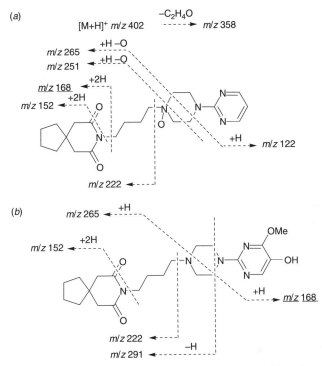

Figure 5.10. *Fragmentation pattern for two buspirone metabolites, (a) buspirone-N-oxide and (b) 5-OH, 4-O-methyl-buspirone, both yielding nominally isobaric fragment ions at m/z 168, which can be correctly assigned with accurate mass measurements.*

becomes essential for successful drug discovery and development. The analytical methods required to achieve this are (i) rapid quantification of the parent drug, (ii) identification of detected metabolites, and (iii) rapid quantification of metabolites of interest. When applied to appropriate in vitro incubations, this suite of analytical experiments affords intrinsic clearance, metabolite soft-spot identification, metabolite kinetic profiles, and reaction phenotyping by loss of parent and metabolite formation. The hydrid LTQ–FTMS instruments are uniquely suited for providing this information in an integrated and automated fashion with little method development time. A full data set, containing both quantitative and qualitative data, can be obtained during an overnight run (Josephs et al., 2008; Shipkova et al., 2004). To ensure linear kinetics for half-life determination of a compound in a metabolic stability incubation or reaction phenotyping experiment, the substrate concentration must be below the Michaelis-Menten constant (K_m) for the compound under investigation/enzyme responsible for clearance. Since the K_m values are not usually known for compounds in discovery, the experiments are typically carried out at 0.1–5 μM substrate concentrations, as the K_m for most drugs cleared by CYP enzymes are >5 μM. Conducting the experiments at lower substrate concentrations increases the likelihood of being below the K_m value. Experiments to determine the route of clearance are typically carried out at much higher concentrations, within the detection range of LC–UV ($\sim$10–30 μM). This choice is made for the purpose of analytical expediency as it becomes easier to detect and elucidate metabolites in the incubation matrix by full-scan MS and MS/MS and the ability to integrate UV peak areas affords a greater degree of relative quantitation for unknown metabolites, which is much better than relying on MS response where differences in the compound ionization potentials may influence the results. Unfortunately, while the relative quantitation of the metabolites may be analytically valid, in most cases relative quantitation does not correctly represent what is taking place in the sub-K_m concentration incubation. In addition, the quantitation by LC–UV requires lengthy chromatography runs which preclude making these determinations at multiple time points. An integrated streamlined process has been developed in our laboratories which allows for the identification of metabolites and the determination of kinetic parameters as well as profiles from a single analytical platform (Josephs et al., 2008).

In an integrated approach, the compounds, under investigation, are first analyzed for their identity and purity by a structural integrity assay (Josephs et al., 2002). A 4-min LC–MS run with a rapid gradient on a short 2.1-mm column employing positive/negative switching with data-dependent MS/MS of anticipated ions for the expected compound is carried out in the electrospray mode in the LTQ section of the LTQ–FTICR–MS. If the ionization method is unsuitable, the process is repeated in the APCI mode. Custom software establishes the most suitable MS/ MS selected reaction monitoring (SRM) transitions for quantitation based on the intensities of product ions obtained in the full-scan MS/MS spectra and builds corresponding quantitation methods for the compounds of interest. Incubations with microsomes (or hepatocytes) are carried out at 1 and 30 μM in parallel on a Tecan liquid handler in duplicate over multiple time points. The 1 μM incubations are analyzed for the parent compound using the established SRM method and a 2-min rapid generic gradient. The half-life ($t_{1/2}$) for each compound is determined, and if the $t_{1/2}$ falls below scientist-defined criteria, the compound is selected for further evaluation of metabolic soft spots. A suitable time point from the 30 μM incubation, typically one where significant metabolism is expected, is analyzed by the LTQ–FTICR–MS using a generic 30-min gradient on a long column (150 mm). UV detection of metabolites allows for the relative quantitation, while full-scan MS and MS/MS in the FTICR provides accurate masses of parent and product ions, used for determination of metabolic modification sites. Custom software determines the five most abundant metabolites based on the UV areas, and by consideration of the nominal parent masses and product ion spectra, unique SRM transitions are generated for each metabolite using similar algorithms to those used to automatically generate SRM methods for the parent compounds. Where no unique transition can be created, isobaric metabolites may still be individually quantitated if they are chromatographically separated on a rapid gradient or, alternatively, they may be quantitated together. The 1 μM time course incubations are then reanalyzed with a rapid 2-min gradient while quantitating the metabolites. In addition, the 30 μM incubation sample used to quantitate the metabolites based on UV response is diluted 30X and run alongside the 1 μM samples and is used as a single point calibration curve for metabolite quantitaion, assuming equal UV response factors at 220 nm. These data are then

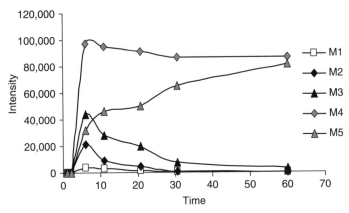

Figure 5.11. Kinetic profiles for top five verapamil metabolites observed in 1 μM microsomal incubations.

represented as kinetic profiles showing appearance and disappearance of metabolites during the incubation time course, as shown in Fig. 5.11 for the five most abundant metabolites of verapamil. Using this automated approach, the half-life and kinetic profiles of metabolites for five drug candidates can be completed in a 24-h period.

The approach described above has been demonstrated on the LTQ–FTICR and further extended to reaction phenotyping (Grubb et al., 2005; Josephs et al., 2005). The SRM quantitation methods for the parent and metabolites can be applied to incubations using human liver microsomes and selective inhibitory antibodies for the individual CYP isoforms. An example of this integrated approach using buspirone as a model compound is shown in Figs 5.12 and 5.13. Figure 5.12 displays the LC–UV–MS chromatogram of a 30 μM incubation of buspirone with human liver microsomes after 30 min. The top panel displays the UV chromatogram (220 nm) used for quantitation of the observed metabolites. The bottom panel is a total ion current plot of the MS response. Buspirone is observed at 12.9 min, together with five of its metabolites (M1–M5), all corresponding to different oxidation products at m/z 402. Unique SRM transitions for the top five metabolites, based on the observed fragment ions, are selected as described above and SRM experiments are performed to quantitate each detected metabolite. Figure 5.13 shows the effect of disappearance of buspirone when using inhibitory monoclonal antibodies, establishing CYP3A4 as the primary enzyme responsible for clearance, with CYP2C19 making a contribution. The table inset shows the corresponding half-life calculated for each CYP. Phenotyping may also be determined by looking at a particular metabolite formation, as shown in Fig. 5.13 for M4, where the contributions of each CYP can be determined over a time course.

Similar experiments have previously been performed separately using a triple-quadrupole or an ion trap mass spectrometer for quantitation followed by accurate mass measurements with a Q-TOF mass spectrometer. With the introduction of the LTQ–FTMS instruments, all the data can be gathered quickly and easily with one experimental setup, and the variations in observed metabolic profiles introduced by

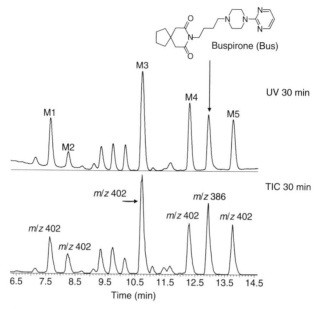

Figure 5.12. *LC–UV (top) and LC–MS (bottom) chromatograms 30 uM human liver microsomal (HLM) incubation samples of buspirone at 30 min.*

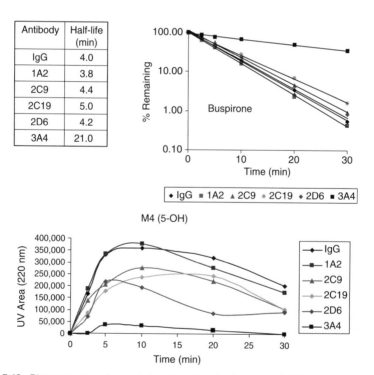

Figure 5.13. *Disappearance of parent drug (buspirone) when using inhibitory monoclonal antibodies and corresponding half-life calculations; individual CYP contributions for M4 metabolite.*

different platforms (i.e., a stand-alone ion trap and a Q-TOF), ionization sources, and so on, are completely eliminated.

5.6 ACCURATE MASS MEASUREMENTS FOR ADDRESSING INTERFERENCES IN COMPLEX MATRICES/FORMULATIONS

Software tools are being developed that take advantage of the high mass resolution and high mass stability that can be obtained with the hybrid FTMS instruments. These tools are providing simple and effective methods for locating the presence of drug metabolites in complex biofluids or the presence of degradants in complex formulations. Drug stability in the formulation vehicle is an important consideration for metabolism studies. Failure to recognize impurities/degradants could lead to

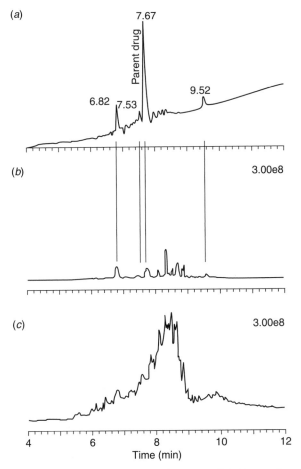

Figure 5.14. LC–MS profiles of PEG formulation of compound that has undergone degradation: (a) UV trace at 230–245 nm; (b) TIC after PEG subtracting process; (c) unprocessed TIC.

incorrect assignment of a detected peak (i.e., degradant) as a metabolite, thus misinterpreting the metabolic pathways involved. Due to the complex chemical nature of some formulations, it is not always easy to determine the drug stability and the task could be very time consuming to fully identify and characterize all possible degradants. Some of the worst matrices for mass spectrometry purposes are those containing polyethylene glycol (PEG) and related substances. Traditional LC–MS analyses are challenged by the fact that the impurities/degradants elute from the HPLC column in the same region as the PEG and the degradant molecular ions are masked by the abundant envelope of PEG ions. For such studies, acquisition of robust and accurate MS data for comparative data analyses (control vs. treated sample) becomes essential. Figure 5.14 shows a drug formulation prepared in PEG 400, where the ion signal is completely overwhelmed by the matrix ions (Fig. 5.14*c*). Utilizing the high resolution and highly stable mass accuracy of LTQ–FTMS, it is possible to subtract all PEG-related ions using control data generated from a PEG 400 sample. After processing with an ion recognition window of only 6 ppm, the major degradants/impurities are revealed as shown in Fig. 5.14*b*.

5.7   MASS DEFECT FILTER

The combination of high resolving power and high-flow liquid chromatography compatibility makes the LTQ–FTMS instruments the platform of choice for utilizing the mass defect filter (MDF) technique for selectively distinguishing drug-related components from endogenous matrix interferences. The ability of MDF to rapidly, and with high confidence, identify drug-related chromatographic peaks greatly facilitates the drug metabolite identification. The MDF is a software-based data filter that interrogates high-resolution LC–MS data for predictable mass defect similarities found in both a drug and its metabolites (Zhang et al., 2003). MDF is an alternative to the existing tools (e.g., precursor ion scans and neutral loss scans on triple-quadrupole instruments) that also allows for selective detection of drug metabolites in complex biological matrices (Zhang et al., 2004). One of the most attractive features of the MDF approach is the simple and straightforward instrument setup. A high-resolution accurate mass LC–MS analysis is all that is needed. Unlike the triple-quadrupole precursor ion or neutral loss scan experiments, where detailed knowledge of the MS/MS fragmentation is required for the compound under study, the MDF approach requires little a priori knowledge in order to collect the appropriate data. This means that as more information is gathered on the metabolism of a compound, the accurate mass LC–MS data set can be reinterrogated using alternate MDF templates to investigate further routes of metabolism without having to reanalyze the sample. More details on the application of MDF are discussed in Chapter 6.

5.8   ENDOGENOUS METABOLITE PROFILING (METABONOMICS)

A related approach which in many ways complements drug metabolism analyses is profiling of endogenous metabolites as markers, often done in conjunction with drug metabolism to gather a complete pharmakokinetic/pharmacodynamic (PK/PD) profile. Endogenous metabolite profiling experiments are often part of metabonomics or metabolomics investigations, which are geared toward understanding how endogenous small-molecule metabolites (or metabolome) are perturbed by an insult (e.g., disease, age, or administration of a drug) (Nicholson et al., 2002). The analytical goal is to perform a comprehensive quantitative screen for all components in a given biofluid or tissue and record any significant changes. The hybrid FTMS instruments have already been shown to be effective for metabonomic analyses, where accurate and robust mass measurements combined with online chromatography are essential (van der Greef et al., 2004; Brown et al., 2005). Analytical approaches in metabonomics are in many ways similar to those used in xenobiotic drug metabolism studies.

An overall metabonomics strategy includes a metabolite ID component, where it is critically important to identify all drug-related metabolites and remove them from the data set, since they could hinder the discovery of dose-dependent changes in the endogenous components. The identification process of both drug-related metabolites and endogenous species are geared towards detecting changes between control and treated samples, and so both could benefit from the same general software and statistical approaches.

Typically, metabolites present in dosed samples but absent in the corresponding controls show up distinctly with a high degree of significance in a metabonomics study (Plumb et al., 2003). The current effort to catalog endogenous components from various biofluids using LC–MS and GC–MS will also serve to help the ADMET scientist to more rapidly distinguish endogenous components from drug-related metabolites. We have shown that in the LC–MS profiling of rat urine, which might contain tens of thousands of ions, a single component will have many ions associated with it beyond the expected isotopic distribution. For hippuric acid we have observed up to 12 related ions in the positive-ion mode and 24 related ions in the negative-ion mode. These contain the expected $[M + Na]^+$ and $[M + K]^+$ adducts as well as gas-phase dimers formed during the ionization process, but as the analysis was being performed with high resolution and mass accuracy, it became clear that more exotic adducts such as $[2M+ CH_3CN + Fe-H]^+$ and $[3M + Ca-H]^+$ were also being formed. The data from the LTQ–FT allowed these adducts to be identified quickly and with high confidence such that time was not wasted trying to identify a new "metabolite." As the small-molecule metabolome and adduct databases are further developed, they will become extremely useful to scientists performing both metabonomics and ADMET studies.

In our hands the statistical approach of principal-component analysis (PCA) for identifying endogenous components that exhibit a dose- or toxicity-related change suffers from a lack of sensitivity. For example, the detection of small (two fold) changes in minor components ($50\,nM$) has not been successful. The ability to detect such changes is desirable since concentration ranges of plasma components are often tightly regulated and potential biomarkers may be present at trace levels. While the traditional metabonomics approaches have posed challenges, adopting a "metabolite ID" strategy to analyze such data has proven to be beneficial (Zhang et al., 2005b). In most metabonomics studies there are groups of samples which can be identified as controls (e.g., nontoxic) and those which correspond to a dose (e.g., toxic). Pooling samples from these groups provides a profile with an "average" response where pronounced changes in the concentration of a metabolite for a group of animals would be reflected in a pooled sample from that group. Comparing the two LC–MS profiles, sample differences can be readily identified using metabolite ID tools such as the Metabolynx software package from Waters. Having detected differences, which may include many false positives, the differential ions are further scrutinized by quantitation in the individual animals. Simple statistical tests can then be applied to the individual animal data to determine if the observed changes were significant. Taking advantage of the mass stability of the LTQ-FT instruments and being able to step through the data in 20-mDa increments have enabled the detection of toxicity markers where other, more traditional, metabonomics approaches have failed.

5.9 SUMMARY

The hybrid LTQ–FTMS mass spectrometers, including the LTQ–FTICR and the LTQ-Orbitrap, have successfully integrated FTMS technology with standard HPLC techniques to provide accurate mass and accurate MS/MS determinations on a chromatographic time scale. These systems can achieve ppb mass accuracy with external calibration which, combined with the ease of use and automation capabilities, greatly enhances and simplifies the metabolite characterization process. Standard HPLC and UPLC chromatography can be employed without flow splitting, and the data-dependent MS^n/and accurate mass capabilities can be routinely utilized to rapidly confirm the identification of expected metabolites, elucidate the structures of unusual or unexpected metabolites, or identify metabolic soft spots. Typically, the value of complex mass spectrometers is determined by the application in which they can perform or the problem they can solve that no other instrument could. With the high-resolution performance characteristics of the hybrid FTMS systems, those situations will most certainly exist, but the real benefit is the productivity gains provided every day on almost every sample. Whether confirming the structure of a known metabolite, assigning the MS/MS spectra of a parent compound, or elucidating the structure of an unknown metabolite, the process is significantly enhanced with utilizing accurate mass for providing an unequivocal assignment of the molecular formula, especially if the data were no more difficult to obtain than from an ion trap mass spectrometer. We are already seeing the emergence of new data-processing approaches that take advantage of the FTMS data. With these instruments, automated data interpretation becomes much more feasible as the data become more precise. The algorithm logic required can be significantly less sophisticated. One can expect the data from these hybrid FT instruments to continue to drive data interpretation and automation software. The LTQ–FTMS instruments have significantly influenced the way in which we approach metabolite characterization, and we anticipate that as this technology is integrated into ADMET laboratories, it will have a profound effect on the way drug metabolism studies are performed throughout the pharmaceutical industry.

REFERENCES

Bateman, K. P., Castro-Perez, J., Wrona, M., Shockcor, J. P., Yu, K., Oballa, R., and Nicoll-Griffith, D. A. (2007). MSE with mass defect filtering for in vitro and in vivo metabolite identification. *Rapid Commun. Mass Spectrom.* **21:**1485–1496.

Bogdanov, B., and Smith, R. D. (2005). Proteomics by FTICR mass spectrometry: Top down and bottom up. *Mass Spectrom. Rev.* **24:**168–200.

Bristow, A. W. (2006). Accurate mass measurement for the determination of elemental formula—A tutorial. *Mass Spectrom. Rev.* **25:**99–111.

Bristow, A. W., and Webb, K. S. (2003). Intercomparison study on accurate mass measurement of small molecules in mass spectrometry. *J. Am. Soc. Mass Spectrom.* **14:**1086–1098.

Brown, S. C., Kruppa, G., and Dasseux, J. L. (2005). Metabolomics applications of FT-ICR mass spectrometry. *Mass Spectrom. Rev.* **24:**223–231.

Burton, R. D., Matuszak, K. P., Watson, C. H., and Eyler, J. R. (1999). Exact mass measurements using a 7 tesla Fourier transform ion cyclotron resonance mass spectrometer in a good laboratory practices-regulated environment. *J. Am. Soc. Mass Spectrom.* **10:**1291–1297.

Chen, G., Khusid, A., Daaro, I., Irish, P., and Pramanik, B. N. (2007). Structural identification of trace level enol tautomer impurity by on-line hydrogen/deuterium exchange HR-LC/MS in a LTQ-Orbitrap hybrid mass spectrometer. *J. Mass Spectrom.* **42:**967–970.

Clarke, N. J., Rindgen, D., Korfmacher, W. A., and Cox, K. A. (2001). Systematic LC/MS metabolite identification in drug discovery. *Anal. Chem.* **73:**430A–439A.

Fandino, A. S., Nagele, E., and Perkins, P. D. (2006). Automated software-guided identification of new buspirone metabolites using capillary LC coupled to ion trap and TOF mass spectrometry. *J. Mass Spectrom.* **41:**248–255.

Fitch, W. L., He, L., Tu, Y.- P., and Alexandrova, L. (2007). Application of polarity switching in the identification of the metabolites of RO9237. *Rapid Commun. Mass Spectrom.* **21:**1661–1668.

Grubb, M. F., Shipkova, P., Langish, R., and Josephs, J. L. (2005). Rapid determination of human liver metabolism reaction phenotype using LC/MS/MS. In *Proceedings of the 53rd ASMS Conference on Mass Spectrometry and Allied Topics*, San Antonio, TX.

Hardman, M., and Makarov, A. A. (2003). Interfacing the orbitrap mass analyzer to an electrospray ion source. *Anal. Chem.* **75:**1699–1705.

Heeren, R. M. A., Kleinnijenhuis, A. J., McDonnell, L. A., and Mize, T. H. (2004). A minireview of mass spectrometry using high-performance FTICR-MS methods. *Anal. Bioanal. Chem.* **378:**1048–1058.

Hopfgartner, G., Chernushevich, I. V., Covey, T., Plomley, J. B., and Bonner, R. (1999). Exact mass measurement of product ions for the structural elucidation of drug metabolites with a tandem quadrupole orthogonal-acceleration time-of-flight mass spectrometer. *J. Am. Soc. Mass Spectrom.* **10:**1305–1314.

Horning, S. R., Lange, O., Wieghaus, A., Malek, R., and Senko, M. W. (2004). Simultaneous acquisition of two ion signals using two detectors in a hybrid ion trap/fourier transform ICR mass spectrometer. In *Proceedings of the 52nd ASMS Conference on Mass Spectrometry and Allied Topics*, Nashville, TN.

Johnson, L., Mollah, S., Garcia, B. A., Muratore, T. L., Shabanowitz, J., Hunt, D. F., and Jacobsen, S. E. (2004). Mass spectrometry analysis of Arabidopsis histone H_3 reveals distinct combinations of post-translational modifications. *Nucleic Acids Res.* **32:**6511–6518.

Josephs, J. L., Grubb, M. F., Yang, Y., and Humphreys, G. W. (2008). A rapid approach to quantitative in vivo metabolite profiling without the need for authentic standards or labeled compounds. In *Proceedings of the 56th ASMS Conference on Mass Spectrometry and Allied Topics*, Denver, Co.

Josephs, J. L., Grubb, M. F., Shipkova, P. S., Langish, A. R., and Sanders, M. (2004a). The hybrid linear ion trap/fourier transform mass spectrometer as a tool for investigating biotransformation pathways. In *Proceedings of the Conference on Small Molecules Science*, Bristol, RI.

Josephs, J. L., Grubb, M. F., Shipkova, P., Langish, R., and Sanders, M. (2005). A comprehensive strategy for the characterization and optimization of metabolic profiles of compounds using a hybrid linear ion trap/FTMS. In *Proceedings of the 53rd ASMS Conference on Mass Spectrometry and Allied Topics*, San Antonio, TX.

Josephs, J. L., and Sanders, M. (2004). Creation and comparison of MS/MS spectral libraries using quadrupole ion trap and triple-quadrupole mass spectrometers. *Rapid Commun. Mass Spectrom.* **18:**743–759.

Josephs, J. L., Sanders, M., Langish, A. R., Hnatyshyn, S. Y., Salyan, M. E., Shipkova, P. S., Drexler, D., Flynn, M., Burdette, H., and Balimane, P. (2002). A high quality high throughput LC/MS strategy for profiling of drug candidates with applications to structural integrity, permeability, stability, and related assays. In *Proceedings of the 50th ASMS Conference on Mass Spectrometry and Allied Topics*, Orlando, FL.

Josephs, J. L., Shipkova, P. A., Langish, R., and Sanders, M. (2004b). A fully automated process for determining compound integrity and building accurate mass MS/MS libraries. In *Proceedings of the 52nd ASMS Conference on Mass Spectrometry and Allied Topics*, Nashville, TN.

Leslie, A. D., and Volmer, D. A. (2007). Dealing with the masses: A tutorial on accurate masses, mass uncertainties, and mass defects. *Spectroscopy* **22:**32, 34–39.

Lim, H. K., Chen, J., Sensenhauser, C., Cook, K., and Subrahmanyam, V. (2007). Metabolite identification by data-dependent accurate mass spectrometric analysis at resolving power of 60,000 in external calibration mode using an LTQ/Orbitrap. *Rapid Commun. Mass Spectrom.* **21:**1821–1832.

Liu, D. Q., and Hop, C. E. (2005). Strategies for characterization of drug metabolites using liquid chromatography-tandem mass spectrometry in conjunction with chemical derivatization and on-line H/D exchange approaches. *J. Pharm. Biomed. Anal.* **37:**1–18.

Ma, L., Zhang, H., Humphreys, G., Sanders, M., and Zhu, M. (2005). Selective and sensitive detection of GSH adducts in complex biological samples using high resolution LC/FTMS with mass defect filtering. In *Proceedings of the 53rd ASMS Conference on Mass Spectrometry and Allied Topics*, San Antonio, TX.

Macek, B., Waanders, L. F., Olsen, J. V., and Mann, M. (2006). Top-down protein sequencing and MS3 on a hybrid linear quadrupole ion trap-orbitrap mass spectrometer. *Mol. Cell. Proteomics* **5:**949–958.

Makarov, A. (2000). Electrostatic axially harmonic orbital trapping: A high-performance technique of mass analysis. *Anal. Chem.* **72:**1156–1162.

Makarov, A., Denisov, E., Kholomeev, A., Balschun, W., Lange, O., Strupat, K., and Horning, S. (2006a). Performance evaluation of a hybrid linear ion trap/orbitrap mass spectrometer. *Anal. Chem.* **78:**2113–2120.

Makarov, A., Denisov, E., Lange, O., and Horning, S. (2006b). Dynamic range of mass accuracy in LTQ Orbitrap hybrid mass spectrometer. *J. Am. Soc. Mass Spectrom.* **17:**977–982.

Marshall, A. G. (2000). Milestones in Fourier transform ion cyclotron resonance mass spectrometry technique development. *Int. J. Mass Spectrom.* **200**:331–356.

Marshall, A. G. (2004). Accurate mass measurement: Taking full advantage of nature's isotopic complexity. *Physica B* **346–347**:503–508.

Marshall, A. G., Hendrickson, C. L., and Jackson, G. S. (1998). Fourier transform ion cyclotron resonance mass spectrometry: A primer. *Mass Spectrom. Rev.* **17**:1–35.

McAlister, G. C., Phanstiel, D., Good, D. M., Berggren, W. T., and Coon, J. J. (2007). Implementation of electron-transfer dissociation on a hybrid linear ion trap-orbitrap mass spectrometer. *Anal. Chem.* **79**:3525–3534.

McDonald, L. A., Barbieri, L. R., Carter, G. T., Kruppa, G., Feng, X., Lotvin, J. A., and Siegel, M. M. (2003). FTMS structure elucidation of natural products: Application to muraymycin antibiotics using ESI multi-CHEF SORI-CID FTMS(n), the top-down/bottom-up approach, and HPLC ESI capillary-skimmer CID FTMS. *Anal. Chem.* **75**:2730–2739.

Nawrocki, J. P., Wigger, M., Watson, C. H., Hayes, T. W., Senko, M. W., Benner, S. A., and Eyler, J. R. (1996). Analysis of combinatorial libraries using electrospray Fourier transform ion cyclotron resonance mass spectrometry. *Rapid Commun. Mass Spectrom.* **10**:1860–1864.

Nicholson, J. K., Connelly, J., Lindon, J. C., and Holmes, E. (2002). Metabonomics: A platform for studying drug toxicity and gene function. *Nat. Rev. Drug. Discov.* **1**:153–161.

Ojanperae, I., Pelander, A., Laks, S., Gergov, M., Vuori, E., and Witt, M. (2005). Application of accurate mass measurement to urine drug screening. *J. Anal. Toxicol.* **29**:34–40.

Peterman, S. M., Duczak, N., Jr., Kalgutkar, A. S., Lame, M. E., and Soglia, J. R. (2006). Application of a linear ion trap/orbitrap mass spectrometer in metabolite characterization studies: Examination of the human liver microsomal metabolism of the non-tricyclic antidepressant nefazodone using data-dependent accurate mass measurements. *J. Am. Soc. Mass Spectrom.* **17**:363–375.

Peterman, S. M., Dufresne, C. P., and Horning, S. (2005). The use of a hybrid linear trap/FT-ICR mass spectrometer for on-line high resolution/high mass accuracy bottom-up sequencing. *J. Biomol. Tech.* **16**:112–124.

Pihakari, K. A. (2007). FT–MS to provide novel insight into complex samples. *Spectroscopy* **22**:18–26.

Plumb, R. S., Stumpf, C. L., Granger, J. H., Castro-Perez, J., Haselden, J. N., and Dear, G. J. (2003). Use of liquid chromatography/time-of-flight mass spectrometry and multivariate statistical analysis shows promise for the detection of drug metabolites in biological fluids. *Rapid Commun. Mass Spectrom.* **17**:2632–2638.

Poulsen, S. A., Gates, P. J., Cousins, G. R., and Sanders, J. K. (2000). Electrospray ionisation Fourier-transform ion cyclotron resonance mass spectrometry of dynamic combinatorial libraries. *Rapid Commun. Mass Spectrom.* **14**:44–48.

Ramanathan, R., Reyderman, L., Kulmatycki, K., Su, A. D., Alvarez, N., Chowdhury, S. K., Alton, K. B., Wirth, M. A., Clement, R. P., Statkevich, P., and Patrick, J. E. (2007a). Disposition of loratadine in healthy volunteers. *Xenobiotica* **37**:753–769.

Ramanathan, R., Reyderman, L., Su, A. D., Alvarez, N., Chowdhury, S. K., Alton, K. B., Wirth, M. A., Clement, R. P., Statkevich, P., and Patrick, J. E. (2007b). Disposition of desloratadine in healthy volunteers. *Xenobiotica* **37**:770–787.

Ramjit, H. G., Kruppa, G. H., Spier, J. P., Ross, C. W., 3rd, and Garsky, V. M. (2000). The significance of monoisotopic and carbon-13 isobars for the identification of a 19-

component dodecapeptide library by positive ion electrospray Fourier transform ion cyclotron resonance mass spectrometry. *Rapid Commun. Mass Spectrom.* **14**:1368–1376.

Saghatelian, A., Trauger, S. A., Want, E. J., Hawkins, E. G., Siuzdak, G., and Cravatt, B. F. (2004). Assignment of endogenous substrates to enzymes by global metabolite profiling. *Biochemistry* **43**:14332–14339.

Sanders, M., Shipkova, P. A., Zhang, H., and Warrack, B. M. (2006). Utility of the hybrid LTQ-FTMS for drug metabolism applications. *Curr. Drug Metab.* **7**:547–555.

Sanders, M., Warrack, B., Lange, O., Strupat, K., and Horning, S. (2005). Optimizing low mass ion transmission on a hybrid linear ion trap FTICR-MS instrument and its application to metabonomic profiling, In *Proceedings of the 53rd ASMS Conference on Mass Spectrometry and Allied Topics*, San Antonio, TX.

Schmid, D. G., Grosche, P., Bandel, H., and Jung, G. (2000). FTICR-mass spectrometry for high-resolution analysis in combinatorial chemistry. *Biotechnol. Bioeng.* **71**:149–161.

Scigelova, M., and Makarov, A. (2006). Orbitrap mass analyzer—Overview and applications in proteomics. *Proteomics* **6** (Suppl. 2): 16–21.

Senko, M., Zabrouskov, V., Lange, O., Wieghaus, A., and Horning, S. (2004). LC/MS with external calibration mass accuracies approaching 100 ppb. In *Proceedings of the 52nd ASMS Conference on Mass Spectrometry and Allied Topics*, Nashville, TN.

Shipkova, P. A., Josephs, J. L., Grubb, M. F., Langish, A. R., Chen, W., and Sanders, M. (2004). Use of a hybrid linear ion trap/Fourier transform mass spectrometer for determining metabolic stability and metabolite characterization. In *Proceedings of the 52nd ASMS Conference on Mass Spectrometry and Allied Topics*, Nashville, TN.

Siegel, M. M. (2005). *Mass-Spectrometry-Based Drug Screening Assays for Early Phases in Drug Discovery.* Wiley, Hoboken, NJ.

Syka, J. E. P., Marto, J. A., Bai, D. L., Horning, S., Senko, M. W., Schwartz, J. C., Ueberheide, B., Garcia, B., Busby, S., Muratore, T., Shabanowitz, J., and Hunt, D. F. (2004). Novel linear quadrupole ion trap/FT mass spectrometer: Performance characterization and use in the comparative analysis of histone H3 post-translational modifications. *J. Proteome Res.* **3**:621–626.

Tozuka, Z., Strupat, W. M., Shiraga, T., Ishimura, R., Hashimoto, T., Kawamura, A., and Kagayama, A. (2005). LTQ FTMS accurate mass and SRM data dependent exclusion MSn measurements for structure determination of FK228 and its metabolites. *J. Mass Spectrom. Soc. Jpn.* **53**:89–99.

Triolo, A., Altamura, M., Dimoulas, T., Guidi, A., Lecci, A., and Tramontana, M. (2005). In vivo metabolite detection and identification in drug discovery via LC-MS/MS with data-dependent scanning and postacquisition data mining. *J. Mass Spectrom.* **40**:1572–1582.

van der Greef, J., van der Heijden, R., and Verheij, E. R. (2004). The role of mass spectrometry in systems biology: Data processing and identification strategies in metabolomics. *Adv. Mass Spectrom.* **16**:145–165.

Wang, Z., Hop, C. E. C. A., Kim, M.- S., Huskey, S.- E. W., Baillie, T. A., and Guan, Z. (2003). The unanticipated loss of SO_2 from sulfonamides in collision-induced dissociation. *Rapid Commun. Mass Spectrom.* **17**:81–86.

Warrack, B., Hnatyshyn, S., Ott, K.- H., Ray, K., Tymiak, A. A., Zhang, H., and Sanders, M. (2004). Application of LTQ-FTMS to metabonomics profiling. In *Proceedings of the 52nd ASMS Conference on Mass Spectrometry and Allied Topics*, Nashville, TN.

Warrack, B., Hnatyshyn, S., Zhang, H., and Sanders, M. (2005). Strategies for the use of mass spectrometry for LC/MS metabonomics profiling: How much resolution is needed and in what dimension? In *Proceedings of the 53rd ASMS Conference on Mass Spectrometry and Allied Topics*, San Antonio, TX.

Williamson, B. L., Marchese, J., and Morrice, N. A. (2006). Automated identification and quantification of protein phosphorylation sites by LC/MS on a hybrid triple quadrupole linear ion trap mass spectrometer. *Mol. Cell. Proteomics* **5**:337–346.

Winger, B. E., and Kemp, C. A. J. (2001). Characterization of pharmaceutical compounds and related substances by using HPLC FTICR-MS and tandem mass spectrometry. *Am. Pharm. Rev.* **4**:55–56, 58, 60, 62–63.

Yates, J. R., Cociorva, D., Liao, L., and Zabrouskov, V. (2006). Performance of a linear ion trap-Orbitrap hybrid for peptide analysis. *Anal. Chem.* **78**:493–500.

Zhang, H., Ray, K., Ma, L., Zhang, D., Drexler, D., and Sanders, M. (2005a). Applicability of mass defect filters (MDF) to drug metabolite detection in biological matrices. In *Proceedings of the 53rd ASMS Conference on Mass Spectrometry and Allied Topics*, San Antonio, TX.

Zhang, H., Ray, K., Zhao, W., Zhang, D., Gozo, S., and Zhu, M. (2004). Selective detection of metabolite ions of non-radiolabeled drugs in complex biological matrices using mass defect filter approach. In *Proceedings of the 52nd ASMS Conference on Mass Spectrometry and Allied Topics*, Nashville, TN.

Zhang, H., Warrack, B., Hnatyshyn, S., Aranibar, N., Friedrechs, M., Ott, K.- H., Ray, K., and Sanders, M. (2005b). A pooled sample strategy for the identification of potential biomarkers utilizing mass defect splitting on high resolution LC/MS data. In *Proceedings of the 53rd ASMS Conference on Mass Spectrometry and Allied Topics*, San Antonio, TX.

Zhang, H., Zhang, D., and Ray, K. (2003). A software filter to remove interference ions from drug metabolites in accurate mass liquid chromatography/mass spectrometric analyses. *J. Mass Spectrom.* **38**:1110–1112.

Zhang, L-K., Rempel, D., Pramanik, B. N., and Gross, M. L. (2005c). Accurate mass measurements by Fourier transform mass spectrometry. *Mass Spectrom. Rev.* **24**:286–309.

Zhang, N., Fountain, S. T., Bi, H., and Rossi, D. T. (2000). Quantification and rapid metabolite identification in drug discovery using API time-of-flight LC/MS. *Anal. Chem.* **72**:800–806.

Zhu, M., Ma, L., Zhang, D., Ray, K., Zhao, W., Humphreys, W. G., Skiles, G., Sanders, M., and Zhang, H. (2006). Detection and characterization of metabolites in biological matrices using mass defect filtering of liquid chromatography/high resolution mass spectrometry data. *Drug Metab. Dispos.* **34**:1722–1733.

Zimmer, J. S., Monroe, M. E., Qian, W. J., and Smith, R. D. (2006). Advances in proteomics data analysis and display using an accurate mass and time tag approach. *Mass Spectrom. Rev.* **25**:450–482.

Ziqiang Guan, J. M. L. (2001). Solvation of acylium fragment ions in electrospray ionization quadrupole ion trap and Fourier transform ion cyclotron resonance mass spectrometry. *J. Mass Spectrom.* **36**:264–276.

6

High-Resolution LC–MS-Based Mass Defect Filter Approach: Basic Concept and Application in Metabolite Detection

Haiying Zhang

Bristol-Myers Squibb Pharmaceutical Research Institute, Department of Biotransformation, Pennington, New Jersey

Donglu Zhang and Mingshe Zhu

Bristol-Myers Squibb Pharmaceutical Research Institute, Department of Biotransformation, Princeton, New Jersey

Kenneth L. Ray

Novatia, Monmouth Junction, New Jersey

Mass Spectrometry in Drug Metabolism and Pharmacokinetics. Edited by Ragu Ramanathan
Copyright © 2009 John Wiley & Sons, Inc.

6.1   INTRODUCTION

A common objective in pharmaceutical research is the detection and identification of drug metabolites. Biological samples or fluids may be subject to liquid chromatography–mass spectrometry (LC–MS) analysis for the purpose of detecting and identifying specific metabolites. The commonly encountered biological samples include plasma, bile, urine, other bodily fluids, tissue extracts, and fecal extracts. Detection and identification of drug metabolites in such biological samples, however, is often difficult due to significant interference from endogenous species. Typically, metabolite ions of interest may be embedded in the ocean of endogenous interference ions from the biological matrices. The total ion chromatogram of an LC–MS experiment may show little resemblance to the corresponding "accepted" metabolite profile, such as a radioactivity chromatogram. As a result, a metabolite of interest may not be easily identifiable even by those skilled in LC–MS analysis. For example, a metabolite ion may be isobaric (i.e., having the same nominal, or integral, mass-to-charge ratio) with a naturally occurring biological substance. In both the theoretical and practical sense, LC–MS data from a high-resolution instrument should assist in distinguishing drug metabolites from endogenous isobaric interferences.

Since determination of pseudomolecular ions is a critical step in identifying drug metabolites in biological fluids, efforts have been focused on that long before the emergence of modern high-resolution LC–MS technology. Over the last 15 years, several MS/MS-based acquisition approaches have been used to discern drug metabolite ions in biological matrices (Vrbanac et al., 1992). The first and the most common approach is the product ion scan, which has been utilized to detect or confirm expected metabolites. Second-tier approaches include precursor ion scans and neutral loss scans, which have been used for selective detection of metabolite ions based on the similarities between the fragment ions of a drug and its metabolites as well as the fragmentation reactions of conjugated metabolites. In addition, several LC–MS data-processing methods have been employed to facilitate the recognition of metabolite ions, including isotope cluster analysis and comparative analysis against control samples (Morand et al., 2001).

The emergence of high-resolution LC–MS technologies has greatly facilitated metabolite identification tasks with enhanced resolution, high mass accuracy, and high sensitivity in a full-scan mode (Meyer et al., 1996; Yoshitsugu et al., 1999;

Corcoran et al., 2000; Hopfgartner and Vilbois, 2000; Zhang et al., 2000; Blum et al., 2001; Nicholson et al., 2001; Marshall et al., 2002; Sundstrom et al., 2002). A major application of high-resolution LC–MS is to assist in distinguishing an ion of interest from its isobaric interferences (Keough et al., 1997; Bahr and Karas, 1999; Hogenboom et al., 1999; Lane and Pipe, 2000; Taylor and Johnson, 2001). With the accurate mass capability of high-resolution LC–MS, one can also perform various MS/MS-based "fishing expeditions" with improved specificity. For example, real-time accurate mass measurements of fragment ions and neutral losses have improved the specificity of data-dependent acquisition of full scan MS/MS data for expected metabolites (Arora et al., 2001; Clarke et al., 2001a, 2002; Castro-Perez et al., 2002; Jemal et al., 2003). Furthermore, the combination of high-resolution LC–MS data with existing data-processing programs (e.g., MetaboLynx from Waters/Micromass and Metabolite ID from Applied Biosystems/MDS Sciex) has provided more selective extracted ion chromatograms for targeted or unknown metabolites (Beaumont et al., 2000; Arora et al., 2001; Clarke et al., 2001a, 2002), although the identification of unexpected metabolites or metabolites with low abundance still remains challenging.

In this chapter, the mass defect filter (MDF) approach, a technique that takes advantage of high-resolution LC–MS data to facilitate the detection of drug metabolite ions in complex biological matrices (Zhang et al., 2003a), is described. MDF is a data-processing approach which works on full-scan LC–MS data to filter out biological interferences. In contrast to the MS/MS-based acquisition techniques that rely on specific knowledge of a drug's fragmentation behavior to identify its metabolites in a sample, a MDF approach relies on a general physical property to segregate drug-related ions from those that are irrelevant (endogenous and/or matrix ions). The aim of a MDF is to render complex data from a biological sample to a level of simplicity that would be equivalent to one that is free of biological matrix and thus to make the drug metabolite ions distinct. A metaphor that is used to depict the MDF approach is to "empty the ocean and get the fish."

In this chapter, the basic concept of MDF is described in detail and various applications of MDF for facile metabolite detection are presented using high-resolution LC–MS data from both time-of-flight (TOF) and Fourier transform mass spectrometry (FTMS) [linear trap quadrupole–Fourier transform ion cyclotron resonance (LTQ–FTICR) and LTQ-Orbitrap] platforms. The effectiveness of MDF in removing endogenous interferences present in plasma, urine, bile, and fecal extracts is also illustrated. In addition, LC–MS data processed using MDF is compared with those from precursor ion scan (PIS) and neutral loss scan (NLS) techniques for the sensitivity and specificity of each technique. Finally, mass defect profiles for various biological matrices are described. Such knowledge gives us an overall picture for the assessment of the applicability of the MDF approach for a given drug and allows for appropriate MDF design for possible drug metabolites.

6.2 WHAT IS MASS DEFECT?

Before getting deeper into the MDF approach, the concept of mass defect is reviewed. Mass defect is a physical phenomenon in nature. At the atomic level, mass defect is

TABLE 6.1. Atomic Weights and Residual Masses of Common Elements

Element	Nominal Mass (Da)	Exact Mass (Da)	Mass Shift (Relative to Nominal Mass) (Da)
H	1	1.0078	0.0078
N	14	14.0031	0.0031
C	12	12.0000	0
O	16	15.9949	−0.0051

the amount by which the mass of an atomic nucleus is less than the sum of the masses of its constituent particles. This amount of mass difference is equivalent to the energy released when the nucleons are bound to form an atomic nucleus. This phenomenon causes the atomic weights of every element in nature to have a unique nonintegral mass when using a relative scale with reference to ^{12}C, whose atomic weight has been arbitrarily defined as 12 Da exactly. Table 6.1 lists the atomic weights of common elements for pharmaceutical compounds. As shown, each element has a characteristic residual mass or nonintegral portion in its atomic weight, which results from the mass defect for that particular element. At a molecular level, the mass defect of a molecule is a simple summation of the mass defects of all of its constituent atoms. This summation of mass defects would, again, result in a characteristic residual, or fractional, mass. For example, Table 6.2 illustrates a classical case where several compounds with the same nominal mass are differentiated by their residual masses. The context here is that the residual mass of a compound results from the summation of the mass defects of its elemental constituents and hence is characteristic of its elemental composition.

To reflect the underlining principle of the approach, it is referred to as a *mass defect filter*, although in practice it works on residual masses of ions. For the simplicity of this chapter, the term "mass defect" is used to refer to the characteristic residue mass of a molecular species.

The mass defect is a unique characteristic associated with the empirical formula of a molecule, which is the basis of accurate mass measurement for molecular characterization. Modern MS technology with high mass-resolving power is capable of differentiating isobaric chemical entities at residual mass levels. For example, FTICR–MS provides the highest mass accuracy and mass-resolving power of the currently available mass spectrometers (Marshall, 2000; Pihakari, 2007). Thanks to the advances in MS technologies, the new generation of mass spectrometers is capable of bridging high-resolution MS with high-performace liquid chromatography (HPLC) technology. For example, with a typical TOF instrument, one can acquire high-resolution LC–MS data and fully resolve isobaric ions that are separated by 50 mDa over a mass range typical for small molecules (Chapter 4) (Castro-Perez et al., 2002).

TABLE 6.2. Molecular Weights and Residual Masses of Example Isobaric Molecules

Compound	Nominal Mass (Da)	Exact Mass (Da)	Mass Shift (Relative to Nominal Mass) (Da)
CO	28	27.9949	−0.0051
N_2	28	28.0061	0.0061
C_2H_4	28	28.0312	0.0312

6.3   CONCEPT OF MDF

A MDF approach, simply put, is a way of segregating ions of interest from interference ions in an entire LC–MS data set by imposing preset criteria in the mass defect dimension. One cannot do mass defect filtering unless the mass defects of interest are defined. This range of values is typically associated with the mass defects of the ions of interest resulting from the expected metabolism or decomposition of a drug. Such a definable range comes from nature. For example, mass defects of a drug and its metabolites are often similar to each other and within predictable narrow ranges (Zhang et al., 2003a). Table 6.3 shows the differences, in terms of mass and mass defects, of a drug and its metabolites resulting from typical metabolic transformations. In the middle columns of Table 6.3, the mass scale is from 2 Da to over 300 Da. Therefore, the metabolites are inevitably mingled with endogenous interferences throughout the mass range. However, the mass defect change shown in the right column of Table 6.3 is typically within a 0.05-Da range. That narrow range of change in the mass defect is the basis for designing a MDF for drug metabolite identification. A comprehensive list of mass defect changes in a wide variety of biotransformations was recently published by Mortishire-Smith et al. (2005). Of course, one can design a MDF based on other types of mass defect similarities for compounds of interest (Zhang and Ray, 2007).

In addition to specifying a predetermined range of mass defect values, a MDF approach also requires high-resolution LC–MS data to detect molecular ions. Mass defect values for the detected ions are obtained from the high-resolution data. Detected ions having mass defect values falling outside the specified range are discarded and those with mass defect values falling within the predetermined range are retained. Metabolites of interest are determined from the retained values. Upon mass defect filtering, the majority of interference ions should be removed. The filtered (and consequently simplified) data can facilitate the identification of drug metabolite ions. In other words, the MDF approach enables the segregation of drug-related components from interference ions of endogenous chemicals based on the unique mass defects of metabolites, so that they can be more selectively detected in the total-ion chromatogram (TIC) or through another data-processing approach. Of course, a MDF will not facilitate metabolite identification when metabolite ions are not

TABLE 6.3. Changes of Masses and Mass Defects of Drug Metabolites Resulting from Typical Biotransformation Reactions

Typical Metabolic Reactions	Nominal Mass Change (Da)	Exact Mass Change (Da)	Mass Defect Change (Da)
+O (hydroxylation)	+16	+15.9949	−0.0051
−H$_2$ (dehydrogenation)	−2	−2.0156	−0.0156
−CH$_2$ (demethylation)	−14	−14.0234	−0.0234
+C$_6$H$_8$O$_6$ (glucuronidation)	+176	+176.0321	+0.0321
+SO$_3$ (sulfation)	+80	+79.9568	−0.0432
+C$_{10}$H$_{15}$N$_3$O$_6$S (glutathione)	+305	+305.0681	+0.0681

manifested in the data due to suppression effects in the sample or the instrument's source or if the species of interest do not ionize to any useful degree since mass defect filtering is strictly a data-processing approach.

Although a mass defect plot is not required to design the filter, the plot helps to visualize how a MDF works. Fundamentally, MS data consist of pairs of intensity and mass values. A mass defect plot basically ignores the intensity portion but separates the mass values into two dimensions: The mass defect is plotted along the *y* axis and the nominal mass is plotted along the *x* axis. For example, for an ion of 400.2 Da, the data point would be plotted at 400 on the *x* axis and at 0.2 on the *y* axis. In this way, each ion in a mass spectrum will be represented as a single point in the mass defect plot. To design a MDF, one or more filters need to be defined in both the *y* (mass defect) and *x* (nominal mass) dimensions to retain possible ions of interest and to reject all other ions whose mass defects are not within the filter ranges. For example, a filter can be designed by defining its range in both *y* and *x* dimensions for a parent drug with a mass of 500 Da and a mass defect around 0.3 Da (see the middle rectangular box in Fig. 6.1 for an example). If glutathione conjugates are expected, a filter around the appropriate mass range can be set (see the rectangular box on the right in Fig. 6.1). Even if a drug molecule is metabolically cleaved into two pieces, thus changing the fragment mass defect to another range, yet another filter can be set for that range (see the rectangular box on the left in Fig. 6.1). By applying the designed MDF, all of the ions outside of the filter ranges, which are expected to be predominantly ions of no interest, will be removed. In our metaphor, it is now much easier for one to catch the fish from a nearly emptied ocean.

In the basic concept of a MDF as illustrated in Fig. 6.1, a range of values for acceptable mass defects is specified. Ions falling outside this range may be discarded while

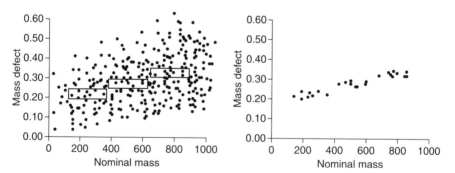

Figure 6.1. *Illustration of MDF concept for removal of interference ions from drug metabolites. Mass defects of ions in mass spectra from high-resolution LC–MS data are represented in the mass defect plots. Left panel: mass defect plot of unprocessed spectrum. The rectangular boxes in the plot illustrate multiple MDFs over distinct mass ranges. The filters can be designed with a narrow width (e.g., ±50 mDa) around the mass defect(s) of a parent drug, of its theoretical glutathione conjugate, or of its anticipated breakdown metabolite. Right panel: mass defect plot of mass spectrum after such filtering, which presents fewer ions and thus facilitates metabolite ion identification.*

ions falling within this range may be retained. The ions of interest falling within this specified filter range (i.e., the retained values) may be considered with greater confidence in analyzing the biological sample. There are multiple ways to design a MDF, depending on the situation. For example, the range of mass defects to be preserved may be constant across the mass range of interest. This is illustrated in Fig. 6.1, where a filter is set between 0.25 and 0.30 Da in the mass defect dimension. That is, both the maximum and minimum values of the mass defects retained by the filter remain constant over the mass ranges of interest.

In addition, as previously illustrated in Fig. 6.1, multiple filters may be established for a single data set, each corresponding to a different mass range, to provide for the examination of a wide range of mass defects resulting from diverse components. (The use of multiple filters may be necessary for such samples because of the metabolic transformations of the molecular species to be investigated.) The parameters for the MDF can be defined prior to examination of the acquired data, based solely on the mass defect of the drug and the expected variation in the mass defects of its related modifications. The MDF parameters can also be defined after examination of the acquired data based on observed trends in the actual acquired data (e.g., the likelihood that interferences will have mass defects similar to that of the metabolites of interest).

A variable window can also be used for a MDF such that the acceptance criteria for the mass defects are more restrictive (or more tolerant) at a given mass than at a different mass. In other words, a specified mass defect range at a particular point over the mass range of interest may be smaller (or larger) than a specified mass defect range at another point over the mass range of interest.

6.4 IMPLEMENTATION OF MDF

Because of the simplicity of the algorithm, the MDF strategy was initially implemented in-house using a scripting language (Zhang et al., 2003a,b). Most recently, all major mass spectrometer vendors have started offering some type of MDF option in their software packages (Leclercq et al., 2004; Mortishire-Smith et al., 2005; Lee et al., 2007; Mortishire-Smith et al., 2007). The in-house script discussed in this chapter was developed using Python (http://www.python.org) with the Scientific Python Module (http://starship.python.net/~hinsen/ScientificPython) to access the NetCDF file format (http://www.unidata.ucar.edu/packages/netcdf/), a universal format to which all of the major types of LC–MS data files can be converted. A centroid LC–MS data file is converted to NetCDF format using either the DataBridge utility in MassLynx for Q-TOF data or the File Converter tool in Xcalibur for LTQ–FTICR data. Through a user interface, one can choose the type of MDF (i.e., either fixed or scalable with respect to mass), define the mass defect ranges for the filter, determine the mass range over which a MDF covers, or implement multiple MDFs over different mass ranges. When applied to an LC–MS data file, the designed filter retains ions whose mass defects fall within its boundaries (which are expected to include drug-related ions) while rejecting ions whose mass defects do not (which are expected to be interference ions). After processing, the output file can be automatically converted from the NetCDF format back to its native data file format using the original file conversion utility to facilitate comparison to the original data. The in-house script described also allows for batch processing of multiple data files using the same MDFs for each data file.

6.5   APPLICATIONS OF MDF IN METABOLITE IDENTIFICATION

The MDF approach has been applied to screen for metabolites in bile, plasma, and fecal extract for drugs with molecular weights between 300 and 600 Da. Both positive- and negative-ionization modes were evaluated. In all cases, the significant removal of interference ions afforded by the MDF resulted in simplified mass spectra that were easier to interpret than the unprocessed spectra. The identification of unusual metabolites and minor metabolite ions under isobaric interferences was considerably enhanced by this approach. The metabolites identified included not only unusual metabolites but also those from expected pathways including hydroxylation $(M + 16)$, dihydroxylation $(M + 32)$, reduction $(M + 2)$, dechlorination $(M - 34)$, O- and N-demethylation $(M - 14)$, O- and N-dealkylation, glucuronidation $(M + 176)$, sulfation $(M + 80)$, glutathione conjugation $(M + 305)$, ester and amide hydrolysis, and combinations of the above. A few examples are given below to illustrate the effectiveness of the MDF approach for the analysis of drug metabolite samples in different biological matrices.

6.5.1   Application to Bile Samples

A bile sample obtained from dogs dosed with razaxaban, a factor Xa inhibitor with an empirical formula of $C_{24}H_{20}N_8O_2F_4$, was analyzed (Zhang et al., 2003a). Positive-ion electrospray LC–MS data of this sample were generated using commercially available Q-TOF Ultima mass spectrometers (Waters Micromass, Manchester, UK). The mass spectrometer resolution was tuned to 18,000 full width at half maximum (FWHM) using a polyalanine tuning solution. LC–MS data, in the centroid mode, were obtained using "all file accurate mass measurement" without lock mass correction. The centroid raw data were processed using a MDF designed to retain ion species with mass defects between 0.006 and 0.106 Da in a mass range of 200–1100 Da. The filter was based on the observed mass of the protonated molecule of razaxaban (529.056 Da) in the centroid raw data file (without lock mass correction).

Figures 6.2a,b illustrate the effectiveness of the MDF in simplifying complicated mass spectra. Figure 6.2a shows a mass spectrum at retention time 35.6 min obtained from unprocessed full-scan MS data containing a metabolite ion embedded among irrelevant endogenous species. It is challenging to recognize the metabolite ion at m/z 503.073 because of the surrounding interference ions. Although by visual examination of individual ions one might be able to identify this metabolite, such an approach is tedious and does not always guarantee satisfactory identification of the ions of interest. In contrast, Fig. 6.2b depicts the same mass spectrum processed

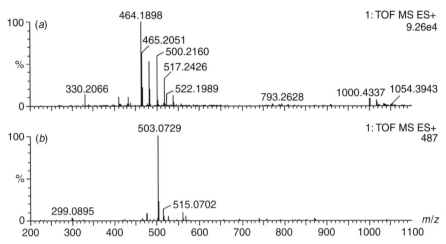

Figure 6.2. *Mass spectra of metabolite peak at retention time 35.6 min obtained from (a) unpro-cessed data (Fig. 6.3a) and (b) MDF-processed data (Fig. 6.3b).*

with mass defect filtering. In Fig. 6.2*b*, the metabolite ion (503.073 Da) is the promi-nent ion, and only very few interference ions are present in comparison to Fig. 6.2*a*.

Figures 6.3*a,b* illustrate the effect of the MDF on making TICs more indicative of metabolite peaks. LC–MS data are essentially a time series of mass spectra along the LC time scale, and the TIC or base-peak chromatogram (BPC) represents either the total mass spectral signals or the most intense ion signal, respectively, acquired at each time point along the time scale. Due to the very large number of interfering ions arising from biological matrices, a TIC or BPC may not reliably indicate the presence of metabolite peaks. As illustrated in Fig. 6.3*a*, the unprocessed TIC shows little indication of the distinct metabolite peaks due to the ocean of interfering ions. However, when a MDF is applied to the whole LC–MS data set, the resulting TIC is a remarkable contrast. As illustrated in Fig. 6.3*b*, the MDF-processed TIC bears a remarkable resemblance to the HPLC radio-chromatographic profile of Fig. 6.3*c*. In this example, all the major metabolite peaks are distinctive in the MDF-processed TIC, and, as illustrated in Fig. 6.2*b*, each pseudomolecular ion is predominant in their respective spectra. The facile identification of potential metabolite ions sub-sequently enables the execution of MS/MS experiments to establish corresponding identities. The metabolites detected in this sample were formed through biotransform-ation pathways including demethylation, reduction, oxygenation, glucuronidation, and combinations thereof. In the example shown in Fig. 6.2*b*, further analysis confirmed that the metabolite ion resulted from reduction and di-demethylation of razaxaban.

6.5.2 Application to Fecal Extracts

The next example is presented to show how a MDF enabled the rapid identification of minor and unexpected metabolites in fecal extracts obtained from human subjects

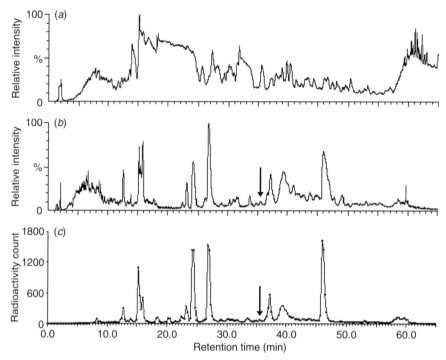

Figure 6.3. *Metabolite profile of razaxaban ($C_{24}H_{20}N_8O_2F_4$) in dog bile: (a) base peak chromatogram of unprocessed LC–MS data; (b) base peak chromatogram of processed LC–MS data; (c) corresponding radioactivity chromatogram. The arrows annotate the retention time of the peak shown in Fig. 6.2. Bile was collected from dogs orally administered with ^{14}C-labeled razaxaban (20 mg/kg). HPLC solvents were 10 mM NH_4HCO_3 (pH 9.0) and acetonitrile. A portion of the HPLC effluent was collected in 15-s fractions for the radioactivity chromatogram. Another portion of the HPLC effluent was directed to a Q-TOF Ultima mass spectrometer.*

orally administered with muraglitazar, a compound with an empirical formula of $C_{29}H_{28}N_2O_7$ (Zhang et al., 2003b). High-resolution LC–MS data of the fecal extract were generated using a commercially available Q-TOF (quadrupole/ time-of-flight) Ultima mass spectrometer (Waters Micromass, Manchester, UK). The mass spectrometer was tuned to 17,000 FWHM and centroid LC–MS data were obtained using "all file accurate mass measurement" without lock spray correction. The data were processed using a MDF designed to retain ion species with mass defects between 0.17 and 0.24 Da for mass range of 490–590 Da. This filter was designed mainly to retain possible oxidative metabolites since the known glucuronide conjugates were largely converted back to the parent drug. The filter setting was based on the observed mass and mass defect of the parent drug (517.21 Da).

The utilization of MDF again allows the TIC to be indicative of metabolite peaks, rendering a sense of simplicity similar to that of working with *in vitro* samples. As shown in Fig. 6.4, the resulting TIC of the MDF-processed data (Fig. 6.4*b*) is comparable to the radioactivity chromatogram (Fig. 6.4*a*). In contrast, the unprocessed TIC (Fig. 6.4*c*) does not show a correlation to the radioactivity chromatogram.

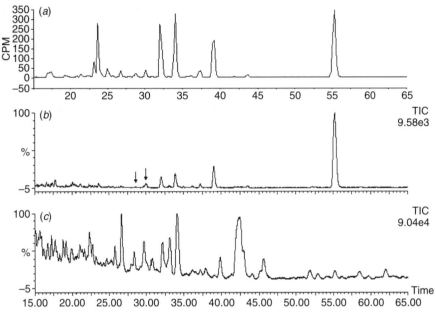

Figure 6.4. *Metabolite profiles of muraglitazar human fecal extracts: (a) radioactivity chromatogram of drug-derived components, (b) TIC of MDF-processed data, and (c) TIC of unprocessed data. The arrows annotate the retention time of the peaks shown in Fig. 6.5. Fecal extracts were obtained from healthy subjects administered with [^{14}C] muraglitazar. HPLC solvents were water and acetonitrile containing 0.06% trifluoroacetic acid. A portion of the LC effluent was collected in 15-s fractions for generating radioactivity chromatogram. Another portion of the LC effluent was directed to a Q-TOF Ultima mass spectrometer equipped with an ESI source operated in the positive- ionization mode. The Q-TOF was tuned to 17,000 mass-resolving power at FWHM and was calibrated using a polyalanine tuning solution. Centroid LC–MS data were obtained using "all file accurate mass measurement" without lock spray correction. The data were then processed by MDF using a decimal window of 0.17–0.24 Da for mass range 490–590 Da. The decimal window was based on the apparent mass of muraglitazar in the data (517.21 Da).*

The most significant benefit of the MDF-simplified LC–MS data, again, resides in the mass spectra themselves. The reduction of interferences in the processed data significantly facilitates the identification of metabolite ions, including minor or unusual ones that otherwise could be missed. For example, the upper panel of Fig. 6.5 shows that a low-intensity metabolite at RT 28.6 (dihydroxylated muraglitazar, m/z 549), not obvious in the unprocessed spectrum in the lower panel, is easily recognizable in the MDF-processed spectrum (middle panel). Because of its low mass spectrometric response, the spectrum of this metabolite was located using the radiochromatographic profile as a reference. In cases where a reference metabolite profile is not available, additional data processing can be applied on LC–MS data simplified by MDF to identify minor metabolites whose peaks are not apparent at the TIC level (detailed in Section 6.6).

The metabolite peaks at 30.0 and 28.6 min (Fig. 6.4) are two muraglitazar metabolites resulting from the opening of the oxazole ring (Zhang et al., 2003b). Both metabolites are unusual not only in their biotransformation pathways but also in

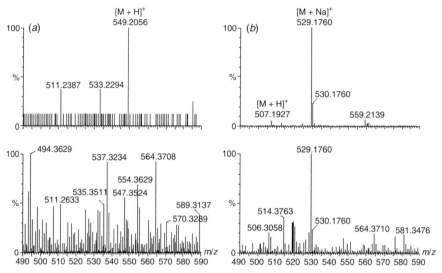

Figure 6.5. *Mass spectra of (a) di-hydroxylated metabolite (RT 28.6 min) and (b) a [M − 10] metabolite of muraglitazar (RT 30.0 min) in human fecal extracts following administration of muraglitazar. Upper panels: obtained from MDF-simplified TIC (Fig. 6.4b) by combining scans over the peak and subtracting scans before and after the peak. Lower panels: from unprocessed TIC (Fig. 6.4c) by combining and subtracting the same scan numbers in the respective upper panel (without MDF).*

the mass spectrometric behavior, as these metabolites produced intense sodium or ammonium adducts instead of protonated molecules. With the notion of proton ionization for all the previously observed muraglitazar metabolites, great difficulty is presented in figuring out which ions are most likely the metabolite ions in the lower panel of Fig. 6.5. But in the upper panel of Fig. 6.5, the unequivocally recognizable $[M + H]^+$ and $[M + Na]^+$ ions in the MDF simplified spectrum facilitated the empirical formula determination of the unusual metabolite of $[M − 10]$ Da at 30 min. Likewise, the predominant $[M + NH_4]^+$ and $[M + Na]^+$ ions led to a quick determination of the acetyl counterpart of the $[M − 10]$ Da metabolite at 43.5 min (data not shown). Once the metabolite ions are identified, several experiments are performed to elucidate the structures of the metabolites. Subsequent MS/MS experiments confirmed that the fecal metabolites at 30.0 and 28.6 min are formed by opening of the muraglitazar oxazole ring. Based on the proposed structures, forced degradation was used to produce a sufficient quantity of the $[M − 10]$ Da metabolite for nuclear magnetic resonance (NMR) confirmation, which fully supported the proposed structure. The proposed structures of both metabolites were ultimately verified by chemical synthesis.

6.5.3 Application to Plasma Extracts

In the next example, human plasma or urine was spiked with a mixture of known metabolites in an effort to investigate how effective the MDF is in removing numerous endogenous interferences (Zhang et al., 2004; Zhu et al., 2006). The metabolite mixture was obtained from a human liver microsomal (HLM) incubation of omeprazole ($C_{17}H_{19}N_3O_3S$) (Hoffmann et al., 1986; Abelo et al., 2000). The omeprazole metabolites were then mixed with human plasma and urine at various ratios. These mixtures and control plasma or urine samples were subjected to high-resolution LC–MS analyses (Waters Micromass Q-TOF Ultima mass spectrometer) followed by MDF processing. The MDF window was set to 50 mDa around the apparent mass defects of the protonated omeprazole ions. MDF processing alone was applied for the detection of omeprazole metabolites in plasma, while a combination of MDF and control sample comparison was used for the identification of metabolite ions in urine (detailed in the next section).

Without MDF processing, the BPC for the LC–MS data showed no distinct indication of metabolite peaks in any of the plasma samples. After MDF processing, the majority of endogenous interference ions were removed. As a result, the BPC of the plasma samples displayed predominantly metabolite peaks. For example, Fig. 6.6*a* shows the BPC of a plasma sample without MDF filtering (1 : 50 dilution of HLM with human plasma, the equivalent of a 1.0-mL plasma injection). Using Fig. 6.6*a*

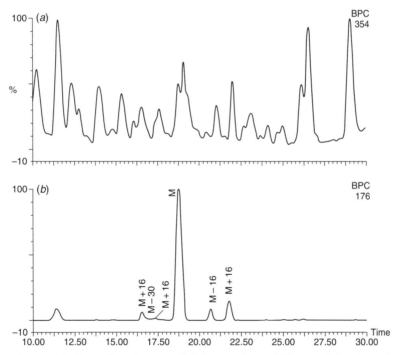

Figure 6.6. Metabolite profiles of omeprazole in human plasma: (a) base peak ion chromatogram of unprocessed data and (b) base peak ion chromatogram of MDF-processed data exhibiting all metabolite peaks present and some endogenous peaks. High-resolution LC–MS data were obtained for a human plasma sample spiked with omeprazole metabolites generated by microsomal incubation (the equivalent of a 1.0-mL plasma injection). A MDF was set at 50 mDa around the apparent mass defect of the omeprazole ion.

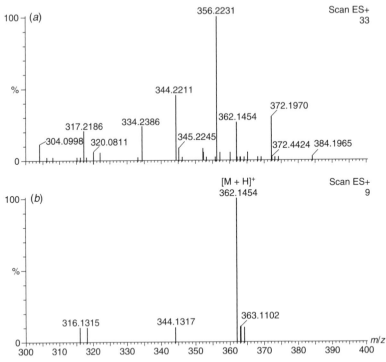

Figure 6.7. *Mass spectra of a circulating human metabolite (RT 16.6 min) of omeprazole obtained from (a) unprocessed data (Fig. 6.6a) and (b) MDF-processed data (Fig. 6.6b). Following MDF processing, predominantly the molecular ions of the omeprazole metabolite was observed.*

no metabolite peak can be immediately identified because of significant endogenous interferences. Furthermore, without prior knowledge of the retention time of each metabolite, obtaining a mass spectrum for each species of interest is nearly impossible. Even if such knowledge were available, the spectrum would have too many endogenous matrix-related ions for one to easily identify the correct pseudomolecular ion (Fig. 6.7a). However, after MDF filtering, the BPC in Fig. 6.6b clearly exhibited all the known metabolite peaks with minimal endogenous peaks. In addition, full-scan mass spectra of the metabolite peaks were greatly simplified. A further examination of the metabolite peaks in Fig. 6.6b shows predominantly the molecular ion for each of the metabolite at its respective retention time (Fig. 6.7b).

6.6   MDF IN CONJUNCTION WITH CONTROL SAMPLE COMPARISON

For drugs with sufficient double-bond equivalents (Chapter 4), MDF tends to work well at the TIC or BPC level to reveal metabolite peaks (of acceptable mass spectrometric response) in plasma, bile, or fecal extract samples. This is because the mass defects of such drugs are typically smaller than those of the endogenous interferences at the same mass range. Urine samples, however, tend to be somewhat more compli-cated, as the mass defects of endogenous components in the urine are typically dis-tributed among those of the drug metabolites. For example, the TIC of a human urine sample containing omeprazole metabolites (a 1 : 50 dilution of a HLM incubation with human urine, the equivalent of a 2.0-mL urine injection) from the MDF-processed data failed to show distinct metabolite peaks (data not shown), even though its LC–MS data were simplified by the mass defect filtering. In such cases where a TIC fails to manifest metabolite peaks and where there is no reference metabolite profile available (e.g., a radioactivity chromatogram) for retrieving metabolite spectra, a technique called control sample comparison is used to help identify metabolite ions from the MDF data.

Control sample comparison traditionally involves the generation of extracted-ion chromatograms (XICs) from a sample and its control matrix for each mass unit over an entire mass range. By comparing the XICs between the sample and the control, one can identify the differences and locate the metabolite peaks. Existing software packages for automatic control sample comparison were somewhat limited in functionality until recent developments in high-resolution LC–MS tech-nology made such an approach more viable. When XICs are compared at unit-mass resolution, significant isobaric interferences from endogenous components can obscure peaks of interest in the sample, resulting in no observable distinction between the sample and the control. Although some software packages provided an option to apply settings to note differences in peak areas between the sample and the control, such an option may generate a significant number of false-positive hits because the levels of endogenous components fluctuate from sample to sample for biological systems. Distinguishing a true hit from many false positives in this situation may be a Herculean task.

MDF, however, can be used in conjunction with control sample comparison to improve the specificity of metabolite ion detection. There are two ways to combine the two techniques. One is first to perform control sample comparison using XICs generated at subunit mass increments (e.g., 0.05 Da) over a mass range and then to invoke a mass defect window at the report-viewing stage to restrict the reported hits to only those whose mass defects are within the expected window. This is essen-tially how Waters has incorporated the MDF functionality into its MetaboLynx package. Another approach is to perform mass defect filtering first and then to perform control sample comparison. As illustrated in the right panel of Fig. 6.1, after mass defect filtering, the majority of isobaric interferences are removed, result-ing in a greatly simplified data set for the generation of XICs. Control sample com-parison can then be performed on such simplified data with minimal isobaric interferences. Typically, comparison criteria can be set to find ions that are present only in the sample and not in the control without invoking complicated settings such as the criteria for differences in peak area.

Using a urine sample spiked with omeprazole and its metabolites as an example, Table 6.4, part (a) shows the control sample comparison results of the MDF data. The urine sample was obtained by a 1 : 50 dilution of the omeprazole HLM mixture with

TABLE 6.4. Results of Control Sample Comparison Between Urine Sample and Urine Sample Spiked with Omeprazole HLM Metabolites

Time	*m/z* Found	Mass Difference	Drug Related?
		(a) With MDF-Processed Data	
16.71	362.1416	15.9968	Yes
17.31	316.1315	−30.0133	Yes
17.56	362.1393	15.9945	Yes
19.07	346.1445	−0.0003	Parent drug
21.16	330.1492	−15.9956	Yes
22.2	362.1417	15.9969	Yes
22.47	313.1263	−33.0185	No
24.1	345.1243	−1.0205	No
26.56	331.1505	−14.9943	No
26.69	315.1541	−30.9907	No
28.75	359.1354	12.9906	No
29.58	315.1517	−30.9931	No
		(b) Without MDF Processing	
10.39	314.1927	−31.9521	No
11.89	327.2493	−18.8955	No
12.45	346.246	0.1012	No
12.72	318.2492	−27.8956	No
16.41	304.2312	−41.9136	No
16.45	302.2156	−43.9292	No

human urine. The injection volume was equivalent to 2 mL of urine. MetaboLynx was used to perform the control sample comparison and to generate reports. With MDF-processed data, all metabolite ions were identified by control sample comparison, along with a few false hits [Table 6.4, part (a)]. A control sample comparison report obtained using the unprocessed data contained several false-positive hits [Table 6.4, part (b)] and no hits for the known metabolites. Even the parent drug was missed in the unprocessed comparison report. These results demonstrate that the selective removal of isobaric interferences and retention of metabolite ions by MDF enhances the specificity of metabolite detection in control sample comparison.

6.7 COMPARISON OF MDF WITH PRECURSOR ION SCAN AND CONSTANT-NEUTRAL-LOSS SCAN

Detection of uncommon metabolites in complex biological matrices can be very challenging. Precursor ion scans (PIS) are often used for selective detection of oxidative metabolites that form one or more product ions identical to (or having similar fragmentation patterns to) those of the parent drug (Clarke et al., 2001b; Kostiainen et al., 2003; Liu and Hop, 2005). Constant-neutral-loss scans (NLS) are usually employed in screening for various conjugative metabolites because these conjugates often undergo common cleavages to generate the loss of specific neutral fragments under collision-induced disassociation (CID) (Jackson et al., 1995; Clarke et al.,

2001b; Kostiainen et al., 2003; Xia et al., 2003; Liu and Hop, 2005). For example, glucuronides usually form a product ion characteristic of the loss of a glucuronic acid moiety (176 Da). Therefore, the neutral loss scan of 176 Da is often used for detecting unknown glucuronides. PIS and NLS are explained in detail in Chapters 1 and 3.

Recently, a study was conducted to compare the MDF method to PIS and NLS methods in finding oxidative metabolites in biological fluids (Zhu et al., 2004, 2006). In this study, diclofenac and clozapine (Fig. 6.8) were incubated with human liver microsomes and the metabolites generated were then spiked into pooled rat urine and bile followed by analyses using either high-resolution LC–MS-based MDF or triple-quadrupole LC–MS with NLS and PIS.

The full-scan BPC obtained from LTQ–FTICR LC–MS data does not display diclofenac or its monohydroxyl metabolite (M1) that were spiked into rat bile sample due to the significant interference from ions of endogenous components in the bile (Fig. 6.9a). In contrast, MDF analysis of the LC–MS data file completely removed these interfering ions, resulting in a BPC that clearly exhibits only the drug and M1 (Fig. 6.9b). The same bile sample was analyzed with NLS and PIS using a triple-quadrupole instrument. Diclofenac lost a neutral fragment (46 Da) to form a major ion at m/z 250 under CID conditions (Fig. 6.8a).

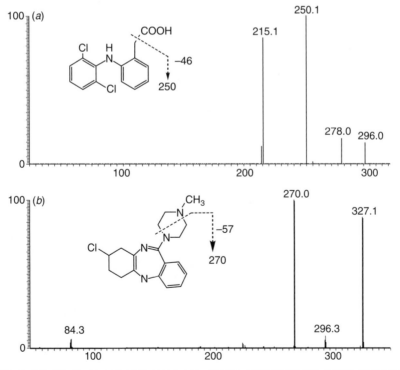

Figure 6.8. Structures and product ion mass spectra of (a) diclofenac and (b) clozapine.

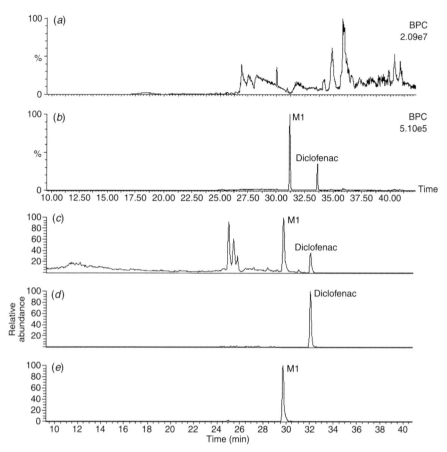

Figure 6.9. Detection of diclofenac metabolite in rat bile by three LC–MS screening methods: (a) full-scan BPC from LTQ–FTICR; (b) MDF-treated full-scan BPC from LTQ–FTICR; (c) ion current of NLS of 46 Da; (d) ion current of PIS of m/z 250; (e) ion current of PIS of m/z 266.

Therefore, the product ion at m/z 250 and a loss of 46 Da were selected for the parent ion scan and the neutral loss scan, respectively. The NLS analysis detected both diclofenac and M1 along with several intense peaks from the bile matrix that were proven as false positives (Fig. 6.9c). The precursor ion scan of m/z 250 revealed the parent drug, but not M1 (Fig. 6.9d). A second PIS was conducted by following a specific product ion at m/z 266 (m/z 250 + oxygen), which was used to detect M1 (Fig. 6.9e).

Another example is the analyses of the clozapine metabolites M2 (the *N*-desmethyl product) and M3 (*N*-oxide) in rat urine. The metabolite peaks of M2 and M3 were hardly observable in the full-scan TIC obtained by LC–LTQ–FTICR (Fig. 6.10a). After removing the majority of endogenous components and chemical noise by MDF, M2, M3, and clozapine became the predominant peaks in the full-scan TIC (Fig. 6.10b). Precursor ion scan analysis of the same spiked

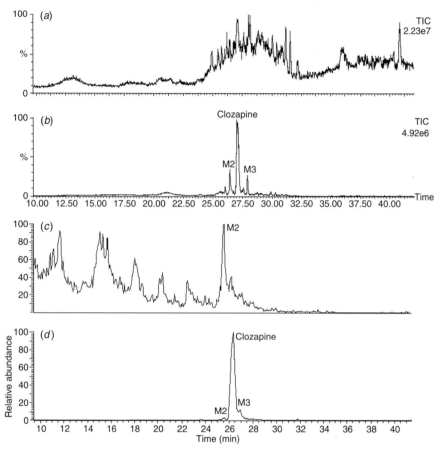

Figure 6.10. Detection of clozapine metabolites in rat urine by three LC–MS screening methods: (a) full-scan TIC from LTQ–FTICR; (b) MDF-treated full-scan TIC from LTQ–FTICR; (c) ion current of NLS of 43 Da; (d) ion current of PIS of m/z 270.

urine using the major product ion at m/z 270 (Fig. 6.8b) of clozapine resulted in detection of both M2 and M3 (Fig. 6.10d). However, a NLS of 57 Da (Fig. 6.8b) did not reveal M2 or M3 (data not shown) because the moiety in these metabolites corresponding to the loss of 57 Da was structurally modified. An additional NLS analysis of 43 Da (57 Da $-$ CH_2), which targeted any metabolites from the N-demethylation, detected M2 along with many false-positive peaks (Fig. 6.10c).

Both of the above examples illustrate that the sensitivity of the MDF method, in general, is comparable to PIS analysis and better than NLS methods (Figs 6.9 and 6.10). The sensitivity of the MDF method is attributed at least in part to the full-scan capability of high-resolution mass spectrometers such as the Q-TOF and LTQ–FTICR, which could be several times more sensitive than full-scan analysis by a triple-quadrupole mass spectrometer (Kostiainen et al., 2003).

The specificity to distinguish drug metabolites from endogenous components varies among the three LC–MS methods, making them complementary to each other. For example, the MDF analysis specifically picked up diclofenac and its metabolite M1, while the NLS ion chromatogram displayed several false-positive peaks (Fig. 6.9*b* vs. 6.9*c*). However, the specificity for clozapine metabolite detection by PI scan analysis was slightly better than by the MDF approach (Fig. 6.10*d* vs. 6.10*b*). The specificity of a MDF method at the TIC level depends on the difference between the mass defects of the metabolites and those of the endogenous components in a biological matrix, as will be discussed in the next section. Since the mass defects of both the diclofenac and the clozapine metabolites were significantly different from those of the interfering components in bile and urine, the ions not related to the drugs can be easily filtered away, resulting in specific detection of these metabolites at the TIC level. If the mass defect of a metabolite is close to those of most endogenous components and background noise, the specificity of a MDF method at the TIC level could be poor.

One of the major advantages of the MDF approach is its capability to extract all possible metabolites from a full-scan LC–MS data file. As demonstrated in the preceding examples, the MDF approach can detect all diclofenac and clozapine metabolites in rat bile or urine without knowing their exact molecular weight or mass defect information. In general, a filter window of ± 40 mDa is set around the mass defect of the parent drug as well as a mass range of ± 50 Da of the drug can be set for screening monohydroxylation, demethylation, and other oxidative metabolites that have structures similar to that of the parent drug. For detecting metabolites from dealkylation, hydrolysis, or other biotransformation reactions that cleave a drug into smaller parts, a filter window is set around the masses and mass defects of these smaller structures (as illustrated in Fig. 6.1). Screening for metabolites by PIS or NLS analyses based on a specific product ion or neutral fragment, however, often misses some metabolites. Therefore, PIS and NLS with multiple ions or multiple LC–MS injections to follow expected product ions or neutral losses are often required to achieve a comprehensive search for metabolites. For instance, a PIS of m/z 266 was performed to detect the monohydroxyl metabolite (M1) of diclofenac after a PIS of m/z 250 failed (Figs 6.9*d* and 6.9*e*). Similarly, to detect clozapine metabolites M2 (demethylation) and M3 (monohydroxylation), specific NLS following the loss of 43 Da $(57 - 14)$ and 73 Da $(57 + 16)$, respectively, are required (refer to Fig. 6.8*b*). The second advantage of the high-resolution LC–MS-based MDF method is the inherent capability to determine the formulas of metabolites (and product ions, at a later stage) based on accurate mass spectral data. The accurate mass aspect is crucial in differentiating a metabolite from other signals that have the same nominal masses and is also helpful in the interpretation of product ion spectra.

In conclusion, the sensitivity and specificity for screening for oxidative metabolites by the three LC–MS methods are variable, compound-dependent, and complementary to each other. The MDF method requires a simple, generic full-scan analysis using a mass spectrometer equipped with high-resolution capability, and the accurate mass information obtained can provide empirical formula information for metabolites and their fragments. Since MDF employs a different detection mechanism and provides some advantages over traditional PIS and NLS analyses performed on a triple-quadrupole mass spectrometer, the high-resolution LC–MS-based MDF approach provides an alternative tool for screening for drug metabolites in biological matrices.

6.8 APPLICABILITY OF MDF FOR DRUGS IN DIFFERENT BIOLOGICAL MATRICES

As one might have guessed from the examples cited thus far, the effectiveness of a MDF depends not only on the similarity of the mass defects of a drug and its metabolites but also on the differences between the mass defects of drug metabolites and those of the biological matrix in which they dwell. To investigate the effects of different biological matrices on the MDF methodology, mass defect profiles of typical biological matrices including plasma, urine, bile, and feces were obtained, and the profiles were compared against those of 115 marketed drugs. The resulting data provide a better picture for understanding the nature of drug metabolite samples in biological matrices with respect to mass defects and allows one to predict the applicability of MDF for selective removal of endogenous interferences for a given drug metabolite sample (Zhang et al., 2005, 2006; Kuehl et al., 2006; Ruan et al., 2006; Zhao et al., 2006; Lee et al., 2007; Mortishire-Smith et al., 2007; Pike et al., 2007; Zhang and Ray, 2007).

High-resolution LC–MS data for plasma, bile, urine, and fecal extract samples from humans, monkeys, and mice were obtained using a ThermoFinnigan LTQ–FTICR mass spectrometer. A Waters Atlantis 2.1×50-mm column was used with a linear gradient of water/acetonitrile solvents containing 0.1% formic acid. The FTICR–MS data were obtained in the positive ESI full-scan acquisition mode, with a mass resolution of 100,000. The plasma samples and fecal homogenates were treated with 3 vol of acetonitrile and 1 vol of methanol for protein precipitation. The bile samples were diluted two-fold with water. The urine samples were injected directly. LC–MS data from each biological matrix were combined into one spectrum. From the resulting spectra, mass defect plots of individual matrices were generated with mass defects along the y axis and nominal masses along the x axis. Similarly, a mass defect plot was generated for 115 marketed drugs based on their mass defects and nominal masses. The mass defect profiles of a particular matrix type across the three species investigated were similar to each other. As representatives, mass defect plots of human plasma, bile, feces, and urine are depicted in Figs 6.11*a* to 6.11*d*. In addition, the mass defects of 115 marketed drugs that were compiled from various public resources are plotted in Fig. 6.11*e*. The circle in Fig. 6.11*e* signifies the dense population of the drugs. The same circle was also overlaid in Figs 6.11*a* to 6.11*d* to show its relationship with these biological matrices.

As shown in Fig. 6.11, the mass defect plot of each biological matrix analyzed revealed certain dense clusters that are roughly monotonic with mass. (The continuous vertical dots at the low-mass end of each plot are from chemical noise above the arbitrary noise cutoff settings.) The dense mass defect clusters of plasma, bile, and

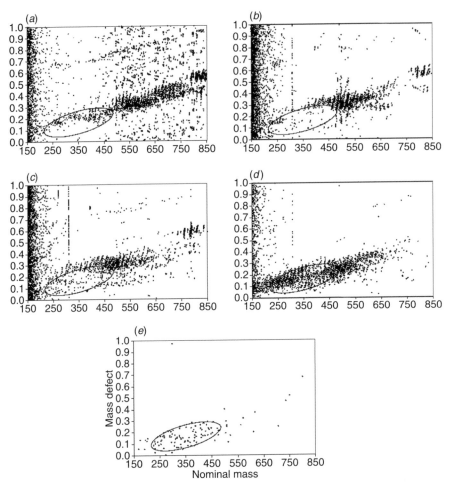

Figure 6.11. *Mass defect plots of human biological matrices obtained from high-resolution LC–MS data: (a) plasma, (b) bile, (c) feces, and (d) urine. LC–MS data from each biological matrix were combined into one spectrum to generate the mass defect plots. (e) Mass defect plot of 115 marketed drugs. The dotted circle in each plot represents the dense population of the mass defects of 115 marketed drugs.*

feces are significantly separated from those of the 115 marketed drugs, suggesting that MDF should be effective for many marketed drugs in these matrices.

The clusters from urine extend toward lower mass and mass defect values than those of plasma, bile, or feces. As a result, the mass defect clusters of urine are overlapped with most of the 115 marketed drugs. Given such a profile, one can expect some of the interference ions to remain with a typical MDF setting for urine. Of course, with MDF processing, any output data are always simplified since interferences outside of the filter range are discarded (as illustrated in Fig. 6.1). However, depending on the amounts of urine injected, the levels of the remaining interference ions may still impact metabolite peak identification if based on a TIC profile.

The profiles of biological interference ions in Fig. 6.11 can serve as a guide to predict the effectiveness of MDF for a drug metabolite sample or to assess whether a stringent filter parameter should be used to facilitate detection of drug metabolite ions. The mass defect window of a MDF and its chosen mass range are set based on the relationship of a

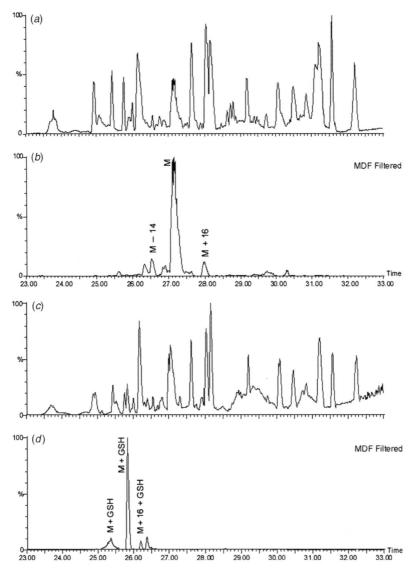

Figure 6.12. Base peak chromatograms obtained from 75-μL urine samples spiked with cloza-pine metabolites using LTQ–FTICR. (a,b) Samples spiked with oxidative metabolites, where (a) was from unprocessed data and (b) from MDF-processed data. (c,d) Samples spiked with GSH metabolites, where (c) was from unprocessed data and (d) from MDF-processed data. The MDF settings were ±0.04 Da in mass defect dimension and ±50 Da around the parent or the parent-plus-GSH mass, respectively.

drug's mass defect or that of its theoretical conjugates to the mass defect profile of the biological matrix in which a drug or its metabolites are observed. For example, in the initial publication of the MDF applications (Zhang et al., 2003a), the compound razaxaban with an empirical formula of $C_{24}H_{20}N_8O_2F_4$ ($[M+H]^+ = 529.172$ Da) falls far outside of the major clusters of the biological matrix studied (dog bile). Indeed, a MDF of ± 0.050 Da around the apparent mass defect of the parent drug in a wide mass range (200–1100 Da) retained all the metabolites in bile with very few unrelated peaks. In the examples shown in this chapter for omeprazole ($C_{17}H_{20}N_3O_3S$, $[M+H]^+ = 346.122$ Da), the results were dramatically different depending on the matrix used. Omeprazole's oxidative metabolite peaks were distinct with 1-mL plasma injection, while with a 2-mL urine injection all the metabolite peaks including the parent peak were masked at the TIC level and necessitated further data-mining techniques (i.e., control sample comparison). This is because the mass defect of omeprazole was outside of the dense clusters of plasma matrix but appeared within the major cluster of urine interferences.

In a mass defect plot, conjugates (e.g., glucuronides, or GSH) would move to the right relative to a parent drug in the nominal mass dimension, without much increase in the mass defect dimension (as opposed to the monotonic increase of matrix clusters), leading the conjugates to move away from the interference clusters. Therefore, the MDF approach is likely more effective for conjugate detection even if a parent drug is within the major interfering clusters. For example, as shown in Fig. 6.11*d*, one should find the relative positions of clozapine ($[M+H]^+ = 327.137$ Da) and its theoretical GSH conjugate ($[M+H]^+ = 632.205$ Da) in urine matrix. Clozapine has a mass defect value similar to that of omeprazole and shares a similar position in the mass defect profile. Its GSH conjugate, however, is essentially in an interference-free region. Figure 6.12 shows the proof-of-principle data obtained from a 75-μL injection of urine samples spiked with oxidative metabolites or GSH conjugates of clozapine. Figures 6.12*a* and 6.12*b* depict that a MDF of ± 0.040 Da around the mass defect of clozapine in a mass range of 277–377 Da could manifest the oxidative metabolites in the base peak ion chromatogram. However, a number of interference peaks that were retained by the filter were also present with levels comparable to those of the metabolite peaks. Figures 6.12*c* and 6.12*d* depict that a MDF of ± 0.040 Da around the mass defect of the theoretical GSH conjugate in a mass range of 580–680 Da revealed GSH conjugates that were more distinct than the oxidative counterparts. Similar studies are ongoing to exploit the advantages of MDF for selective conjugate metabolites detection (Ma et al., 2005).

6.9 SUMMARY

In summary, a simple and novel data-processing strategy using a MDF has been applied to identify drug-related species in complex environments. A MDF can be designed provided that there is a similarity in the mass defects among the species of interest. This is often the case for drug metabolites in biological matrices. By selective removal of all ions that fall outside of the filter window, the MDF-processed data facilitate the identification of drug-related ion species of interest. In addition, the MDF-processed data can also be used in conjunction with other techniques such as control sample comparisons to further facilitate the identification of ion species of interest.

The significance of the MDF approach for drug metabolite identification is a valuable alternative for identifying minor or unknown metabolites. This is because MDF approach works by a different mechanism from the more conventional techniques by utilizing the general physical property of mass defect to quickly screen a pool of possible metabolite ion candidates. Most other "fishing expedition" techniques, including parent ion scan, daughter ion scan, and neutral loss scan, require knowledge of structural information to fish out a particular type of metabolite.

Of course, any technique has its limitations. What are the limitations of the MDF approach? The mass defect and the probable mass defect similarity among drug-related species are all gifts from nature and will depend upon how large the difference between the mass defects and those of the endogenous components will be. If nature does not allow a compound to be differentiated from matrices, then it will not be useful to implement a MDF. Also, if the similarity of the mass defects of the drug-related species is not predictable and the mass defects of some metabolites are not retained by the filter, the MDF approach will provide little, if any, information. It is important nonetheless to bear in mind that MDF is a postacquisition approach. The filtering method can always be refined as more information becomes available regarding a compound's metabolite profile. Finally, it is critical to remember that a good data-processing technique may reveal what is hidden, but it should not lead one to imagine something in the data that is not really there. With this perspective, it is apparent that the MDF is a worthwhile approach to simplify complex LC–MS data in an easily implemented manner.

ACKNOWLEDGMENTS

We would like to thank Dr. W. Griffith Humphreys and Dr. Mark Sanders for their support and expert advice. We would also like to thank Ming Yao, Li Ma, Nirmala Raghavan, and Weiping Zhao for their assistance in various experiments.

REFERENCES

Abelo, A., Andersson, T. B., Antonsson, M., Naudot, A. K., Skanberg, I., and Weidolf, L. (2000). Stereoselective metabolism of omeprazole by human cytochrome P450 enzymes. *Drug Metab. Dispos.* **28**:966–972.

Arora, V. K., Li, Y., Philip, T., Yeola, S., and Mayol, R. F. (2001). Automated high-throughput metabolite identification using quadrupole time-of-flight (QTOF 2) mass spectrometer. In *Proceeding of the 49th ASMS Conference on Mass Spectrometry and Allied Topics*, Chicago, IL.

Bahr, U., and Karas, M. (1999). Differentiation of "isobaric" peptides and human milk oligosaccharides by exact mass measurements using electrospray ionization orthogonal time-of-flight analysis. *Rapid Commun. Mass Spectrom.* **13**:1052–1058.

Beaumont, C., Kelly, P. J., and Castro-Perez, J. (2000). Structure elucidation of drug metabolites in a complex matrix using a hybrid quadrupole orthogonal time of flight mass spectrometer and automated data processing. In *Proceeding of the 49th ASMS Conference on Mass Spectrometry and Allied Topics*, Chicago, IL.

Blum, W., Aichholz, R., Ramstein, P., Kuhnol, J., Bruggen, J., O'Reilly, T., and Florsheimer, A. (2001). In vivo metabolism of epothilone B in tumor-bearing nude mice: Identification of three new epothilone B metabolites by capillary high-pressure liquid chromatography/mass spectrometry/tandem mass spectrometry. *Rapid Commun. Mass Spectrom.* **15**:41–49.

Castro-Perez, J., Hoyes, J., Major, H., and Preece, S. (2002). Advances in MS-based approaches for drug and metabolism studies. *Chromatographia* **55**:S59–S63.

Clarke, N., Rindgen, D., Cox, K., Korfmacher, W., Castro-Perez, J., and Preece, S. (2001a). Use of new technologies for metabolic characterization in plasma. In *Proceeding of the 49th ASMS Conference on Mass Spectrometry and Allied Topics*, Chicago, IL.

Clarke, N., Rindgen, D., Grotz, D., and Cox, K. (2002). Detection and identification of a toxic metabolite from monkey plasma. In *Proceeding of the 50th ASMS Conference on Mass Spectrometry and Allied Topics*, Orlando, FL.

Clarke, N. J., Rindgen, D., Korfmacher, W. A., and Cox, K. A. (2001b). Systematic LC/MS metabolite identification in drug discovery. *Anal. Chem.* **73**:430A–439A.

Corcoran, O., Nicholson, J. K., Lenz, E. M., Abou-Shakra, F., Castro-Perez, J., Sage, A. B., and Wilson, I. D. (2000). Directly coupled liquid chromatography with inductively coupled plasma mass spectrometry and orthogonal acceleration time-of-flight mass spectrometry for the identification of drug metabolites in urine: Application to diclofenac using chlorine and sulfur detection. *Rapid Commun. Mass Spectrom.* **14**:2377–2384.

Hoffmann, K. J., Renberg, L., and Olovson, S. G. (1986). Comparative metabolic disposition of oral doses of omeprazole in the dog, rat, and mouse. *Drug Metab. Dispos.* **14**:336–340.

Hogenboom, A. C., Niessen, W. M. A., Little, D., and Brinkman, U. A. T. (1999). Accurate mass determinations for the confirmation and identification of organic microcontaminants in surface water using online solid-phase extraction liquid chromatography electrospray orthogonal-acceleration time-of-flight mass spectrometry. *Rapid Commun. Mass Spectrom.* **13**:125–133.

Hopfgartner, G., and Vilbois, F. (2000). The impact of accurate mass measurements using quadrupole/time-of-flight mass spectrometry on the characterisation and screening of drug metabolites: Structure elucidation by LC–MS. *Analysis* **28**:906–914.

Jackson, P. J., Brownsill, R. D., Taylor, A. R., and Walther, B. (1995). Use of electrospray ionization and neutral loss liquid chromatography/tandem mass spectrometry in drug metabolism studies. *J. Mass Spectrom.* **30**:446–451.

Jemal, M., Ouyang, Z., Zhao, W., Zhu, M., and Wu, W. W. (2003). A strategy for metabolite identification using triple-quadrupole mass spectrometry with enhanced resolution and accurate mass capability. *Rapid Commun. Mass Spectrom.* **17**:2732–2740.

Keough, T., Lacey, M. P., Ketcha, M. M., Bateman, R. H., and Green, M. R. (1997). Orthogonal acceleration single-pass time-of-flight mass spectrometry for determination of the exact masses of product ions formed in tandem mass spectrometry experiments. *Rapid Commun. Mass Spectrom.* **11**:1702–1708.

Kostiainen, R., Kotiaho, T., Kuuranne, T., and Auriola, S. (2003). Liquid chromatography/ atmospheric pressure ionization-mass spectrometry in drug metabolism studies. *J. Mass Spectrom.* **38**:357–372.

Kuehl, D., Gu, M., and Wang, Y. (2006). Comparing mass defect filtering and accurate mass profile extracted ion chromatogram (AMPXIC) for metabolism studies. In *Proceedings of the 54th ASMS Conference on Mass Spectrometry and Allied Topics*, Seattle, WA.

Lane, S. J., and Pipe, A. (2000). Single bead and hard tag decoding using accurate isotopic difference target analysis-encoded combinatorial libraries. *Rapid Commun. Mass Spectrom.* **14**:782–793.

Leclercq, L. L., Delatour, C., Marlot, E., Brunelle, F., McCullagh, M., Castro-Perez, J., Harland, G., and Preece, S. (2004), A new concept for fully automated metabolite identification by LC/MS. In *Proceedings of the 52nd ASMS Conference on Mass Spectrometry and Allied Topics*, Nashville, TN.

Lee, L., McLaughlin, T., Chen, Y., Ma, J., Cho, R., Le, H., and Miao, S. (2007). Re-interrogation of verapamil metabolites from bile using time-based data-dependent mass lists and mass defect filter with the LTQ Orbitrap. In *Proceedings of the 55th ASMS Conference on Mass Spectrometry and Allied Topics*, Indianapolis, IN.

Liu, D. Q., and Hop, C. E. (2005). Strategies for characterization of drug metabolites using liquid chromatography-tandem mass spectrometry in conjunction with chemical derivatization and on-line H/D exchange approaches. *J. Pharm. Biomed. Anal.* **37**:1–18.

Ma, L., Zhang, H., Humphreys, G., Sanders, M., and Zhu, M. (2005). Selective and sensitive detection of GSH adducts in complex biological samples using high resolution LC/FTMS with mass defect filtering. In *Proceedings of the 53rd ASMS Conference on Mass Spectrometry and Allied Topics*, San Antonio, TX.

Marshall, A. G. (2000). Milestones in Fourier transform ion cyclotron resonance mass spectrometry technique development. *Int. J. Mass Spectrom.* **200**:331–356.

Marshall, P., Heudi, O., McKeown, S., Amour, A., and Abou-Shakra, F. (2002). Study of bradykinin metabolism in human and rat plasma by liquid chromatography with inductively coupled plasma mass spectrometry and orthogonal acceleration time-of-flight mass spectrometry. *Rapid Commun. Mass Spectrom.* **16**:220–228.

Meyer, K., Fobker, M., Christians, U., Erren, M., Sewing, K. F., Assmann, G., and Benninghoven, A. (1996). Characterization of glucuronidated phase II metabolites of the immunosuppressant cyclosporine in urine of transplant patients using time-of-flight secondary-ion mass spectrometry. *Drug Metab. Dispos.* **24**:1151–1154.

Morand, K. L., Burt, T. M., Regg, B. T., and Tirey, D. A. (2001). Advances in high-throughput mass spectrometry. *Curr. Opin. Drug Discov. Devel.* **4**:729–735.

Mortishire-Smith, R. J., Hill, A., and Castro-Perez, J. M. (2007). Generic dealkylation: A tool for increasing the hit-rate of metabolite identification, and customizing mass defect filters. In *Proceedings of the 55th ASMS Conference on Mass Spectrometry and Allied Topics*, Indianapolis, IN.

Mortishire-Smith, R. J., O'Connor, D., Castro-Perez, J. M., and Kirby, J. (2005). Accelerated throughput metabolic route screening in early drug discovery using high-resolution liquid chromatography/quadrupole time-of-flight mass spectrometry and automated data analysis. *Rapid Commun. Mass Spectrom.* **19**:2659–2670.

Nicholson, J. K., Lindon, J. C., Scarfe, G. B., Wilson, I. D., Abou-Shakra, F., Sage, A. B., and Castro-Perez, J. (2001). High-performance liquid chromatography linked to inductively coupled plasma mass spectrometry and orthogonal acceleration time-of-flight mass spectrometry for the simultaneous detection and identification of metabolites of 2-bromo-4-trifluoromethyl. *Anal. Chem.* **73:**1491–1494.

Pihakari, K. A. (2007). FT–MS to provide novel insight into complex samples. *Spectroscopy* July (special issue):18–26.

Pike, A., Naegele, E., and Bilsborough, S. (2007). Discovery and identification of low level drug metabolites using a novel searching method combined with exact mass measurement. In *Proceedings of the 55th ASMS Conference on Mass Spectrometry and Allied Topics*, Indianapolis, IN.

Ruan, Q., Peterman, S., Szewc, M. A., Ma, L., Cui, D., Humphreys, W. G., and Zhu, M. (2006). Metabolite detection and characterization by high-resolution orbitrap mass spectrometry: Application of mass defect and product ion filtering techniques. In *Proceedings of the 54th ASMS Conference on Mass Spectrometry and Allied Topics*, Seattle, WA.

Sundstrom, I., Hedeland, M., Bondesson, U., and Andren, P. E. (2002). Identification of glucuronide conjugates of ketobemidone and its phase I metabolites in human urine utilizing accurate mass and tandem time-of-flight mass spectrometry. *J. Mass Spectrom.* **37:**414–420.

Taylor, J. A., and Johnson, R. S. (2001). Implementation and uses of automated de novo peptide sequencing by tandem mass spectrometry. *Anal. Chem.* **73:**2594–2604.

Vrbanac, J. J., O'Leary, I. A., and Baczynskyj, L. (1992). Utility of the parent-neutral loss scan screening technique: Partial characterization of urinary metabolites of U-78875 in monkey urine. *Biol. Mass Spectrom.* **21:**517–522.

Xia, Y. Q., Miller, J. D., Bakhtiar, R., Franklin, R. B., and Liu, D. Q. (2003). Use of a quadrupole linear ion trap mass spectrometer in metabolite identification and bioanalysis. *Rapid Commun. Mass Spectrom.* **17:**1137–1145.

Yoshitsugu, H., Fukuhara, T., Ishibashi, M., Nanbo, T., and Kagi, N. (1999). Key fragments for identification of positional isomer pair in glucuronides from the hydroxylated metabolites of RT-3003 (Vintoperol) by liquid chromatography/electrospray ionization mass spectrometry. *J. Mass Spectrom.* **34:**1063–1068.

Zhang, D., Wang, L., Raghavan, N., Zhang, H., Li, W., Cheng, P. T., Yao, M., Zhang, L., Zhu, M., Bonacorsi, S., Yeola, S., Mitroka, J., Hariharan, N., Hosagrahara, V., Chandrasena, G., Shyu, W. C., and Humphreys, W. G. (2007). Comparative metabolism of radiolabeled muraglitazar in animals and humans by quantitative and qualitative metabolite profiling. *Drug Metab. Dispos.* **35:**150–167.

Zhang, H., Heinig, K., and Henion, J. (2000). Atmospheric pressure ionization time-of-flight mass spectrometry coupled with fast liquid chromatography for quantitation and accurate mass measurement of five pharmaceutical drugs in human plasma. *J. Mass Spectrom.* **35:**423–431.

Zhang, H., and Ray, K. (2007). Profiling impurities in a taxol formulation: Application of a mass-dependent mass defect filter to remove polymeric excipient interferences. In *Proceedings of the 55th ASMS Conference on Mass Spectrometry and Allied Topics*, Indianapolis, IN.

Zhang, H., Ray, K., Ma, L., Zhang, D., Drexler, D., Zhu, M., and Sanders, M. (2005). Applicability of mass defect filters (MDF) to drug metabolite detection in biological matrices. In *Proceedings of the 53rd ASMS Conference on Mass Spectrometry and Allied Topics*, San Antonio, TX.

Zhang, H., Ray, K., Zhao, W., Zhang, D., Gozo, S., and Zhu, M. (2004). Selective detection of metabolite ions of non-radiolabeled drugs in complex biological matrices using mass defect filter approach. In *Proceedings of the 52nd ASMS Conference on Mass Spectrometry and Allied Topics*, Nashville, TN.

Zhang, H., Zhang, D., and Ray, K. (2003a). A software filter to remove interference ions from drug metabolites in accurate mass liquid chromatography/mass spectrometric analyses. *J. Mass Spectrom.* **38**:1110–1112.

Zhang, H., Zhang, D., Ray, K., Gozo, S., Kleintop, B., and Mitroka, J. (2003b). Leveraging high resolution LC/MS data to identify drug metabolite ions in samples of human excreta. In *Proceeding of the 51st ASMS Conference on Mass Spectrometry and Allied Topics*, Montreal, Canada.

Zhang, H., Zhu, M., Ma, L., He, H., Humphreys, W. G., and Sanders, M. (2006). Combining MDF and peak match techniques for comprehensive and selective detection of drug metabolites in vivo. In *Proceedings of the 54th ASMS Conference on Mass Spectrometry and Allied Topics*, Seattle, WA.

Zhao, W., Zhang, H., Zhu, M., Warrack, B., Ma, L., Humphreys, W. G., and Sanders, M. (2006). An integrated method for quantification and identification of radiolabeled metabolites: Application of chip-based nanoelectrospray and mass defect filter techniques. In *Proceedings of the 54th ASMS Conference on Mass Spectrometry and Allied Topics*, Seattle, WA.

Zhu, M., Ma, L., Zhang, D., Ray, K., Zhao, W., Humphreys, W. G., Skiles, G., Sanders, M., and Zhang, H. (2006). Detection and characterization of metabolites in biological matrices using mass defect filtering of liquid chromatography/high-resolution mass spectrometry data. *Drug Metab. Dispos.* **34**:1722–1733.

Zhu, M., Zhang, H., Yao, M., Zhang, D., Ray, K., and Skiles, G. L. (2004). Detection of metabolites in plasma and urine using a high resolution LC/MS-based mass defect filter approach: Comparison with precursor ion and neutral loss scan analyses. *Drug Metab. Rev.* **36** (Suppl. 1):43.

7

Applications of High-Sensitivity Mass Spectrometry and Radioactivity Detection Techniques in Drug Metabolism Studies

Wing W. Lam

Johnson and Johnson Pharmaceutical Research and Development, Department of Drug Metabolism and Pharmacokinetics, Raritan, New Jersey

Cho-Ming Loi

Pfizer Global Research and Development, Department of Pharmacokinetics, Dynamics and Metabolism, San Diego, California

Angus Nedderman and Don Walker

Pfizer Global Research and Development, Department of Pharmacokinetics, Dynamics and Metabolism, Sandwich, Kent, United Kingdom

Mass Spectrometry in Drug Metabolism and Pharmacokinetics. Edited by Ragu Ramanathan
Copyright © 2009 John Wiley & Sons, Inc.

7.1 INTRODUCTION

During the discovery and development process for a new chemical entity (NCE), it is common to use [^{3}H]- or [^{14}C]-radiolabeled compounds to evaluate the in vivo metabolism (Chang et al., 1998; Zhang et al., 2007). The nature of drug-related material present in the systemic circulation is clearly of high relevance to the pharmacological and toxicological action of the NCE because the tissues of the body are exposed to the chemical components present in the systemic circulation (blood/plasma). However, in the past, due to the limitations of analytical technologies, it was only possible to derive extensive characterization of drug-related components present in the excreta (urine, bile, feces, etc.) due to the higher concentrations and amounts of chemical materials available. Concentrations of drug-derived material present in plasma are generally in the range of parts per billion and the amount of blood/plasma available for analysis is limited to 1–10 mL. On the other hand, in the excreta, drug-derived material is present in the range of parts per million and hundreds of milliliters of material is available for analysis. The challenge, to move from detailed characterization in excreta to plasma, is therefore in the magnitude of four orders or more. This task is clearly assisted by state-of-the-art mass spectrometry (MS) techniques available for structural elucidation, but semiquantitative assessment of the drug-related material remains a challenge. Furthermore, drug development programs present even greater challenges when unusually lower systemic exposure of drug-related materials is encountered from dermal or inhaled routes of administration.

Conventionally, a flow-through radiochemical detector is used in-line to detect the presence of metabolites in plasma and excreta (Egnash and Ramanathan, 2002). However, in-line radiochemical detection is generally of insufficient sensitivity for profiling plasma samples collected from typical clinical studies involving lower doses. For this reason, profiling of plasma-derived radioactive components has historically required laborious fraction collection procedures with offline scintillation counting. The mixing, of high-performance liquid chromatography (HPLC) eluent fractions with scintillation fluid in this process, prevents subsequent mass spectrometric analysis of the collected fractions and adds more complexity to the metabolite profiling and identification process.

Recently, newer techniques have been developed to overcome these challenges. In this chapter, a discussion of two relatively new methods, namely liquid chromatography–accurate radioisotope counting–mass spectrometry (LC–ARC–MS) and the use of 96-well microtiter plates containing solid scintillants, is presented. These new methods are of higher sensitivity for detecting low levels of radiolabeled metabolites. In addition, the use of accelerator mass spectrometry (AMS), which detects individual ions of elemental isotopes, is discussed as an alternative to radiochemical detection techniques (Lappin and Garner, 2005; Sarapa et al., 2005). These high-sensitivity techniques, when applied to circulating metabolite semiquantitation and identification, permit a more detailed evaluation of the exposure of drug-related materials to animals and humans.

7.2 LIQUID CHROMATOGRAPHY–ACCURATE RADIOISOTOPE COUNTING–MASS SPECTROMETRY

LC–ARC–MS is a novel radiochemical detection system that is designed to detect samples with very low levels of radioactivity (Lee et al., 2000; Lee, 2003). Similar to the conventional flow through radiochemical detection method, LC–ARC is an in-line detection technique that allows real-time display of metabolite peaks. LC–ARC can either be set up as a stand-alone system or be coupled with a mass spectrometer to become LC–ARC–MS. Thus, the combination of ARC and MS enhances the sensitivity of peak detection and also provides mass spectral information for structural elucidation of metabolite(s). Other interfaces, coupled with the LC–ARC system, are also available; for example, LC–ARC/RD–MS/FC is a system of LC–ARC which couples with a radiochemical detector (RD), a mass spectrometer, and fraction collector (FC) (Lu et al., 2002).

7.2.1 LC–ARC–MS System

7.2.1.1 System Setup The LC–ARC–MS system consists of a liquid chromatograph, a RD, a StopFlow™ pump, an ARC™ control and data acquisition software system, and a mass spectrometer (Fig. 7.1). The schematic of the system is

Figure 7.1. Photograph of typical setup of LC–ARC–MS system.

shown in Fig. 7.2. Various LC systems are compatible with the ARC system, for example, Agilent 1100 systems, Waters 2695 series, and Perkin Elmer systems. Radiochemical detectors, like beta-RAM from IN/US and Radiomatric 600 series from Packard, are also compatible with the ARC system.

7.2.1.2 System Operation LC–ARC operates in both the stop-flow and non-stop-flow modes. Each of these is described below.

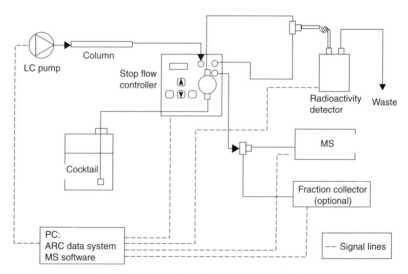

Figure 7.2. Schematic of LC–ARC–MS setup.

Stop-Flow Mode   The LC–ARC StopFlow controller performs two major functions. First, the controller maintains the back pressure during stop-flow data acquisition. This function is crucial for maintaining the shape and resolution of a peak. The second function is to deliver a liquid scintillation cocktail necessary for peak detection by the radiochemical detector.

There are two ways to operate the stop-flow mode. The by-level mode, which is based upon the defined count zone(s), operates by signaling the instrument to continue running until a radioactive peak above a designated level is detected. The instrument will then stop and count the predefined fraction (in seconds), which is defined in the count zone. A second way to operate the stop-flow mode is the by-fraction mode. The instrument will stop and count every fraction (in seconds) within the predefined count zone, regardless of whether any radioactivity is detected or not. The fraction size in the stop-flow mode and the volume of liquid scintillant used for counting can be automatically calculated by the instrument or the parameters can be defined by the user.

Non-Stop-Flow Mode   In the non-stop-flow mode, the LC–ARC system is operated in a similar manner to the conventional continuous-flow analysis. If a mass spectrometer is coupled to the LC–ARC system, the LC effluent is split postcolumn to deliver a fraction to the radiochemical detector and the balance to a mass spectrometer.

7.2.1.3 *Data System*   The ARC data system is designed to control the LC, RD, ARC, and partly the MS system. The data system consists of two modules: (1) the run module and (2) the evaluate module. The run module is used to set up the run conditions and instrument control, initiate the run sequence, and monitor the run in real time. The evaluate module is used for postrun data evaluation and processing.

Run Module   The run module is used to set up the LC conditions, including the pump (e.g., gradient method), UV methods (e.g., wavelength), and autosampler (e.g., sequence setup). Apart from the LC, the method for radiochemical detection and the volume of liquid scintillant for radioactivity counting is also programmed in this module. The autosampler (as set up by the run module) is used to trigger the Start Run for all other instruments through external contact closures.

If the MS is incorporated as part of the run, the autosampler will also trigger the mass spectrometer to run and collect MS data. The run module, however, does not set up the mass spectrometer conditions, which are determined using the MS software. Thus, the run time used in both the ARC and MS must be consistent in order for the acquisition to be compatible. Prior to initiating a run, the sequence identifier of the MS run is entered into the ARC data system to initiate the correct sequences simultaneously in the MS and ARC systems.

Evaluate Module   The functions of this module are to process, integrate, and report radiochemical, UV, and/or MS data. The MS data are best processed using this module, especially when the data are acquired using the stop-flow method. When operating in this mode, the acquired MS profile appears discontinuous. The ARC software is capable of removing the gaps observed during all the ARC stop periods to display the MS profile similar to that obtained using the MS data system. This feature, however, is currently available only to mass spectrometers manufactured by Thermo Electron Corporation.

7.2.2 Utilities of LC–ARC in Drug Discovery and Development

LC–ARC can be used to study samples containing either ^{14}C or [3]H (Nassar et al., 2003; Zhao et al., 2004; Lam et al., 2007; Nassar and Lee, 2007a,b). The following examples demonstrate the sensitivity and utility of the LC–ARC technique as applied to metabolite profiling and characterization involving samples with low amounts of radioactivity. The enhanced sensitivity of the LC–ARC in detecting compound-related radioactivity is illustrated in Fig. 7.3 (Lee, 2003). A rat urine sample, containing ^{14}C-derived components with a total radioactivity count of 2855 dpm, was analyzed using the StopFlow technique with the by-level mode and a counting time of 120 s. The limit of detection in this experiment was determined to be 12 cpm. Based on this limit of detection, a peak with 32 cpm is detected without ambiguity. This result is consistent with those reported by others, indicating that a limit of detection of 10–20 cpm is routinely achievable (Nassar et al., 2003; Gaddamidi et al., 2004). Although the counting efficiency is lower with ^{3}H compared to ^{14}C, the limit of detection for samples containing ^{3}H-derived components has been reported with ranges of 10–40 dpm (Lee et al., 2000). Taken together, these data suggest that the LC–ARC technique potentially offers an enhanced sensitivity by at least 10- to 20-fold compared to conventional flow through detection methods.

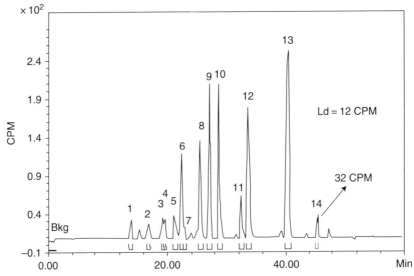

Figure 7.3. LC–ARC chromatogram of rat urine samples (2855 dpm) showing limit of detection of 12 cpm. Column recovery is 91%. The total run time is 177 min, which is approximately 3 times the original run time of 60 min.

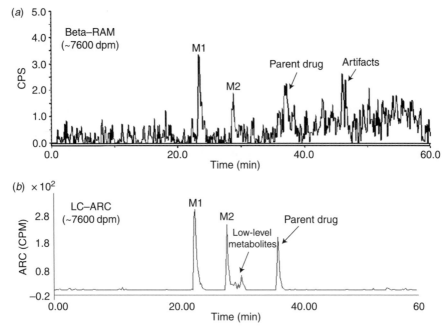

Figure 7.4. *Radiochemical detection of parent drug and its metabolites using (a) conventional beta-RAM method and (b) LC–ARC.*

The enhanced sensitivity of this technique can improve the ability to perform metabolite characterization in support of drug discovery and development so long as radiolabeled material is available. For instance, using the conventional flow-through beta-RAM detection, analysis of a plasma sample containing ^{3}H-derived components with a total radioactivity count of ~7600 dpm revealed that the peaks of interest (M1, M2, and parent) had low signal-to-noise ratio (Fig. 7.4a). Furthermore, the presence of artifactual peaks may also confound the assignment of peaks of interest, for example in Fig. 7.4, M2 and parent peaks are obscured among the background noise. When an identical amount of ^{3}H radioactivity from the same plasma sample was analyzed using the LC–ARC method, the chromatograms clearly indicated the presence of three peaks corresponding to the parent and its two major metabolites (Fig. 7.4b). In addition, several minor metabolites were also detected. This technique enables the peaks of interest to be detected so that structural elucidation can be performed to characterize the metabolites corresponding to each peak.

7.2.3 Summary

The LC–ARC system, with the stop-flow and the non-stop-flow modes, provides a convenient platform for analysis of samples with a wide range of radioactivity. A major advantage of this technique, compared to other conventional in-line flow-through detection methods, is derived from its enhanced sensitivity. Detection of radioactivity as low as ~6 cpm has been reported using the LC–ARC stop-flow technique (Nassar et al., 2003), which is approximately 20 times more sensitive than the conventional flow through radiochemical detector. This capability renders the LC–ARC technology an important tool in metabolite profiling and characterization when the level of radioactivity associated with the metabolite(s) of interest is low. When coupled with MS, this online LC–ARC–MS system provides real-time analysis and monitoring.

Although the StopFlow method provides enhanced sensitivity to detect low levels of radioactivity in biological matrices, the technique requires longer run time with stop-flow operation and therefore is less desirable for coupling with LC–MS due to the flow interruption. Most recently, to circumvent the flow interruption limitation, DynamicFlow™ LC–ARC was introduced (Lee, 2007; Nassar and Lee, 2007a). This newer system controls HPLC flow and downstream flow cell residence time dynamically so that the radioactivity detection sensitivity is maximized. The detection limit of this system is down to 20 cpm for ^{14}C-labeled compound without stopping the HPLC flow (Lee, 2007; Nassar and Lee, 2007b). The run time of this method is similar to the conventional radio-HPLC analysis and thus offers an advantage over the StopFlow method. An additional advantage of the newer system is the ability to process the data using the software supplied by the MS vendors. However, a potential drawback of the current version of this technology is related to peak broadening, especially those peaks containing very low levels of radioactivity.

Currently, optimization of background and counting efficiency for the LC–ARC system are being determined semimanually. As accurate determination of these parameters is paramount to the quality of the chromatograms and the data acquired, automation of these determinations by the software will render greater efficiency and convenience in utilizing this technique for metabolite profiling.

7.3   96-WELL MICROTITER PLATES CONTAINING SOLID SCINTILLANT

An alternative strategy to LC–ARC, in terms of generating semiquantitative metabolism data from samples containing low levels of radioactivity, is to collect fractions into 96-well plates during HPLC profiling for offline scintillation counting. Similar to LC–ARC and other conventional radiochemical detection techniques, increasing the counting time improves the sensitivity for [^{14}C] and [^{3}H] detection. Two common types of plates are available for this approach, the Lumaplate (Fig. 7.5a) and the Scintiplate (Fig. 7.5b), both manufactured by PerkinElmer (Meriden, CT). The former consists of polystyrene wells embedded with a solid scintillant and is counted on the Microbeta scintillation counter (Yin et al., 2000), while the latter incorporates a solid layer of yttrium silicate at the bottom of each well and is analyzed on a TopCount® instrument (Floeckher, 1991a,b; Boernsen, 2000; Boernsen et al., 2000; Zhu et al., 2005). Since the 96-well plate approach involves drying of the plates and offline counting, throughput is diminished compared to conventional

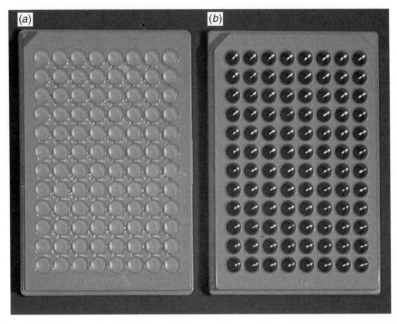

Figure 7.5. *Photograph of (a) a 96-well deep-well Lumaplate and (b) a 96-well Scintiplate.*

online methodologies and LC–ARC. Nevertheless, the approach is clearly less labor intensive and time consuming than using scintillation vials for fraction collection and offline counting and therefore offers an attractive method for metabolite profiling where levels of radioactivity are low.

7.3.1 Characteristics of 96-Well Approach

7.3.1.1 Counting Efficiency and Background Noise   The counting efficiency for [^{14}C]-labeled samples is as high as 91% when a dried 96-well Lumaplate is used with the TopCount instrument (Boernsen, 2000). In contrast, the efficiency is markedly reduced for [^{3}H]-labeled material (40%) due to the low energy output of this isotope. In both cases, however, the values are comparable to those achievable using conventional liquid scintillation counting techniques. The efficiency of the Scintiplate approach, using dried plates, is somewhat lower than that of the Lumaplate for both isotopes (80 and 18% for [^{14}C] and [^{3}H], respectively) (Yin et al., 2000). Counting efficiencies close to those of the Lumaplate method can be achieved with Scintiplate by the addition of liquid scintillant to the Scintiplate wells (Yin et al., 2000). It is clear that static counting of radioactive samples using either 96-well plate approach yields a clear advantage in counting efficiency over online radiochemical flow detection. Background noise levels, achievable with Lumaplate and Scintiplate approaches, are respectively at approximately 1–3 and 5–8 cpm for both isotopes (Boernsen, 2000). For comparison, background values in the range 12–16 cpm are typical for conventional liquid scintillation counters using vials, although improved shielding with modern instruments can result in background levels of approximately 3–5 cpm.

7.3.1.2 Limits of Detection

The theoretical limit of detection (LOD) using the 96-well plate approaches was calculated using the methodology of Currie (1968). Using typical background and efficiency values and a counting time of 5 min, the LOD for [^{14}C] detection using dried Lumaplate and Scintiplate is approximately 5 and 7 dpm, respectively. However, in view of the fact that a radiolabeled component is likely to be collected into more than one well under typical chromatographic conditions, a realistic limit of detection is approximately 10–15 dpm. The detection and quantitation of [^{14}C]-labeled drug-related components, accounting for approximately 25 dpm using Scintiplate, have been reported (Nedderman et al., 2004) (Fig. 7.6). Due to the lower counting efficiency, the theoretical LOD for [^{3}H] is somewhat higher at approximately 9 and 16 dpm for the Lumaplate and Scintiplate methods, respectively. Metabolite profiling studies, with rat urine following administration of a [^{3}H]-labeled compound using the Lumaplate approach, have shown that drug-related components with approximately 25 cpm (equivalent to approximately 60 dpm) are readily detectable and quantifiable (Boernsen et al., 2000) (Fig. 7.7).

7.3.1.3 Quenching Effects

Chemical and color quenching of the radiochemical signal by endogenous matrix components will vary during HPLC profiling of biological samples. While fraction collection into vials followed by the addition of liquid scintillant for offline counting involves external standardization, such that the quenching effects are corrected, the 96-well plate approach typically employs no quench correction, such that the validity of the quantitative data may be compromised. A previous study using the Scintiplate (Nedderman et al., 2004) showed

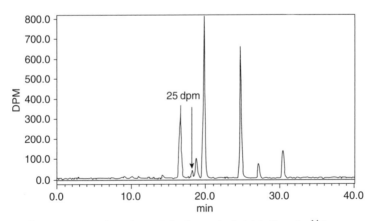

Figure 7.6. *Chromatogram of rat plasma following oral administration of a ^{14}C compound using Scintiplate approach.*

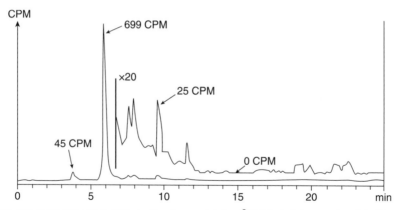

Figure 7.7. *Chromatogram of 5 μL rat urine containing [³H]-labeled drug and metabolites after injection of 5500 dpm, counted offline with the TopCount system. Inset: 20-fold expansion of y axis (Boernsen, 2000).*

that quenching caused by the endogenous components in extracted fecal homogenates and urine samples was significant at sample amounts typically used in metabolism studies, regardless of the levels of radioactivity in the samples. Furthermore, the effects were markedly different for [³H] than [¹⁴C] due to the lower energy of the [³H] isotope. However, profiling of a plasma sample resulting from precipitation of 5 mL of rat plasma showed that significant quenching of the [¹⁴C] signal only occurred at the solvent front, where metabolite elution is unlikely (Fig. 7.8a). In contrast, the lower energy of the [³H] isotope resulted in significant quenching of the signal by endogenous plasma components (Fig. 7.8b). Although formal studies to investigate quenching effects using the Lumaplate have not been reported, it is likely that large quantities of endogenous material will have a similar effect on the radiochemical response using this approach.

7.3.2 Combination of 96-Well Plate Approach with Metabolite Identification

The results discussed above point out that both the Scintiplate and Lumaplate can readily be used in an online mode by splitting the HPLC flow during HPLC–MS analysis and diverting an appropriate proportion to the fraction collector and the rest to the mass spectrometer. Although a reduction in radiochemical signal-to-noise ratio is inevitable with this approach, this reduction is most likely insignificant. This online approach in combination with data-dependent acquisition software yields an abundance of structural information. However, acquisition of sufficient data across a chromatographic peak remains a potential issue for metabolite identification studies. In such cases, the 96-well plate approach enables elution of the drug-related components from the plates after radiochemical counting for additional analysis, if required (Nedderman et al., 2004). In this respect, the Scintiplate has the advantage due to the embedded nature of the solid scintillant, such that elution is possible

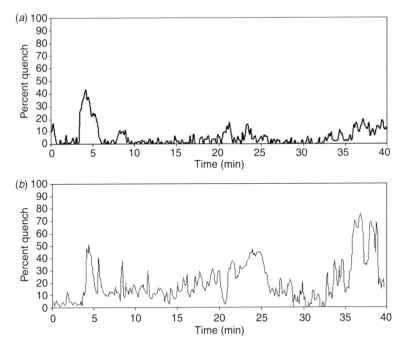

Figure 7.8. *Quenching effect of 5 mL rat plasma, following protein precipitation, on radiochemical signal using Scintiplate approach: (a) [^{14}C]; (b) [^{3}H].*

without contamination of the reconstituted sample with particulate matter. In contrast, the yttrium silicate coating of the Lumaplate is not robust and is likely to be disturbed during reconstitution of samples on the plates. In many cases, direct infusion of drug-related components is preferable to the online approach, enabling sufficient experimentation time on selected samples in an effort to maximize the opportunity for structural characterization. If the offline approach is to be pursued, then care must be taken to choose relatively benign solvents. An appropriate reconstituting solvent ensures minimal ion suppression effects from solubilization of plate material.

7.3.3 Summary

The use of either the Scintiplate in combination with the Microbeta scintillation counter or the Lumaplate with the TopCount instrument offers significant advantages over conventional online radiochemical detection in terms of counting efficiency and associated detection limits. Therefore, both methodologies are well suited to low-level metabolite identification and profiling studies (Boernsen et al., 2000; Kiffe et al., 2003; Nedderman et al., 2004). Furthermore, both approaches are significantly less time consuming and labor intensive than fraction collection into scintillation vials and liquid scintillation counting, such that sample throughput is dramatically improved. Quenching effects of endogenous matrix components may affect the validity of the counting data, especially when profiling excreta samples containing [^{3}H]-derived material. However, the quenching effects are not as severe for plasma samples. As a result of higher counting efficiency and lower backgrounds for both [^{3}H] and [^{14}C], the LOD is lower using the Lumaplate approach. The magnitude of difference between the metabolism data derived from the Lumaplate and the 96-well plate approaches is less likely to have a significant differences. When metabolite identification data are required in addition to quantitative data, the 96-well plate approaches can be used online with a mass spectrometer by splitting the flow. Alternatively, metabolite identification data can be obtained offline with subsequent elution and further analysis of drug-related components from the plates. In the latter mode, the Scintiplate offers an advantage due to the embedded nature of the scintillant.

7.4 FUTURE TRENDS IN LOW-LEVEL METABOLITE CHARACTERIZATION IN COMBINATION WITH LC–ARC–MS AND 96-WELL PLATE APPROACH

Both LC–ARC and the 96-well plate approaches have the potential to be used in combination with a number of additional methodologies to optimize qualitative and quantitative metabolism data. In conventional radiolabeled metabolism studies, plasma samples are typically prepared by protein precipitation prior to HPLC profiling. However, the offline sample preparation step can be removed by incorporating online sample extraction techniques involving turbulent-flow liquid chromatography (TFLC, Chapter 10) (Oberhauser et al., 2000; Chassaing et al., 2001, 2005; Ynddal and Hansen, 2003), restricted-access media (RAM) columns (Souverain et al., 2004; Veuthey et al., 2004), or solid-phase extraction (SPE) (Calderoli et al., 2003). Improved sample cleanup results in reduced radiochemical quenching and ion suppression effects from endogenous components as well as enhanced analytical chromatography and yields further improvements in mass spectrometric and radiochemical signal-to-noise ratio. In addition, online extraction will inevitably improve sample throughput over offline approaches.

Although the advantages associated with online plasma extraction are attractive, care must be taken to monitor the recovery of drug-related material during the extraction process. Unlike quantitative plasma analysis, where poor recovery only affects the limit of quantitation, the recovery of all drug-related components in metabolite profiling studies must be high in order to ensure that the quantitative data are meaningful.

Alternative approaches to improve chromatography and thereby enhance signal-to-noise ratio include the use of ultrapressure liquid chromatography (UPLC) (Dear et al., 2006), where chromatographic resolution can be dramatically increased (Castro-Perez et al., 2004, 2005; Plumb et al., 2004), or nano-HPLC (Schmidt et al., 2003), which may serve to reduce ion suppression effects. For offline use, the 96-well plate approach (typically the Scintiplate) will normally be used in combination with direct mass spectrometric infusion at conventional flow rates (5–10 µL/min). This approach enables multiple experiments to be performed in order to achieve metabolite structural elucidation. Alternatively, automated chip-based infusion approaches such as the TriVersa Nanomate from Advion Biosciences (Dethy et al., 2003a,b; Kapron et al., 2003; Leuthold et al., 2004; Ackermann and Dethy, 2005) can be used, which employs low-flow-rate infusion (50–200 nL/min) through a chip containing 400 separate electrospray nozzles. The low flow rate has the potential to increase MS signal-to-noise ratio in three ways: infusion times may be increased, such that signal averaging can be used to enhance the mass spectrometry signal; samples may be concentrated without adversely affecting analysis time; and ion suppression effects may be reduced (Schmidt et al., 2003).

7.5 ACCELERATOR MASS SPECTROMETRY

Accelerator mass spectrometry is a highly sensitive method predominantly used for the detection of [^{14}C] ions, although the analysis of many other isotopes, including [^{3}H], is also possible (Cupid and Garner, 1998; Barker and Garner, 1999; Garner, 2000; Garner et al., 2000). In brief, the approach for [^{14}C] detection involves combustion of the sample and subsequent reduction of the evolved CO_2 to form a graphite pellet (Vogel, 1992). Ionization of the pellet, usually by cesium ion sputtering, forms negative ions of [^{12}C], [^{13}C], and [^{14}C] which are focused into the accelerator, whereupon electrons are stripped from the ions, either by collision with an inert gas or a foil, to form a range of positively charged species. These ions are transferred out of the accelerator, separated, and focused prior to detection (Fig. 7.9) (additional details in Chapter 2).

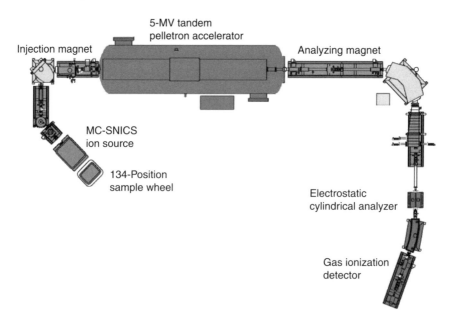

Figure 7.9. Schematic of AMS instrument.

Because the detection is at the level of individual ions, the sensitivity is dramatically improved compared to scintillation counting, which relies upon relatively rare disintegration events. Originally used for radiocarbon dating (Bennett et al., 1977; Nelson et al., 1977), the methodology was later used for a range of biomedical applications, including drug metabolism and pharmacokinetic studies (Litherland, 1980; Elmore and Phillips, 1987; Elmore et al., 1990). Early work in the field (Kaye et al., 1997) assessed the utility of AMS for excretion balance or mass balance and concluded that equivalent data to conventional studies could be achieved with dramatically reduced doses of radioactivity, thus minimizing issues such as safety and disposal. Subsequent studies (Gilman et al., 1998; Garner et al., 2000, 2002; Young et al., 2001) demonstrated that excretion balance study data were routinely achievable using the approach. However, for plasma analysis, the sensitivity of AMS is limited by the endogenous levels of radioactivity found in both human and animal plasma samples.

Metabolic profiles for excreta samples (urine, feces, etc) were generated using fraction collection and AMS analysis of individual fractions (Young et al., 2001; Garner et al., 2002). In this respect, AMS is not as vulnerable to radiochemical quenching effects caused by endogenous excreta components and therefore offers an additional advantage over alternative low-level profiling methodologies. The AMS approach can also be applied to the generation of circulating metabolite data for extremely low level samples, which would be impossible using conventional methodologies. Such data have been reported for low-level oral administrations (Garner et al., 2002) and also generated following the dermal application of a conventional radioactive dose (Fig. 7.10). For profiling circulating metabolites, the limitations of AMS for plasma analysis are alleviated by sample preparation, which removes the endogenous radioactivity in the samples, such that quantitative data with good signal-to-noise ratio were achieved for circulating metabolites accounting for approximately 0.05 dpm in the dermal study.

In terms of facilitating radiolabeled safety studies, AMS has generated considerable interest in the field of human microdosing studies, which are designed to generate human pharmacokinetic data earlier in the drug discovery/development process

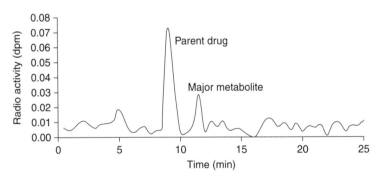

Figure 7.10. Radiochromatogram of mini-pig plasma following dermal application (6 µCi) of radiolabeled compound.

than current practice (Lappin and Garner, 2007; Smith et al., 2003; Sarapa et al., 2005; Lappin et al., 2006a,b). By administering an extremely low dose of radioactivity, plasma samples can be analyzed for drug-related material even when levels of total drug in plasma are very low. Although the approach will not distinguish between parent compound and metabolites unless additional fractionation is undertaken, the approach may potentially yield useful data to facilitate compound selection. While AMS is a very sensitive technique for metabolite analysis following fraction collection, the process is labor intensive and the sample throughput is slow and the cost of analysis is high. In recent years, attempts have been made to make the technique more accessible by developing smaller, less expensive instruments, although this strategy will inevitably result in a reduction in sensitivity (Ognibene et al., 2003; Ognibene and Vogel, 2004). In addition, new interfaces are currently under development in an attempt to facilitate direct analysis of biological samples (Liberman et al., 2004), ultimately leading to an HPLC–AMS system (Liberman et al., 2004). Although the potential to improve the throughput of the process using this approach is clear, achieving acceptable sensitivity remains problematic.

7.6 SUMMARY

High-sensitivity techniques for the quantitation and characterization of circulating metabolites following administration of radiolabeled compounds are of critical importance to understand the safety and efficacy profiles of novel drug candidates. AMS is one of the most sensitive techniques for the detection of radiolabeled components. However, the high cost and slow throughput of AMS analysis preclude the routine use of the techniques for metabolism studies.

The use of LC–ARC or 96-well microtiter plates containing solid scintillant, in contrast, represents low-cost methodologies to improve the sensitivity of the semi-quantitative analysis of metabolites. The LC–ARC and 96-well microtiter plate approaches have similar LOD for ^{14}C and ^{3}H and both can be used in combination with MS to generate quantitative and structural data. Both approaches are particularly useful in cases where the sample is limited. The throughput of LC–ARC is higher than for the plate approach due to the need to dry and count the plates offline. However, the 96-well plate approach is nondestructive, such that drug-related components can be subsequently eluted from the plates for further analysis, if necessary.

All the techniques described in this chapter contribute to improved metabolite characterization. Notably, these techniques will enable the generation of extended time profiles for plasma radioactivity, the assessment of free metabolite concentrations following plasma dialysis, and the characterization of metabolites at very low levels of exposure, for example, following inhaled or dermal routes of administration. Furthermore, the use of high-sensitivity techniques could impact the design of radiolabel studies, for example, by enabling a lower radiolabel dose to be administered to human subjects without compromising the ability to characterize the metabolic fate of the compound of interest.

ACKNOWLEDGMENT

The authors would like to thank Dr. Dian Lee for helpful discussion on LC–ARC.

REFERENCES

Ackermann, B. L., and Dethy, J. M. (2005). Understanding the role and potential of infusion nanoelectrospray ionization for pharmaceutical bioanalysis. In *Using Mass Spectrometry for Drug Metabolism Studies* (Korfmacher, W. A., Ed.). CRC Press, Boca Raton, pp. 329–356.

Barker, J., and Garner, R. C. (1999). Biomedical applications of accelerator mass spectrometry-isotope measurements at the level of the atom. *Rapid Commun. Mass Spectrom.* **13**:285–293.

Bennett, C. L., Beukens, R. P., Clover, M. R., Gove, H. E., Liebert, R. B., Litherland, A. E., Purser, K. H., and Sondheim, W. E. (1977). Radiocarbon dating using electrostatic accelerators: Negative ions provide the key. *Science* **198**:508–510.

Boernsen, K. O. (2000). Using the TopCount microplate scintillation and luminescence counter and deep-well lumaplate microplates in combination with micro-separation techniques for metabolic studies. in Application note, AN004-TC. Packard Instrument Co. Available at http://www.perkinelmer.com.

Boernsen, K. O., Floeckher, J. M., and Bruin, G. J. M. (2000). Use of a microplate scintillation counter as a radioactivity detector for miniaturized separation techniques in drug metabolism. *Anal. Chem.* **72**:3956–3959.

Calderoli, S., Colombo, E., Frigerio, E., James, C. A., and Sibum, M. (2003). LC-MS-MS determination of brostallicin in human plasma following automated on-line, SPE. *J. Pharm. Biomed. Analy.* **32**:601–607.

Castro-Perez, J., Plumb, R., and Granger, J. (2004). Maximizing chromatographic resolution using ultra performance liquid chromatography. *LCGC North Am.* **23**:59.

Castro-Perez, J., Plumb, R., Liang, L., and Yang, E. (2005). A high-throughput liquid chromatography/tandem mass spectrometry method for screening glutathione conjugates using exact mass neutral loss acquisition. *Rapid Commun. Mass Spectrom.* **19**:798–804.

Chang, M., Sood, V. K., Wilson, G. J., Kloosterman, D. A., Sanders, P. E., Schuette, M. R., Judy, R. W., Voorman, R. L., Maio, S. M., and Slatter, J. G. (1998). Absorption, distribution, metabolism, and excretion of atevirdine in the rat. *Drug Metab. Dispos.* **26**:1008–1018.

Chassaing, C., Luckwell, J., Macrae, P., Saunders, K., Wright, P., and Venn, R. F. (2001). Direct analysis of crude plasma samples by turbulent flow chromatography/tandem mass spectrometry. *Chromatographia* **53**:122–130.

Chassaing, C., Stafford, H., Luckwell, J., Wright, A., and Edgington, A. (2005). A parallel micro turbulent flow chromatography-tandem mass spectrometry method for the analysis of a pharmaceutical compound in plasma. *Chromatographia* **62**:17–24.

Cupid, B. C., and Garner, R. C. (1998). Accelerator mass spectrometry—A new tool for drug metabolism studies. In *Drug Metabolism: Towards the Next Millennium* (Gooderham, N. J., Ed.). Ion Press, Amsterdam, pp. 175–187.

Currie, L. A. (1968). Limits for qualitative detection and quantitative determination. *Anal. Chem.* **40:**568–593.

Dear, G. J., Patel, N., Kelly, P. J., Webber, L., and Yung, M. (2006). TopCount coupled to ultra-performance liquid chromatography for the profiling of radiolabeled drug metabolites in complex biological samples. *J. Chromatogr. B Anal. Technol. Biomed. Life Sci.* **844:**96–103.

Dethy, J. M., Ackermann, B. L., Delatour, C., Henion, J. D., and Schultz, G. A. (2003a). Demonstration of direct bioanalysis of drugs in plasma using nanoelectrospray infusion from a silicon chip coupled with tandem mass spectrometry. *Anal. Chem.* **75:**805–811.

Dethy, J. M., Ackermann, B. L., Delatour, C., Henion, J. D., and Schultz, G. A. (2003b). Demonstration of direct bioanalysis of drugs in plasma using nanoelectrospray infusion from a silicon chip coupled with tandem mass spectrometry. *Anal. Chem.* **75:**805–811.

Egnash, L. A., and Ramanathan, R. (2002). Comparison of heterogeneous and homogeneous radioactivity flow detectors for simultaneous profiling and LC-MS/MS characterization of metabolites. *J. Pharm. Biomed. Anal.* **27:**271–284.

Elmore, D., Bhattacharyya, M. H., Sacco-Gibson, N., and Peterson, D. P. (1990). Calcium-41 as a long-term biological tracer for bone resorption. *Nucl. Instrum. Methods Phys. Res.* **B52:**531–535.

Elmore, D., and Phillips, F. M. (1987). Accelerator mass spectrometry for measurement of long-lived radioisotopes. *Science* **236:**543–550.

Floeckher, J. M. (1991a). Solid scintillation counting. Application note TCA-002. Packard Instrument Co. Available at http://www.perkinelmer.com.

Floeckher, J. M. (1991b). Theory of TopCount operation. Application note TCA-003. Packard Instrument Co. Available at http://www.perkinelmer.com.

Gaddamidi, V., Scott, M. T., Swain, S. R., Brown, A. M., Young, G., Hashinger, B. M., Lee, D., and Bookhart, W. (2004). Sensitive on-line detection of radioisotopes using, LC–ARC applications in agrochemical research. *Drug Metab. Rev.* **36:**256.

Garner, R. C. (2000). Accelerator mass spectrometry in pharmaceutical research and development—A new ultrasensitive analytical method for isotope measurement. *Curr. Drug Metab.* **1:**205–213.

Garner, R. C., Barker, J., Flavell, C., Garner, J. V., Whattam, M., Young, G. C., Cussans, N., Jezequel, S., and Leong, D. (2000). A validation study comparing accelerator, M. S., and liquid scintillation counting for analysis of 14C-labelled drugs in plasma, urine and faecal extracts. *J. Pharm. Biomed. Anal.* **24:**197–209.

Garner, R. C., Goris, I., Laenen, A. A., Vanhoutte, E., Meuldermans, W., Gregory, S., Garner, J. V., Leong, D., Whattam, M., Calam, A., and Snel, C. A. (2002). Evaluation of accelerator mass spectrometry in a human mass balance and pharmacokinetic study—Experience with 14C-labeled (*R*)-6-[amino(4-chlorophenyl)(1-methyl-1*H*-imidazol-5-yl)methyl]-4-(3-chlorophenyl)-1-methyl-2(1*H*)-quinolinone (R115777), a farnesyl transferase inhibitor. *Drug Metab. Dispos.* **30:**823–830.

Gilman, S. D., Gee, S. J., Hammock, B. D., Vogel, J. S., Haack, K., Buchholz, B. A., Freeman, S. P. H. T., Wester, R. C., Hui, X., and Maibach, H. I. (1998). Analytical performance of accelerator mass spectrometry and liquid scintillation counting for detection of 14C-labeled atrazine metabolites in human urine. *Anal. Chem.* **70:**3463–3469.

Kapron, J. T., Pace, E., Van Pelt, C. K., and Henion, J. (2003). Quantitation of midazolam in human plasma by automated chip-based infusion nanoelectrospray tandem mass spectrometry. *Rapid Commun. Mass Spectrom.* **17:**2019–2026.

Kaye, B., Garner, R. C., Mauthe, R. J., Freeman, S. P., and Turteltaub, K. W. (1997). A preliminary evaluation of accelerator mass spectrometry in the biomedical field. *J. Pharm. Biomed. Anal.* **16:**541–543.

Kiffe, M., Jehle, A., and Ruembeli, R. (2003). Combination of high-performance liquid chromatography and microplate scintillation counting for crop and animal metabolism studies: A comparison with classical on-line and thin-layer chromatography radioactivity detection. *Anal. Chem.* **75:**723–730.

Lam, W., Loi, C. M., Atherton, J., Stolle, W., Easter, J., and Mutlib, A. (2007). Application of in-line liquid chromatography–accurate radioisotope counting-mass spectrometry (LC–ARC–MS) to evaluate metabolic profile of [³H]-Mefanamic acid in rat plasma. *Drug Metab. Lett.* **1:**179–188.

Lappin, G., and Garner, R. C. (2003). Big physics, small doses: The use of AMS and PET in human microdosing of development drugs. *Nat. Rev. Drug Discov.* **2:**233–240.

Lappin, G., and Garner, R. C. (2005). The use of accelerator mass spectrometry to obtain early human ADME/PK data. *Expert Opin. Drug Metab. Toxicol.* **1:**23–31.

Lappin, G., Kuhnz, W., Jochemsen, R., Kneer, J., Chaudhary, A., Oosterhuis, B., Drijfhout, W. J., Rowland, M., and Garner, R. C. (2006a). Use of microdosing to predict pharmacokinetics at the therapeutic dose: Experience with 5 drugs. *Clin. Pharmacol. Ther.* **80:**203–215.

Lappin, G., Rowland, M., and Garner, R. C. (2006b). The use of isotopes in the determination of absolute bioavailability of drugs in humans. *Expert Opin. Drug Metab. Toxicol.* **2:**419–427.

Lee, D. Y., Anderson, J. J., and Ryan, D. L. (2000). LC–ARC: A novel sensitive in-line detection system/method for radio-HPLC. Paper presented at the 7th International Symposium of IIS, Dresden, Germany.

Leuthold, L. A., Grivet, C., Allen, M., Baumert, M., and Hopfgartner, G. (2004). Simultaneous selected reaction monitoring, MS/MS and MS3 quantitation for the analysis of pharmaceutical compounds in human plasma using chip-based infusion. *Rapid Commun. Mass Spectrom.* **18:**1995–2000.

Liberman, R. G., Tannenbaum, S. R., Hughey, B. J., Shefer, R. E., Klinkowstein, R. E., Prakash, C., Harriman, S. P., and Skipper, P. L. (2004). An interface for direct analysis of (14)c in nonvolatile samples by accelerator mass spectrometry. *Anal. Chem.* **76:**328–334.

Litherland, A. E. (1980). Ultrasensitive mass spectrometry with accelerators. *Ann. Rev. Nucl. Part Sci.* **30:**437–473.

Lu, W., Yu, C. P., and Lee, D. Y. (2002). A novel hyphenated LC/ARC/RD/MS/FC system for identification of drug metabolites. In *Proceedings of the 50th ASMS Conference on Mass Spectrometry and Allied Topics*, Orlando, FL.

Nassar, A. E., Bjorge, S. M., and Lee, D. Y. (2003). On-line liquid chromatography-accurate radioisotope counting coupled with a radioactivity detector and mass spectrometer for metabolite identification in drug discovery and development. *Anal. Chem.* **75:**785–790.

Nassar, A. E., Martine, M., Parmentier, Y., and Lee, D. Y. (2003). Comparison between liquid chromatography–accurate radioisotope counting and microplate scintillation counter technologies in drug metabolism studies. *Drug Metab. Rev.* **35**(Suppl. 2):79.

Nassar, A. E., and Lee, D. Y. (2007a). Novel approach to performing metabolite identification in drug metabolism. *J. Chromatogr. Sci.* **45**:113–119.

Nassar, A. E., and Lee, D. Y. (2007b). Novel radio-HPLC detector for sensitive metabolite profiling and structural elucidation in support of drug metabolism studies. *Drug Metab. Rev.* **39**(Suppl. 1):73.

Nedderman, A. N., Savage, M. E., White, K. L., and Walker, D. K. (2004). The use of 96-well Scintiplates to facilitate definitive metabolism studies for drug candidates. *J. Pharm. Biomed. Anal.* **34**:607–617.

Nelson, D. E., Korteling, R. G., and Scott, W. R. (1977). Carbon-14: Direct detection at natural concentrations. *Science* **198**:507–508.

Oberhauser, C. J., Niggebrugge, A. E., Lachance, J. D., Takarewski, J., and Quinn, H. M. (2000). Directions in discovery: Turbulent-flow LC for LC-MS and LC-MS-MS bioanalysis. *LC-GC* **18**(7):716–724.

Ognibene, T. J., Bench, G., Vogel, J. S., Peaslee, G. F., and Murov, S. (2003). A high-throughput method for the conversion of CO_2 obtained from biochemical samples to graphite in septa-sealed vials for quantification of 14C via accelerator mass spectrometry. *Anal. Chem.* **75**:2192–2196.

Ognibene, T. J., and Vogel, J. S. (2004). Highly sensitive 14C and 3H quantification of biochemical samples using accelerator mass spectrometry. In *Synthesis and Applications of Isotopically Labelled Compounds* (Dean, D. C., Filer, C. N., and McCarthy, K. E., Eds.). Wiley, Hoboken, NJ, pp. 293–295.

Plumb, R., Castro-Perez, J., Granger, J., Beattie, I., Joncour, K., and Wright, A. (2004). Ultra-performance liquid chromatography coupled to quadrupole-orthogonal time-of-flight mass spectrometry. *Rapid Commun. Mass. Spectrom.* **18**:2331–2337.

Sarapa, N., Hsyu, P. H., Lappin, G., and Garner, R. C. (2005). The application of accelerator mass spectrometry to absolute bioavailability studies in humans: simultaneous administration of an intravenous microdose of 14C-nelfinavir mesylate solution and oral nelfinavir to healthy volunteers. *J. Clin. Pharmacol.* **45**:1198–1205.

Schmidt, A., Karas, M., and Dulcks, T. (2003). Effect of different solution flow rates on analyte ion signals in nano-ESI MS, or: when does ESI turn into nano-ESI? *J. Am. Soc. Mass Spectrom.* **14**:492–500.

Smith, D. A., Johnson, D. E., and Park, B. K. (2003). Editorial overview: use of microdosing to probe pharmacokinetics in humans—Is it too much for too little? *Curr. Opin. Drug Discov. Devel.* **6**:39–40.

Souverain, S., Rudaz, S., and Veuthey, J. L. (2004). Matrix effect in LC-ESI-MS and LC-APCI-MS with off-line and on-line extraction procedures. *J. Chromatogr. A.* **1058**:61–66.

Veuthey, J. L., Souverain, S., and Rudaz, S. (2004). Column-switching procedures for the fast analysis of drugs in biologic samples. *Ther. Drug. Monit.* **26**:161–166.

Vogel, J. S. (1992). Rapid production of graphite without contamination for biomedical AMS. *Radiocarbon* **34**:344–350.

Yin, H., Greenberg, G. E., and Fischer, V. (2000). Application of Wallac Microbeta radio-activity counter and Wallac Scintiplate in metabolite profiling and identification studies. Application note. PerkinElmer Life Sciences, accessed at http://www.perkinelmer.com.

Ynddal, L., and Hansen, S. H. (2003). On-line turbulent-flow chromatography-high-performance liquid chromatography-mass spectrometry for fast sample preparation and quantitation. *J. Chromatogr. A.* **1020:**59–67.

Young, G., Ellis, W., Ayrton, J., Hussey, E., and Adamkiewicz, B. (2001). Accelerator mass spectrometry (AMS): Recent experience of its use in a clinical study and the potential future of the technique. *Xenobiotica* **31:**619–632.

Zhang, D., Wang, L., Raghavan, N., Zhang, H., Li, W., Cheng, P. T., Yao, M., Zhang, L., Zhu, M., Bonacorsi, S., Yeola, S., Mitroka, J., Hariharan, N., Hosagrahara, V., Chandrasena, G., Shyu, W. C., and Humphreys, W. G. (2007). Comparative metabolism of radiolabeled mur-aglitazar in animals and humans by quantitative and qualitative metabolite profiling. *Drug Metab. Dispos.* **35:**150–167.

Zhao, W., Wang, L., Zhang, D., and Zhu, M. (2004). Rapid and sensitive determination of enzyme kinetics of drug metabolism using HPLC coupled with a stop-flow radioactivity flow detector. *Drug Metab. Rev.* **36:**257.

Zhu, M., Zhang, D., and Skiles, G. L. (2005). Quantification and structural elucidation of low quantities of radiolabeled metabolites using microplate scintillation counting techniques in conjunction with LC-MS. In: *Identification and Quantification of Drugs, Metabolites and Metabolizing Enzymes by LC-MS* (Chowdhury, S. K., Ed.). Elsevier, Amsterdam, pp. 195–223.

8

Online Electrochemical– LC–MS Techniques for Profiling and Characterizing Metabolites and Degradants

Paul H. Gamache, David F. Meyer, Michael C. Granger, and Ian N. Acworth

ESA Biosciences, Chelmsford, Massachusetts

8.1   INTRODUCTION

There has been a steady increase over the last decade in the number of publications that describe online combination of electrochemical (EC) techniques with liquid chromatography–mass spectrometry (EC–LC–MS) (Kertesz and Van Berkel, 2004; King et al., 2004; Van Berkel, 2004; Carr et al., 2006; Nozaki et al., 2006; Madsen et al., 2007). These techniques have been shown to possess unique capabilities to both utilize and study oxidative and reductive (redox) reactions. EC techniques provide analytical utility (e.g., LC detection) as well as insight that can be beneficial to drug discovery and development processes. This latter capability is largely based on the relevance of redox processes to many chemical and biochemical aspects of drug development, including chemical oxidative degradation (Waterman et al., 2002), oxidative metabolism and formation of reactive species (Hill et al., 1993; Eyer, 1994), and the relevance of oxidative stress in disease and toxicity. The incorporation of rugged commercially available EC cells within existing LC–MS systems is typically inexpensive and readily implemented. These and other factors have led to an increased use of these techniques at strategic times within the drug development process.

This chapter is an overview of several EC–LC–MS approaches, specifically those that employ three-electrode EC flow cells for controlled potential electrolysis. The basic theory and applications, including those that use EC cells as reaction devices in series with MS and those involving parallel EC and MS for high-performance liquid chromatography (HPLC) detection, are discussed. Serial EC–MS applications include generating and characterizing products that often correspond to metabolites, degradants, or short-lived species of interest. The quantitative and qualitative aspects of parallel EC-Array–MS are also described for studying chemical oxidative stability, for profiling metabolites and degradants, and for metabolomics. To facilitate expanded adoption of EC–LC–MS techniques, emphasis will be placed on basic methodology and practical considerations.

8.2   EC OVERVIEW

Common uses of three-electrode EC flow cells include HPLC detection (LC–EC) (Acworth and Gamache, 1996; Riis, 2002; Gonzalez de la Huebra et al., 2003; Sabbioni et al., 2004) and hydrodynamic voltammetry (Nagels et al., 1989). There are many excellent sources of electrochemical information for detailed description of theory and applications (Lund and Baizer, 1991; Chen et al., 1996; Kissinger and Heineman, 1996; Lacourse, 2001). Briefly, the cell functions by establishing a specific applied potential between a reference (RE) and working electrode (WE). The auxiliary electrode (AE), through electrical feedback via a potentiostat, provides the energy necessary to maintain the potential and relies on solution conductivity (typically mobile phase with ≥ 20 mM buffer) to complete the circuit. This energy drives electrolysis of solution-phase species (i.e., analyte, reactant) at the WE surface. The majority of the applications described below involves recording

EC cell current and associated mass spectra as a function of and as a result of the applied potential.

EC cells with carbon-based WE are the most widely used for LC–EC and, as such, are the primary focus of this discussion. Other WE materials (e.g., Au, Ag, and Pt), while more advantageous for certain applications (Rocklin, 1984; Bowers, 1991), are typically more prone to surface passivation and are generally less practical for many solvent and pH conditions (Rocklin, 1984). There are several possible EC flow cell designs, but only three basic geometries are in widespread use: thin layer, wall jet and, porous flow-through. Thin-layer and wall-jet amperometric cells have small surface area WEs and, when using typical HPLC flow rates (i.e., 0.2–2.0 mL/min), only a small percentage (typically <5%) of analyte comes into close enough proximity to the WE for either oxidation or reduction. Response is therefore affected by changes in flow rate, and the overall yield of reaction products is typically low. Also, electrode maintenance is frequently required due to WE passivation, particularly when analyzing complex matrices or a large number of samples (Catarino et al., 2003). A third class of EC flow cell utilizes high-surface-area microporous WEs to achieve higher electrolysis efficiencies (typically >95% at 2.0 mL/min). This design is termed coulometric since the integrated response (peak area) represents the charge realized (coulombs) from nearly complete electrolysis of analyte as described by Faraday's law. A disadvantage to the coulometric cell design is that the WE is not accessible for mechanical resurfacing. However, in comparison to thin-layer and wall-jet cells, coulometric cells provide higher yield and more reproducible reactions (i.e., response) and require much less maintenance (Riis, 2002; Gonzalez de la Huebra et al., 2003; Sabbioni et al., 2004). These advantages have led to widespread use of coulometric cells for analysis of complex matrices and for higher throughput approaches including those that utilize gradient elution HPLC. Further discussion will focus on two basic implementations of coulometric EC: (1) a single cell upstream and in series with MS and (2) multiple cells (EC-Array) in parallel with MS.

8.3 SERIAL EC–LC–MS

8.3.1 General Considerations

Several laboratories have used single EC cells as a simple "add-on" to a typical LC–MS setup. Common experimental configurations (Fig. 8.1) include flow injection, direct infusion, and both pre- and postcolumn EC–MS. Several design aspects must be considered when adding an EC cell to an LC–MS system, such as solution conductivity, electrical grounding, and cell volume. Supporting electrolyte is required to provide solution conductivity. Common LC–MS mobile phases containing, for example, 0.1% formic acid or 5–10 mM ammonium acetate may have insufficient conductivity and the resulting IR drop must be compensated through the use of higher applied potentials. While this may be acceptable for some experiments, the use of more conductive solutions is generally preferred. This is particularly

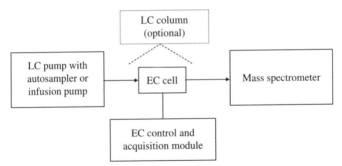

Figure 8.1. *Representative serial EC–LC–MS configurations using either ESA model 5021A or 5030 coulometric EC cell controlled with Coulochem III for MS detector (ESA Biosciences, Chelmsford, MA).*

important for LC–EC detection and for generating hydrodynamic voltammograms (HDVs) where low solution conductivity may lead to poor response, peak tailing, and loss of voltammetric resolution. Volatile buffers (e.g., ammonium acetate or ammonium formate) of at least 20 mM concentration typically provide sufficient conductivity. While higher buffering capacity may provide improved control of chromatography and detection, it is important to consider potential adverse effects on MS detection (e.g., ionization suppression, adduct formation, background noise, etc.).

A further consideration is related to the use of conductive fluids and high-voltage ion sources. The liquid inlet to the ion source (e.g., electrospray) of most LC–MS instruments is electrically grounded. However, in some instruments there exists the possibility that current flow may occur from the high-voltage ion source through a conductive fluid. Since many LC effluents are somewhat conductive, grounding the fluid line is typically recommended, irrespective of the use of EC. Current flow through the fluid line may compromise analyte integrity since it can affect the interfacial potential of distal, upstream, wetted components (e.g., injector and analytical column), which can lead to unwanted redox reactions (Liu et al., 2003). A simple way to ground the fluid line is to use a stainless steel fluidic union connected to the ground of the MS high-voltage power supply as previously described (Zhou and Van Berkel, 1995).

8.3.2 Applications

Serial EC–LC–MS configurations are used in drug discovery and development to study the susceptibility of a compound to oxidation (or reduction), the nature of products and short-lived intermediates, and the associated reaction pathways. These data are used to provide insight to chemical stability, metabolism, and toxicity in a variety of contexts from rapid, early-stage screening to more in-depth studies. A common approach is to use flow injection analysis (FIA) for medium-throughput analysis of a series of compounds (Gamache et al., 2003b). Each compound is typically analyzed at several EC potentials (e.g., 0, 400, 800, 1200 mV vs. Pd) to quickly (e.g., <30 s/injection) generate voltammetric data which are indicative of relative ease of EC oxidation. The corresponding MS data are then used to provide information on the likely chemical sites of oxidation and the nature of products. These data are typically generated for analogous series to provide input to lead optimization strategies or to provide structural alerts for oxidatively unstable compounds. One should note that instantaneous voltammetric data may also be generated from a single injection by using several EC cells arranged in series, each successive cell maintained at a higher potential than the preceding cell. This technique, typically conducted in early-stage discovery for studying oxidative stability, is discussed in Section 8.4.

A logical extension of FIA with serial EC–LC–MS is to incorporate a separation technique to further characterize the products. Figure 8.2 is an example of precolumn EC oxidation of tamoxifen. With the EC cell at 500 mV, the total ion chromatogram (TIC) shows two primary peaks corresponding to starting material (tamoxifen) and, at an earlier retention time, an N-demethylated product. With the precolumn EC cell at 1000 mV (Fig. 8.3), additional peaks are evident, including possible aromatic

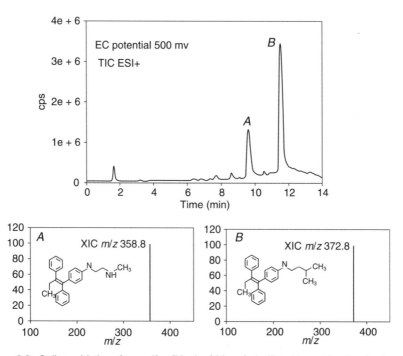

Figure 8.2. Online oxidation of tamoxifen (50 μL of 20 μg/mL diluted in mobile phase) using precolumn ESA model 5021 EC cell (500 mV vs. Pd). Binary gradient from 16 to 80% acetonitrile over 10 min, 20 mM ammonium acetate as supporting electrolyte. Shiseido C18 MG 3μ 75 × 4.6 mm i.d. HPLC column; HPLC flow: 1 mL/min, 200 μL/min split to MS. Mass spectrometer: Agilent 1100 single quadrupole; ionization mode: positive electrospray ionization (ESI); scan range: 80–500 Da; Fragmentor: 70 V; gain: 1.0; threshold: 150; step: 0.25; drying gas flow: 12 L/min; nebulizer pressure: 35 psi; drying gas temperature: 350°C; capillary voltage: 3500 V.

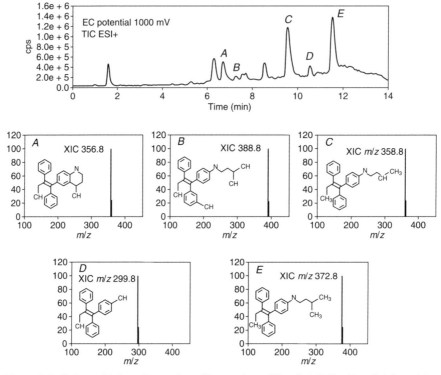

Figure 8.3. *Online oxidation of tamoxifen with precolumn EC cell at 1000 mV vs. Pd. Remaining conditions as in Fig. 8.2.*

hydroxylation and dealkylation products. This example shows that product formation is, to some extent, controlled by simple control of EC cell potential and that several of these EC oxidation products correspond to biological metabolites or degradants of tamoxifen (Desta et al., 2004).

There are many examples, as cited above, which have demonstrated that the EC oxidation products for a given compound often coincide with metabolites observed in biological assays (Volk et al., 1988, 1989; Getek et al., 1989; Deng and Van Berkel, 1999; Jurva et al., 2000; van Leeuwen et al., 2005). These and other EC oxidation products may also be observed in purposeful degradation or stability studies. One must recognize, however, that enzyme-catalyzed, chemical and EC oxidations are often very complex and involve different phenomena.

Studies conducted by Jurva and colleagues (2003) provide very useful insight to the mechanistic aspects of EC and enzymatic oxidative reaction pathways, particularly in the context of cytochrome P450 (CYP450) metabolism. These studies have shown that EC oxidation using porous carbon WE generally leads to the formation of similar products for those enzyme-catalyzed reactions that are supposed to proceed through a mechanism initiated by a one-electron transfer oxidation. Examples of CYP450 reactions mimicked by carbon-based WE include dehydrogenation,

N-deacetylation, N-dealkylation, S-oxidation, and aromatic O-dealkylation. Reactions initiated by H-atom abstraction, such as aliphatic C-oxidation and hydroxylation of aromatic rings without electron-donating groups, are, however, generally not mimicked by EC oxidation. There are several examples of alternative or modified WE materials and reaction conditions to effect specific reactions. This includes the immobilization of enzymes (Joseph et al., 2003) and biomimetic redox indicators (Shiryaeva et al., 2003) on solid electrodes and solution-phase conditions such as those described for EC-assisted Fenton reactions (Jurva et al., 2002). While these approaches are extremely useful for many applications, EC is generally considered complementary to, rather than a substitute for, a given in vitro assay. A similar argument applies to chemical oxidative degradation studies, which are discussed in Section 8.4.2.

The example in Figs 8.2 and 8.3 illustrates online generation and analysis of EC products using LC–MS conditions that are typical of drug metabolism studies (e.g., in vitro microsomal analysis). Serial EC–LC–MS can thus be used with "neat" parent compound solutions for preliminary optimization of LC and MS conditions for subsequent metabolite analysis in biological samples. By using identical conditions, EC data are then used as input to automated metabolite identification software to aid in finding metabolites present in more complex biological matrices (Kieser et al., 2004). Furthermore, when the data from an EC-generated product correspond to that of a biological metabolite, the EC technique is viewed as a selective and rapid synthetic route to small quantities of this metabolite. This technique is then used to facilitate structural elucidation or, more generally, to produce additional chemical entities. Recent studies have demonstrated the feasibility of scaling up online EC techniques to produce sufficient quantities for structural confirmation by nuclear magnetic resonance (NMR) (Gamache et al., 2004a). The simplicity and speed of online EC–LC–MS may thus provide an effective means of characterizing some of the many unknown metabolites encountered in multivariate profiling studies.

Many studies suggest that redox metabolism of a wide range of chemical structures leads to formation of reactive electrophiles which participate in a diverse array of toxic processes that typically involve covalent binding or other modifications to small and large molecules (e.g., DNA, proteins, peptides, lipids), redox cycling, antioxidant/scavenger depletion, and other elements of oxidative stress (Hill et al., 1993; Eyer, 1994; Halliwell and Gutteridge, 1999; Dieckhaus et al., 2002). The propensity of compounds to undergo redox-based metabolic activation to form reactive electrophilic species is therefore a major consideration in pharmaceutical development (Evans and Baillie, 2005; Evans et al., 2004). Several reports have shown that EC–LC–MS is useful in the study of reactive intermediate metabolites (Getek et al., 1989; Deng and Van Berkel, 1999; Gamache et al., 2003b; Kertesz and Van Berkel, 2004; King et al., 2004; van Leeuwen et al., 2005; Carr et al., 2006). A variety of experimental configurations have been described that incorporate a nucleophilic trapping agent (typically glutathione, GSH), by either coinjection or infusion, with FIA and pre- or postcolumn EC. FIA and precolumn EC have been used to analyze neat drug solutions in order to investigate the susceptibility of a parent drug to EC oxidation, to characterize the reactive intermediates formed, and to study the chromatographic and MS behavior of trapped species. Postcolumn

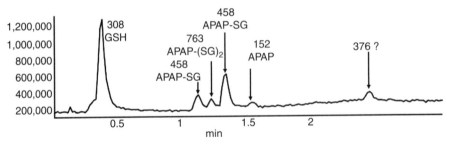

Figure 8.4. *Precolumn oxidation of APAP with GSH as trapping agent: 10 μL of 20 μg/mL APAP, 1 mM GSH mixture with precolumn ESA model 5021 cell at 500 mV vs. Pd. Binary gradient elution from 1 to 80% acetonitrile in 5 min, 20 mM ammonium acetate as supporting electrolyte. Shiseido C18 MG 3μ 50 × 4.6 mm i.d. HPLC column; HPLC flow: 1 mL/min, 200 μL/min split to MS. MS conditions as in Fig. 8.3 except the scan range was changed to 80–1000 Da.*

configurations have also been used to analyze biological (microsomal) preparations to investigate individual metabolite susceptibility to EC oxidation and to identify which metabolites form reactive species.

Figure 8.4 provides a basic example of precolumn EC oxidation using the widely studied compound acetaminophen (APAP). Oxidative metabolic activation of APAP to form *N*-acetyl-*p*-benzoquinoneimine (NAPQI) is widely regarded as an essential component of its hepatotoxic effects in humans (Eyer, 1994). MS data indicate that EC oxidation of APAP in the presence of GSH (coinjected) resulted in two separate peaks corresponding to monoglutathionyl conjugates and one peak indicative of a di-glutathionyl conjugate. These results are comparable to those reported nearly two decades ago by Getek and colleagues (1989) using coulometric EC cells with thermospray MS.

Several other "model" compounds also have been studied using the EC techniques, including 4-aminophenol, BHT, estradiol and metabolites, phenacetin (Gamache et al., 2003b), and clozapine (van Leeuwen et al., 2005). While the products expected from the known systems are generally formed and can be interrogated to reveal structural information, the electrochemical oxidation sometimes results in a wide array of oxidation products not always seen in biological systems (King et al., 2004). Recognizing that both nonmicrosomal and nonenzymatic redox reactions may play a role in toxic processes (Smith, 2003), further study is required to assess the significance of these observations. These data do highlight the usefulness of EC–LC–MS as a rapid and efficient means of generating trapped products of reactive intermediates for structural characterization by LC–MS/MS from both parent drug and in vitro metabolites.

Figure 8.5 illustrates the precolumn oxidation of a phenacetin solution with LC–EC-Array detection (discussed in more detail in the following section). A peak, which eluted at 3.8 min, shows a characteristic voltammetric profile (i.e., reduction followed by oxidation) of a quinone species. Based on EC-Array and MS data (not shown), this peak has been identified as NAPQI, the expected reactive intermediate. This peak was not evident in a microsomal incubate of phenacetin analyzed using the same conditions (not shown). A possible explanation for this is that

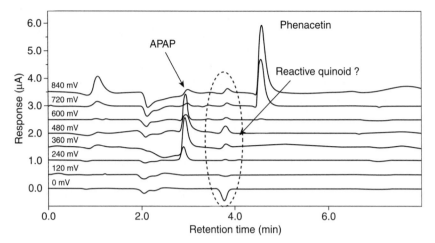

Figure 8.5. *Precolumn oxidation of phenacetin. Conditions as in Figure 8.2 except LC detector was an eight-channel EC-Array (CoulArray, ESA Biosciences, Chelmsford, MA) with model 6210 EC cell potentials of 0–840 mV vs. Pd in 120-mV increments.*

nonspecific binding of this reactive species occurred in the biological preparation. Thus, an additional advantage to the use of online EC–LC–MS includes the ability to examine short-lived species that, while potentially relevant, may not as readily be observed in biological systems. Many other pharmaceutically relevant applications of EC–MS have been described that are beyond the scope of this discussion. These include EC derivatization to enhance MS ionization, mass tagging, and protein cleavage techniques, several of which are discussed in a recent focus issue on electrochemistry combined with MS (Van Berkel, 2004).

8.4 EC-ARRAY

8.4.1 General Considerations

The general concepts of EC-Array are described in detail elsewhere (Acworth and Gamache, 1996). Briefly this technique employs up to 16 series EC cells (Fig. 8.6) with porous graphitic carbon-based WEs, similar in concept to those described in

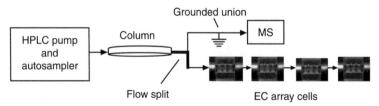

Figure 8.6. *Representative configuration of LC with postcolumn flow passively split to EC-Array and MS. Model 6210 EC cells were controlled with a 16-channel CoulArray (ESA Biosciences, Chelmsford, MA). (Reprinted with permission from Gamache et al., 2004b.)*

the previous section. Each cell is typically poised at a different fixed potential, thereby spanning a wide potential window to allow detection of a broad scope of redox active analytes during the transit of a single injected aliquot through the cells. Efficient electrolysis obtained with high-surface-area coulometric WEs allows selective detection and resolution of coeluting analytes, based on differences in the relative ease of oxidation and/or reduction. Stated differently, each analyte, within a coeluting band, will demonstrate signal dominance on a different WE based on their relative ease of oxidation or reduction.

In general, EC reactions are typically observed according to the following general rank order (by relative ease of oxidation): *o,p*-quinol and *o,p*-aminophenol > tertiary amine > *m*-quinol ≈ phenol ≈ arylamine > secondary amine ≈ thiol > thioether ≠ primary amines, aliphatic alcohols. (HDVs) each redox active metabolite are obtained from the response across adjacent EC-Array sensors. These data are a reflection of the kinetic and thermodynamic components of electron transfer reactions. Since chemical structure is a critical determinant of an analyte's redox behavior, the intrinsic generation of an HDV with EC-Array provides qualitative information for each species.

8.4.2 Applications

The use of FIA with a single EC cell in series with MS was described above to study compound susceptibility to oxidation. This requires multiple injections for each compound. An alternative approach is to use EC-Array (without MS) to generate voltammetric data from a single injection. EC-Array has the advantage of providing higher throughput but lacks the structure–activity and product information provided by MS. MS is not used downstream of EC-Array because the products that are selectively formed at upstream electrodes may further react at downstream electrodes and elude detection by MS. Many laboratories use EC-Array to generate voltammetric data to study oxidative stability, largely based on the pioneering work of Lombardo and Campos (2004). Chemical oxidation, a common mode of degradation for active compounds and drug products (Waterman et al., 2002), is a significant concern at all stages of drug discovery and development. The relative tendency for compounds to undergo chemical oxidative degradation via electron transfer mechanisms is closely related to their EC redox potentials (Waterman et al., 2002). As described in the previous section, oxidation reactions are often complex phenomena with a variety of mechanisms not necessarily modeled by a given EC technique. EC is typically used as part of a suite of techniques (e.g., oxygen/radical initiator, photolytic) for comprehensive study.

Representative data from FIA with 16-channel EC-Array detection is shown in Fig. 8.7. Using the described conditions, voltammetric data obtained for several model compounds demonstrated that the most stable compounds oxidized at the highest potentials while the least stable compounds oxidized at the lowest potentials (Gamache et al., 2003a). While FIA was suitable for this pilot study, the approach was found to be inadequate for high-throughput library stage screening, particularly when considering the possible presence of electroactive impurities. To overcome the

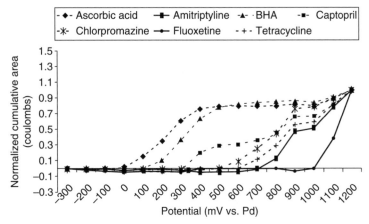

Figure 8.7. *Voltammetric plots for seven model compounds representing normalized cumulative peak area. Dashed lines correspond to those compounds reported as oxidatively unstable; solid lines are indicative of oxidative stability.*

limitation, Lombardo and Campos (2004) developed a method that used a short (3-cm) C18 column and isocratic elution in a strong (90% organic) eluent. This technique, together with the inherent selectivity of EC-Array and the appropriate choice of signal threshold, has allowed the generation of voltammetric data with a typical throughput of eight 96-well plates per week per instrument. This work, now applied to more than 30,000 compounds, was used to develop alerts for oxidatively unstable compounds, thus reducing the potential for later stage issues. Furthermore, there are several additional potential uses of this approach that are currently under investigation. These include the ability to assess the potential for degradation upon library storage, to develop predictive structure–stability relationships, and to examine excipient compatibility.

The more traditional use of EC-Array is with LC separation for multicomponent quantitative and qualitative analysis. The primary advantages to this technique include two-dimensional (i.e., chromatographic and voltammetric) resolution, femtomole sensitivity, and data-dependent acquisition (autoranging) which facilitates use with gradient elution and provides a 10^5 dynamic response range (Ferruzzi et al., 1998). LC–EC-Array has been widely used for routine analysis of redox active substances with primary in vivo application to clinical chemistry (Cheng et al., 1991; Gamache et al., 1993), neurochemistry (Volicer et al., 1985; Beal et al., 1992; LeWitt et al., 1992), and redox biochemistry (Sofic et al., 1992; Hensley et al., 1997, 2000; Collins et al., 1998; Acworth et al., 1999; Yanagawa et al., 2001; Christen et al., 2002). Many pharmaceutical laboratories have adopted EC-Array for LC analysis of degradants and metabolites (i.e., related substances). A majority [estimated >85% (Jane et al., 1985)] of pharmacologically active small molecules possess one or more "EC-active" moieties (see above). The sensitivity (typically 10- to 1000-fold greater than UV–Vis absorbance detection) (Ferruzzi et al., 1998), selectivity, and qualitative information obtained can provide significant advantages for a variety of applications. There are many examples of the use of EC-Array specifically for bioavailability and pharmacokinetic studies (Achilli et al., 1996; Chow et al., 2001; Graefe et al., 2001; Roy et al., 2002, Muldrew, 2002; Penalvo et al., 2004; Zhang et al., 2004; Wittemer et al., 2005). The coupling of EC-Array in parallel with MS is a logical extension of these approaches and is discussed below.

A major aspect of recent studies has been to couple EC-Array detection in parallel with MS for metabolomic studies (Gamache et al., 2004b). The comprehensive study of small molecules in living systems, metabolomics, has several challenges, including analyte number and diversity, range of concentrations, and complexity of sample matrices. A variety of techniques exists that are effective in studying certain classes of molecules within well-delineated concentration ranges; however, no one particular technique can overcome all of the field's inherent challenges. For instance, NMR is able to detect any molecule that contains an active nuclide (e.g., ^{1}H and ^{13}C); however, its detection is limited to microgram quantities and does not apply to certain functional groups such as amines and sulfates. MS is very versatile but relies on the ionizability of the compound. EC techniques, specifically EC-Array, are extremely sensitive detectors of redox active compounds—compounds that, in many instances, are not observed with MS. The parallel use of EC and MS provides the capability to obtain more information from a given metabolite and to extend the number of metabolites and range of chemical classes that can be effectively studied.

A representative chromatogram of rat urine analyzed by parallel EC-Array–MS is shown in Fig. 8.8. As expected, the MS base peak chromatogram shows relatively few, directly visible metabolite peaks. Mass spectral data from full-scan exploratory studies are typically processed by extracting discrete signals each defined by a particular retention time and m/z and using algorithms to help distinguish analytical signal from background noise (Plumb et al., 2003). The resulting data, often consisting of hundreds of discrete signals, are then typically processed by chemometric techniques such as principal-component analysis (PCA). Chromatographic variability, ionization suppression, adduct formation and in-source oxidation with MS, and electrode adsorptive and other non-Faradaic processes with EC are important factors to consider in these multivariate analyses. Current data show that the concurrent acquisition of EC-Array and MS data for each metabolite peak helps to address these potential issues. For example, the observation of a particular redox active metabolite peak allows the analyst to conduct a more informed and targeted interrogation of corresponding MS data. Likewise, specific MS data are useful to normalize for retention time variability observed with both MS and EC-Array data. In addition, the results indicate that many redox active urinary metabolites exist as solution-phase neutral species under a variety of reversed-phase chromatographic conditions. For example, peaks annotated in Fig. 8.8 are ascorbic acid (AA), uric acid (UA), 5-hydroxyindoleacetic acid (5HIAA), and homovanillic acid (HVA). Of these metabolites, only uric acid was detected from extracted ion chromatograms (above baseline noise) as its protonated or adducted molecule [e.g., $M + X$ $(X = H^{+}, Na^{+}, K^{+}, NH_4^{+})$]. The combined detection scope of MS and EC-Array

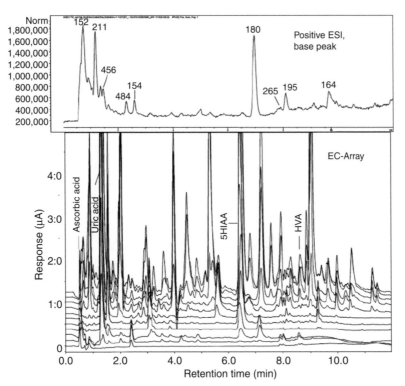

Figure 8.8. *MS base peak chromatogram labeled with base peak m/z (top) and EC-Array multichannel chromatogram (11 of 16 channels shown for clarity). Analytical conditions: 20 µL of 10-fold diluted urine. Gradient elution 1–100% aqueous acetonitrile with 10 mM ammonium formate and 50 mM formic acid; flow rate: 1.5 mL/min; Shiseido C18, 3 mm, 75 mm × 4.6 mm i.d. HPLC column; 4 : 1 passive postcolumn flow split to EC-Array and MS, respectively. EC-Array potentials: 0–1050 mV in increments of 70 mV; ESI: positive mode; capillary voltage: 3500 V, fragmentor voltage: 70 V; scan range: 50–850 Da; scan speed: 1.2 s/cycle. (Reprinted with permission from Gamache et al., 2004b.)*

thus provided higher coverage in a single analysis of the wide dynamic range and broad chemical diversity of urinary metabolites.

In a model study of APAP-induced hepatotoxicity, results from PCA of EC-Array data showed consistent differentiation of high-dose APAP (200 and 300 mg/kg, 0–8 h collection) from control, low-dose (20 mg/kg) APAP, and high-dose (200 mg/kg) acetylsalicylic acid. This differentiation was observed both exclusive and inclusive of xenobiotic metabolite variables. From analysis of eigenvector loadings, possible endogenous (E1–E4) marker peaks and xenobiotic metabolites (M1, M3, M4) are shown in Fig. 8.9. These data further demonstrate the complementary nature of EC and MS detection. For example, MS data, along with prior knowledge of a parent compound's analytical behavior and informed prediction of biotransformations, were used to remove xenobiotic data. This process allowed a more direct study of changes in endogenous metabolite profiles by PCA. Also, EC-Array data showed

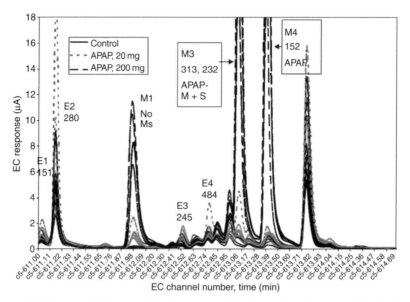

Figure 8.9. *Overlay of EC-Array data from urine following intraperitoneal administration of vehicle, 20 mg/kg, or 300 mg/kg APAP to rats (n = 5 each group). Peaks labeled* E *indicate endogenous metabolites while peaks labeled* M *indicate drug metabolites. Base peak* m/z *ratios as determined from corresponding MS data are shown. APAP-M + S indicates coelution of sulfate and mercapturate metabolites of APAP. (Reprinted with permission from Gamache et al., 2004b.)*

clear evidence of an APAP metabolite that was not detected by MS (peak M1). Furthermore, MS data indicated that peak M3 consisted of two major components having m/z 313 and 232. These m/z values are consistent with the commonly observed O-sulfated metabolite and the less frequently observed mercapturic acid metabolite of APAP, respectively. This observation is supported by EC-Array voltammetric data where the putative O-sulfated metabolite oxidized at a higher potential (840 mV) than parental APAP (600 mV), which can be expected based on phenol substitution. Furthermore, the similarity in voltammetric response between parental APAP and the peak corresponding to m/z 232 suggests that the easily oxidized 4-amidophenol group is intact. These data are consistent with the commonly observed ring thioether metabolites associated with high-dose APAP (Nelson, 1990). Since APAP–mercapturate is a recognized marker of the apparent toxic pathway of APAP, the described approach of exploratory multivariate analysis and targeted interrogation of EC and MS data was an efficient and effective way to determine relevant changes associated with high-dose APAP toxicity. These results demonstrate that the combined use of EC-Array and MS provided useful insight to these metabolic profiling studies by expanding the scope of detection, selectivity, and qualitative information obtained in a single analysis.

8.5 SUMMARY

Electrochemical cells may be coupled with LC–MS in a variety of experimental configurations to provide analytical utility and insight to drug discovery and development. These include the use of EC as reaction devices in series with MS for studies relevant to oxidative metabolism, degradation, and reactive species formation. EC-Array, in parallel with MS, provides the capability to increase the detection scope of LC-based multivariate profiling with high sensitivity, selectivity, wide dynamic range, and complementary qualitative data. The usefulness of these hyphenated techniques and the data generated for drug discovery and development are largely dependent on the challenges of a particular program and should be enhanced by informed strategic implementation.

REFERENCES

Achilli, G., Cellerino, G. P., Melzi d'Eril, G. V., and Tagliaro, F. (1996). Determination of illicit drugs and related substances by high-performance liquid chromatography with an electrochemical coulometric-array detector. *J. Chromatogr. A* **729:**273–277.

Acworth, I. N., Bogdanov, M. B., McCabe, D. R., and Beal, M. F. (1999). Estimation of hydroxyl free radical levels in vivo based on liquid chromatography with electrochemical detection. *Methods Enzymol.* **300:**297–313.

Acworth, I. N., and Gamache, P. H. (1996). The coulometric electrode array for use in HPLC analysis: Part 1: Theory. *Amer. Lab.* **28:**33–38.

Beal, M. F., Matson, W. R., Storey, E., Milbury, P., Ryan, E. A., Ogawa, T., and Bird, E. D. (1992). Kynurenic acid concentrations are reduced in Huntington's disease cerebral cortex. *J. Neurol. Sci.* **108:**80–87.

Bowers, M. L. (1991). A new analytical cell for carbohydrate analysis with a maintenance-free reference electrode. *J. Pharm. Biomed. Anal.* **9:**1133–1137.

Carr, R., Feng, J., Kolodsick, K., Wang, D., Zhong, M., Holliman, C., and Kingsmill, C. A. (2006). Electrochemical assisted electrospray ionization tandem mass spectrometry: Improving the detection limit of electrochemically active species. Poster presented at *54th ASMS Conference on Mass Spectrometry and Allied Topics*, Seattle, WA.

Catarino, R. I., Conceicao, A. C., Garcia, M. B., Goncalves, M. L., Lima, J. L., and dos Santos, M. M. (2003). Flow amperometric determination of pharmaceuticals with on-line electrode surface renewal. *J. Pharm. Biomed. Anal.* **33:**571–580.

Chen, J. G., Woltman, S. J., and Weber, S. G. (1996). Electrochemical detection of biomolecules in liquid chromatography and capillary electrophoresis. *Adv. Chromatogr.* **36:**273–313.

Cheng, M. H., Lipsey, A. I., Lee, J., and Gamache, P. H. (1991). Automated analysis of urinary VMA, HVA, and 5-HIAA by gradient HPLC using an array of eight coulometric electrochemical detectors. *Lab. Robot. Auto.* **4:**297–303.

Chow, H. H., Cai, Y., Alberts, D. S., Hakim, I., Dorr, R., Shahi, F., Crowell, J. A., Yang, C. S., and Hara, Y. (2001). Phase I pharmacokinetic study of tea polyphenols following single-dose administration of epigallocatechin gallate and polyphenon E. *Cancer Epidemiol. Biomarkers Prev.* **10**:53–58.

Christen, S., Jiang, Q., Shigenaga, M. K., and Ames, B. N. (2002). Analysis of plasma tocopherols alpha, gamma, and 5-nitro-gamma in rats with inflammation by HPLC coulometric detection. *J. Lipid Res.* **43**:1978–1985.

Collins, A. R., Gedik, C. M., Olmedilla, B., Southon, S., and Bellizzi, M. (1998). Oxidative DNA damage measured in human lymphocytes: Large differences between sexes and between countries, and correlations with heart disease mortality rates. *FASEB J.* **12**:1397–1400.

Deng, H., and Van Berkel, G. J. (1999). A thin-layer electrochemical flow cell coupled online with electrospray-mass spectrometry for the study of biological redox reactions. *Electroanalysis* **11**:857–865.

Desta, Z., Ward, B. A., Soukhova, N. V., and Flockhart, D. A. (2004). Comprehensive evaluation of tamoxifen sequential biotransformation by the human cytochrome P450 system in vitro: Prominent roles for CYP3A and CYP2D6. *J. Pharmacol. Exp. Ther.* **310**:1062–1075.

Dieckhaus, C. M., Thompson, C. D., Roller, S. G., and Macdonald, T. L. (2002). Mechanisms of idiosyncratic drug reactions: The case of felbamate. *Chem. Biol. Interact.* **142**:99–117.

Evans, D. C., and Baillie, T. A. (2005). Minimizing the potential for metabolic activation as an integral part of drug design. *Curr. Opin. Drug Discov. Devel.* **8**:44–50.

Evans, D. C., Watt, A. P., Nicoll-Griffith, D. A., and Baillie, T. A. (2004). Drug-protein adducts: An industry perspective on minimizing the potential for drug bioactivation in drug discovery and development. *Chem. Res. Toxicol.* **17**:3–16.

Eyer, P. (1994). Reactions of oxidatively activated arylamines with thiols: Reaction mechanisms and biologic implications. An overview. *Environ. Health Perspect.* **102** (Suppl. 6): 123–132.

Ferruzzi, M. G., Sander, L. C., Rock, C. L., and Schwartz, S. J. (1998). Carotenoid determination in biological microsamples using liquid chromatography with a coulometric electrochemical array detector. *Anal. Biochem.* **256**:74–81.

Gamache, P., McCarthy, R., Waraska, J., and Acworth, I. N. (2003a). Pharmaceutical oxidative stability profiling with high-throughput voltammetry. *Am. Lab.* **35**:21–25.

Gamache, P., Smith, R., McCarthy, R., Waraska, J., and Acworth, I. N. (2003b). ADME/Tox profiling using coulometric electrochemistry and electrospray ionization mass spectrometry. *Spectroscopy* **18**:14–21.

Gamache, P., Solomon, M., Acworth, I. N., and Cole, R. (2004a). Rapid on-line electrochemical synthesis of pharmaceutical degradants and metabolites for profiling, identification and quantitation. Poster presented at *Pittcon*, Chicago, IL.

Gamache, P. H., Kingery, M. L., and Acworth, I. N. (1993). Urinary metanephrine and normetanephrine determined without extraction by using liquid chromatography and coulometric array detection. *Clin. Chem.* **39**:1825–1830.

Gamache, P. H., Meyer, D. F., Granger, M. C., and Acworth, I. N. (2004b). Metabolomic applications of electrochemistry/mass spectrometry. *J. Am. Soc. Mass Spectrom.* **15**:1717–1726.

Getek, T. A., Korfmacher, W. A., McRae, T. A., and Hinson, J. A. (1989). Utility of solution electrochemistry mass spectrometry for investigating the formation and detection of biologically important conjugates of acetaminophen. *J. Chromatogr.* **474**:245–256.

Gonzalez de la Huebra, M. J., Bordin, G., and Rodriguez, A. R. (2003). Comparative study of coulometric and amperometric detection for the determination of macrolides in human urine using high-performance liquid chromatography. *Anal. Bioanal. Chem.* **375**:1031–1037.

Graefe, E. U., Wittig, J., Mueller, S., Riethling, A. K., Uehleke, B., Drewelow, B., Pforte, H., Jacobasch, G., Derendorf, H., and Veit, M. (2001). Pharmacokinetics and bioavailability of quercetin glycosides in humans. *J. Clin. Pharmacol.* **41**:492–499.

Halliwell, B., and Gutteridge, J. M. C. (1999). *Free Radicals in Biology and Medicine*. Oxford University Press, New York.

Hensley, K., Maidt, M. L., Pye, Q. N., Stewart, C. A., Wack, M., Tabatabaie, T., and Floyd, R. A. (1997). Quantitation of protein-bound 3-nitrotyrosine and 3,4-dihydroxy-phenylalanine by high-performance liquid chromatography with electrochemical array detection. *Anal. Biochem.* **251**:187–195.

Hensley, K., Williamson, K. S., and Floyd, R. A. (2000). Measurement of 3-nitrotyrosine and 5-nitro-gamma-tocopherol by high-performance liquid chromatography with electrochemical detection. *Free Radic. Biol. Med.* **28**:520–528.

Hill, B. A., Kleiner, H. E., Ryan, E. A., Dulik, D. M., Monks, T. J., and Lau, S. S. (1993). Identification of multi-S-substituted conjugates of hydroquinone by HPLC-coulometric electrode array analysis and mass spectroscopy. *Chem. Res. Toxicol.* **6**:459–469.

Jane, I., McKinnon, A., and Flanagan, R. J. (1985). High-performance liquid chromatographic analysis of basic drugs on silica columns using non-aqueous ionic eluents. II. Application of UV, fluorescence and electrochemical oxidation detection. *J. Chromatogr.* **323**:191–225.

Joseph, S., Rusling, J. F., Lvov, Y. M., Friedberg, T., and Fuhr, U. (2003). An amperometric biosensor with human CYP3A4 as a novel drug screening tool. *Biochem. Pharmacol.* **65**:1817–1826.

Jurva, U., Wikstrom, H. V., and Bruins, A. P. (2000). In vitro mimicry of metabolic oxidation reactions by electrochemistry/mass spectrometry. *Rapid Commun. Mass Spectrom.* **14**:529–533.

Jurva, U., Wikstrom, H. V., and Bruins, A. P. (2002). Electrochemically assisted Fenton reaction: Reaction of hydroxyl radicals with xenobiotics followed by on-line analysis with high-performance liquid chromatography/tandem mass spectrometry. *Rapid Commun. Mass Spectrom.* **16**:1934–1940.

Jurva, U., Wikstrom, H. V., Weidolf, L., and Bruins, A. P. (2003). Comparison between electrochemistry/mass spectrometry and cytochrome P450 catalyzed oxidation reactions. *Rapid Commun. Mass Spectrom.* **17**:800–810.

Kertesz, V., and Van Berkel, G. J. (2004). Investigation of reserpine oxidation using on-line electrochemistry/electrospray mass spectrometry. Poster presented at *52nd ASMS Conference on Mass Spectrometry and Allied Topics*, Nashville, TN.

Kieser, B., Impey, G., Meyer, D. F., Caraiman, D., and Gamache, P. (2004). Metabolite profiling utilizing an in-line electrochemical system for LC/MS. Poster presented at *52nd ASMS Conference on Mass Spectrometry and Allied Topics*, Nashville, TN.

King, R., Dieckhaus, C., Nitkowski, N., and Gamache, P. H. (2004). On-line electrochemical oxidation used with HPLC-MS for the study of reactive drug intermediates. In *Proceedings of the 52nd ASMS Conference on Mass Spectrometry and Allied Topics*, Nashville, TN.

Kissinger, P. T., and Heineman, W. R. (1996). *Laboratory Techniques in Electroanalytical Chemistry*. Marcel Dekker, New York.

Lacourse, W. R. (2001). Electrochemical detectors: Functional group analysis. *Enantiomer* **6**:141–152.

LeWitt, P. A., Galloway, M. P., Matson, W., Milbury, P., McDermott, M., Srivastava, D. K., and Oakes, D. (1992). Markers of dopamine metabolism in Parkinson's disease. The Parkinson Study Group. *Neurology* **42**:2111–2117.

Liu, S., Griffiths, W. J., and Sjovall, J. (2003). On-column electrochemical reactions accompanying the electrospray process. *Anal. Chem.* **75**:1022–1030.

Lombardo, F., and Campos, G. (2004). How do we study oxidative chemical stability in discovery? Some ideas, trials, and outcomes. In *Pharmaceutical Profiling in Drug Discovery for Lead Selection* (Borchardt, R. T., Kerns, E. H., Lipinski, C. A., Thakker, D. R., and Wang, B. Eds.). AAPS Press, Arlington, VA, pp. 183–194.

Lund, H., and Baizer, M. M. (1991). *Organic Electrochemistry, an Introduction and Guide*. Marcel Dekker, New York.

Madsen, K. G., Olsen, J., Skonberg, C., Hansen, S. H., and Jurva, U. (2007). Development and evaluation of an electrochemical method for studying reactive phase-I metabolites: Correlation to in vitro drug metabolism. *Chem. Res. Toxicol.* **20**:821–831.

Muldrew, K. L., James, L. P., Coop, L., McCullough, S. S., Hendrickson, H. P., Hinson, J. A., and Mayeux, P. R. (2002). Determination of acetaminophen-protein adducts in mouse liver and serum and human serum after hepatotoxic doses of acetaminophen using high-performance liquid chromatography with electrochemical detection. *Drug Metab. Dispos.* **30**:446–451.

Nagels, L. J., Mush, G., and Massart, D. L. (1989). Rapid-scan hydrodynamic voltammetry and cyclic voltammetry of pharmaceuticals in flow injection analysis conditions. *J. Pharm. Biomed. Anal.* **7**:1479–1483.

Nelson, S. D. (1990). Molecular mechanisms of the hepatotoxicity caused by acetaminophen. *Semin. Liver Dis.* **10**:267–278.

Nozaki, K., Kitagawa, H., Kimura, S., Kagayama, A., and Arakawa, R. (2006). Investigation of the electrochemical oxidation products of zotepine and their fragmentation using on-line electrochemistry/electrospray ionization mass spectrometry. *J. Mass Spectrom.* **41**:606–612.

Penalvo, J. L., Nurmi, T., Haajanen, K., Al-Maharik, N., Botting, N., and Adlercreutz, H. (2004). Determination of lignans in human plasma by liquid chromatography with coulometric electrode array detection. *Anal. Biochem.* **332**:384–393.

Plumb, R., Granger, J., Stumpf, C., Wilson, I. D., Evans, J. A., and Lenz, E. M. (2003). Metabonomic analysis of mouse urine by liquid-chromatography-time of flight mass spectrometry (LC-TOFMS): Detection of strain, diurnal and gender differences. *Analyst* **128**:819–823.

Riis, B. (2002). Comparison of results from different laboratories in measuring 8-oxo-2′-deoxyguanosine in synthetic oligonucleotides. *Free Radic. Res.* **36**:649–659.

Rocklin, R. D. (1984). Working-electrode materials. *Liquid Chromatogr.* **2**:588–594.

Roy, S., Venojarvi, M., Khanna, S., and Sen, C. K. (2002). Simultaneous detection of tocopherols and tocotrienols in biological samples using HPLC-coulometric electrode array. *Methods Enzymol.* **352**:326–332.

Sabbioni, C., Saracino, M. A., Mandrioli, R., Pinzauti, S., Furlanetto, S., Gerra, G., and Raggi, M. A. (2004). Simultaneous liquid chromatographic analysis of catecholamines and 4-hydroxy-3-methoxyphenylethylene glycol in human plasma. Comparison of amperometric and coulometric detection. *J. Chromatogr. A* **1032:**65–71.

Shiryaeva, I. M., Collman, J. P., Boulatov, R., and Sunderland, C. J. (2003). Nonideal electrochemical behavior of biomimetic iron porphyrins: Interfacial potential distribution across multilayer films. *Anal. Chem.* **75:**494–502.

Smith, M. T. (2003). Mechanisms of troglitazone hepatotoxicity. *Chem. Res. Toxicol.* **16:**679–687.

Sofic, E., Lange, K. W., Jellinger, K., and Riederer, P. (1992). Reduced and oxidized glutathione in the substantia nigra of patients with Parkinson's disease. *Neurosci. Lett.* **142:**128–130.

Van Berkel, G. J. (2004). Focus issue on electrochemistry combined with mass spectrometry. *J. Am. Soc. Mass Spectrom.* **15:**1691–1692.

van Leeuwen, S. M., Blankert, B., Kauffmann, J. M., and Karst, U. (2005). Prediction of clozapine metabolism by on-line electrochemistry/liquid chromatography/mass spectrometry. *Anal. Bioanal. Chem.* **382:**742–750.

Volicer, L., Langlais, P. J., Matson, W. R., Mark, K. A., and Gamache, P. H. (1985). Serotoninergic system in dementia of the Alzheimer type. Abnormal forms of 5-hydroxytryptophan and serotonin in cerebrospinal fluid. *Arch. Neurol.* **42:**1158–1161.

Volk, K. J., Lee, M. S., Yost, R. A., and Brajter-Toth, A. (1988). Electrochemistry/thermospray/tandem mass spectrometry in the study of biooxidation of purines. *Anal. Chem.* **60:**720–722.

Volk, K. J., Yost, R. A., and Brajter-Toth, A. (1989). On-line electrochemistry/thermospray/tandem mass spectrometry as a new approach to the study of redox reactions: The oxidation of uric acid. *Anal. Chem.* **61:**1709–1717.

Waterman, K. C., Adami, R. C., Alsante, K. M., Hong, J., Landis, M. S., Lombardo, F., and Roberts, C. J. (2002). Stabilization of pharmaceuticals to oxidative degradation. *Pharm. Dev. Technol.* **7:**1–32.

Wittemer, S. M., Ploch, M., Windeck, T., Muller, S. C., Drewelow, B., Derendorf, H., and Veit, M. (2005). Bioavailability and pharmacokinetics of caffeoylquinic acids and flavonoids after oral administration of artichoke leaf extracts in humans. *Phytomedicine* **12:**28–38.

Yanagawa, K., Takeda, H., Egashira, T., Matsumiya, T., Shibuya, T., and Takasaki, M. (2001). Changes in antioxidative mechanisms in elderly patients with non-insulin-dependent diabetes mellitus. Investigation of the redox dynamics of alpha-tocopherol in erythrocyte membranes. *Gerontology* **47:**150–157.

Zhang, X. Z., Gan, Y. R., and Zhao, F. N. (2004). Determination of salbutamol in human plasma and urine by high-performance liquid chromatography with a coulometric electrode array system. *J. Chromatogr. Sci.* **42:**263–267.

Zhou, F., and Van Berkel, G. J. (1995). Electrochemistry combined online with electrospray mass spectrometry. *Anal. Chem.* **67:**3643–3649.

9

LC–MS Methods with Hydrogen/Deuterium Exchange for Identification of Hydroxylamine, N-Oxide, and Hydroxylated Analogs of Desloratadine

Natalia A. Penner, Joanna Zgoda-Pols, Ragu Ramanathan, Swapan K. Chowdhury, and Kevin B. Alton

Schering-Plough Research Institute, Department of Drug Metabolism and Pharmacokinetics, Kenilworth, New Jersey

Mass Spectrometry in Drug Metabolism and Pharmacokinetics. Edited by Ragu Ramanathan
Copyright © 2009 John Wiley & Sons, Inc.

9.1   INTRODUCTION

Following administration to animals and humans, drugs are typically biotransformed into more polar species, which facilitates clearance from the body. This is not necessarily a benign process since metabolites can be toxic, more potent, or more reactive than the administered drug. Biotransformation can lead to other undesirable drug characteristics such as poor oral bioavailability due to substantial first-pass metabolism and drug–drug interactions due to induction and/or inhibition of metabolizing enzymes (Kiese, 1966; Sugimura et al., 1966; Bickel, 1969; Mattocks, 1971; Cook et al., 1983; Mattocks and Bird, 1983; Reilly et al., 1998; Gibson and Skett, 2001; Kostiainen et al., 2003; Lin et al., 2003; Miao et al., 2003). These are the reasons why, in order to create a molecule with better safety and pharmacokinetic (PK) characteristics, pharmaceutical companies invest significant effort into metabolite profiling during lead compound optimization when the main focus is on the identification of metabolic hot spots and reactive and active metabolites. After the development candidate is selected, a more thorough examination of its metabolic profile is conducted, including characterization of major human metabolites and the extent of exposure to preclinical species.

Superior sensitivity, efficiency, and specificity have made high-performance liquid chromatography coupled with tandem mass spectrometry (HPLC–MS/MS), the predominant analytical technique for characterization and quantitative analysis of metabolites (Kostiainen et al., 2003; Ma et al., 2006; Prakash et al., 2007). Ion trap, triple-quadrupole, and quadrupole time-of-flight (Q-TOF) mass spectrometers are routinely used to profile and characterize metabolites in plasma and excreta (Ma et al., 2006). The combination of scan types and features available on mass spectrometers of different design (product ion, MS^n, neutral loss, precursor ion scans, accurate mass measurements) allows identification and characterization of putative and unexpected metabolites with or without little prior knowledge of biotransformation pathways of a given drug molecule.

Nevertheless, even with accurate mass measurement by Q-TOF, LTQ-Orbitrap, or LTQ–Fourier transform ion cyclotron resonance (LTQ–FTICR) MS, it is not always possible to fully characterize certain metabolites based solely on mass spectrometric data. A good example is oxidation, which is one of the most common metabolic pathways. This process involves incorporation of one or more oxygen atoms into a drug molecule resulting in formation of hydroxylated metabolites (oxidation of carbon atom), *N*-oxides, or hydroxylamines. Distinguishing C- and N-oxidation by LC–MS without other spectroscopic data [e.g., nuclear magnetic resonance (NMR)] is a challenging task since a large number of isomers which differ from the parent drug by +16 amu can be formed. Previous studies have shown that for some compounds this problem can be successfully solved by using a combination of common hyphenated mass spectrometric techniques. For example, LC–MS and LC–MS/MS coupled with atmospheric pressure chemical ionization (APCI) can be used to distinguish N-oxidation from hydroxylated metabolites based on the thermal instability of *N*-oxides and hydroxylamines relative to other oxygenated metabolites (Ramanathan et al., 2000; Tong et al., 2001; Penner et al., 2003; Zgoda-Pols et al., 2004).

Another approach useful for LC–MS-based metabolite profiling work employs hydrogen/deuterium (H/D) exchange (HDX) to determine the number of exchangeable protons in a molecule and therefore assist in the structural elucidation (Ohashi et al., 1998; Olsen et al., 2000; Liu et al., 2001, 2003; Lam and Ramanathan, 2002; Nassar, 2003; Eichhorn et al., 2005; Liu and Hop, 2005). Deuterium exchange of small molecules is usually fast and can be performed on line using deuterated mobile phase (Ohashi et al., 1998; Olsen et al., 2000; Liu et al., 2001, 2003; Lam and Ramanathan, 2002; Nassar, 2003; Eichhorn et al., 2005; Liu and Hop, 2005). Data obtained by HDX in combination with the diagnostic fragmentation patterns from different isomers under various mass spectrometric conditions can provide enough evidence to differentiate among the isobaric oxidative metabolites in question.

In this chapter, the utility of LC–electrospray ionization (ESI)–MS, LC–ESI–MS/MS, LC–APCI–MS, LC–APCI–MS/MS, and HDX–LC–MS for structural identification of metabolites and derivatives of loratadine and desloratadine is discussed.

9.2 EXPERIMENTAL METHOD

9.2.1 Test Compounds

Loratadine (**LOR**, Claritin®) and desloratadine (**DL**, Clarinex®) are potent, long-acting, nonsedating antihistamines available over the counter or by prescription. Previously, LC–MS methods to differentiate between aliphatic and aromatic hydroxylated compounds as well as *N*-oxides and *N*-hydroxylamines of DL and LOR without HDX were proposed (Ramanathan et al., 2000; Penner et al., 2003).

As shown in the scheme (see next page), oxidation of DL results in formation of a number of isobaric monooxygenated products (m/z 327) including 3-, 5-, or 6-hydroxy-DL as well as *N*-oxide and *N*-hydroxylamine of DL. The LC–MS behaviors of four isomers with a molecular weight of 326 Da were previously evaluated (Ramanathan et al., 2000; Penner et al., 2003). A model system included the parent drug desloratadine (DL), its aliphatic (6-OH-DL) and aromatic (3-OH-DL) hydroxylated metabolites, along with *N*-oxide (1-pyridine-*N*-oxide-DL), and hydroxylamine (*N*-OH-DL) derivatives. All compounds used in this study were synthesized and purified at Schering-Plough Research Institute. A mixture containing all five compounds (20 ng/μL each) was prepared in methanol and a 10-μL aliquot was injected for analysis.

Loratadine (Lor)
m/z 383

N-Oxide of DL
m/z 323

N-Oxide of loratadine
(*N*-Oxide-Lor)
m/z 399

N-oxidation
(derivatives)
m/z 327

Desloratadine(DL)
m/z 311

C-oxidation
(metabolites)
m/z 327

3-Hydroxy-DL
(3-OH-DL)

1-Pyridine *N*-oxide of DL
(1-pyridine-*N*-oxide-DL)

N-Hydroxylamine of DL
(*N*-OH-DL)

5-Hydroxy-DL
(5-OH-DL)

6-Hydroxy-DL
(6-OH-DL)

9.2.2   Mass Spectrometry Conditions

LC–MS experiments were performed using an ion trap (LCQ Classic, Thermo Electron Corp., San Jose, CA) or triple-quadrupole (TSQ Quantum, Thermo Electron Corp., San Jose, CA) mass spectrometers. The experimental conditions are summarized below:

Parameter	LCQ Classic	TSQ Quantum Classic
Ionization source	ESI and APCI	ESI and APCI
Ionization mode	Positive	Positive
Scan range	100–1100 Da	100–1100 Da
Precursor isolation window	1.0 Th	0.7 Th
Spray voltage	4500 V	4500 V
Heated capillary/ion transfer tube temperature	150–350°C	200–390°C
Vaporizer temperature (APCI)	250–550°C	200–590°C
Discharge current (APCI)	7 μA	10 μA
Sheath gas	Nitrogen 80	Nitrogen 10–50
Auxiliary gas	Nitrogen 20	Nitrogen 5–25
Collision gas	Helium 0.015 mtorr	Argon 1.2–1.5 mtorr

9.2.3 Chromatographic Conditions

Separation of five compounds (DL, 6-OH-DL, 3-OH-DL, *N*-OH-DL, and 1-pyridine-*N*-oxide-DL) was achieved using an Alliance HPLC system (Waters Corp., Milford, CA) equipped with a 2690 model pump, an autoinjector, a Polaris C18-A guard column (Varian Inc., Lake Forest, CA), and a Luna Phenyl-Hexyl analytical column (Phenomenex, Inc., Torrance, CA) maintained at 40°C. For robust characterization of each isomeric compound, an online HDX LC–MS method was developed. The composition of regular and deuterated mobile phases is summarized below:

Mobile Phase	Regular	Deuterated
A	20 m*M* ammonium acetate adjusted to pH 6 with acetic acid, mixed with acetonitrile (95 : 5, v/v)	20 m*M* ammonium-d_4-acetate-d_3 in D_2O adjusted to pH 6 with acetic-d_3 acid-*d*, mixed with acetonitrile (95 : 5, v/v)
B	Acetonitrile	Acetonitrile

A linear gradient elution was used with A/B ratios of 90/10, 90/10, 60/40, 30/70, 30/70, 90/10, and 90/10 at 0, 2, 25, 35, 40, 40.5, and 45 min, respectively, at a flow rate of 1 mL/min. As shown in Fig. 9.1 base line resolution of all five analytes was achieved. The concentrations of ammonium acetate and pH were critical for obtaining optimum separation. The column effluent was split to divert 15–25% or

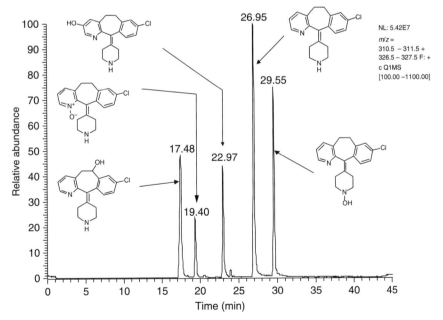

Figure 9.1. Extracted ion chromatograms for m/z 311 (DL) and 327 (6-OH-DL, 3-OH-DL, N-OH-DL, and 1-pyridine-N-oxide-DL) generated using a TSQ Quantum mass spectrometer (LC–APCI–MS).

40–50% of the flow into the mass spectrometer for HPLC–ESI–MS or HPLC–APCI–MS experiments, respectively.

9.3 APPLICATIONS

9.3.1 LC–ESI–MS and LC–APCI–MS Experiments

9.3.1.1 Regular Mobile Phase With the exception of minor water loss observed for *N*-OH-DL, no significant in-source fragmentation occurred for all five compounds under LC–ESI–MS conditions (Fig. 9.2). In the LC–APCI–MS spectra, $[M + H]^+$ ions were observed for all the compounds (Fig. 9.3). In agreement with our previously reported results (Zgoda-Pols et al., 2004), aliphatic hydroxylation can be easily differentiated from aromatic hydroxylation due to formation of an $[M + H - H_2O]^+$ ion at m/z 309 for 6-OH-DL. A corresponding ion for 3-OH-DL was not observed. In addition to a water loss, elimination of an oxygen atom was observed for both 1-pyridine-*N*-oxide-DL and *N*-OH-DL. This combination makes

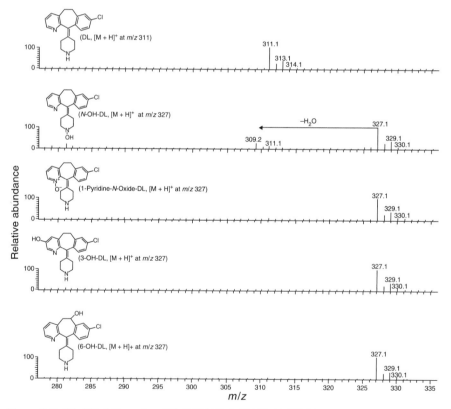

Figure 9.2. *LC–ESI–MS spectra of DL, 6-OH-DL, 3-OH-DL, N-OH-DL, and 1-pyridine-N-oxide-DL generated using a TSQ Quantum triple-quadrupole mass spectrometer.*

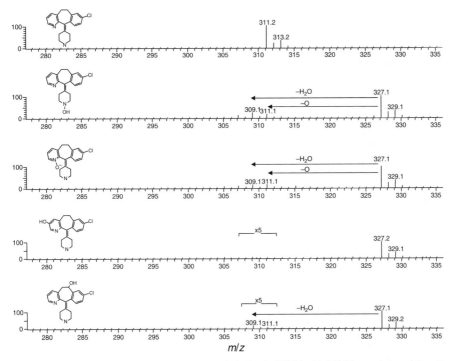

Figure 9.3. *LC–APCI–MS spectra of DL, 6-OH-DL, 3-OH-DL, N-OH-DL, and 1-pyridine-N-oxide-DL generated using a TSQ Quantum triple-quadrupole mass spectrometer. Vaporizer temperature 590°C.*

it possible to differentiate between N- and C-hydroxylated analogs of DL. It should be noted that the intensity of fragment ions relative to the molecular ion peaks increased significantly when the vaporizer temperature was gradually elevated from 200 to 590°C. Therefore, where possible, LC–APCI–MS experiments should be carried out at higher temperatures, where the loss of water and atomic oxygen is more pronounced.

9.3.1.2 Deuterated Mobile Phase

In the absence of other modifications, hydroxylated compounds can be easily differentiated from *N*-oxides and hydroxylamines in either LC–ESI–MS or LC–APCI–MS mode by HDX. This is a well-established technique for studying structure, stability, folding dynamics, and intermolecular interactions of proteins in solution. During HDX, hydrogens on heteroatoms (O, N, S) undergo exchange to deuterium, which results in the increase of molecular mass according to a number of exchangeable protons present in a given molecule. In most cases, exchange occurs very quickly on column during separation, so no special handling (drying and reconstitution in a deuterated solvent) is required for the sample to be analyzed.

LC–ESI–MS and LC–APCI–MS experiments involving HDX were carried out on a TSQ Quantum mass spectrometer. All labile protons, in DL, 6-OH-DL, 3-OH-DL, *N*-OH-DL, and 1-pyridine-*N*-oxide-DL, underwent complete deuterium exchange. C-Hydroxylated compounds (6-OH-DL, 3-OH-DL) underwent a total of three HDXs, while *N*-oxide and the hydroxylamine exchanged only two protons.

| Compound | *m/z* | | Mass Increase | Number of Labile Hydrogens in Molecule |
	Regular Mobile Phase	Deuterated Mobile Phase		
DL	311	313	2	1
DL-*N*-OH	327	329	2	1
DL-*N*-oxide	327	329	2	1
3-OH-DL	327	330	3	2
6-OH-DL	327	330	3	2

In all experiments, the charge-providing species were deuterium and $[M + D]^+$ ions were observed for all compounds under ESI and APCI conditions.

Similar to experiments conducted with conventional mobile phase, no significant in-source fragmentation was observed in the ESI–MS spectra (spectra not shown). With the *N*-oxide and hydroxylamine derivatives, fragment ions at *m/z* 313 corresponding to a loss of one oxygen atom were observed in the LC–APCI–MS mass spectra (Fig. 9.4). These results confirm that deoxygenation of *N*-oxides is a result of thermal degradation rather than collision activation (Tong et al., 2001). Since different numbers of exchangeable protons and deoxygenation pattern were observed for *N*-OH-DL and 1-pyridine-*N*-oxide-DL under LC–APCI–MS conditions, it is possible to distinguish products of N- and C-oxidation eventhough the $[M + D - O]^+$ ions were not very intense. Before performing LC–MS and LC–MS/MS experiments, one should take into consideration that the background level observed in MS spectra following HDX is significantly higher and the signal is lower compared to separation conducted with regular mobile phase.

None of the LC–MS experiments, under deuterated or nondeuterated conditions, provided a means for distinguishing hydroxylamine *N*-OH-DL from 1-pyridine-*N*-oxide-DL, and therefore these compounds were further investigated by LC–MS/MS.

9.3.2 LC–MS/MS Experiments

The same standard mixture, containing DL, 6-OH-DL, 3-OH-DL, *N*-OH-DL, and 1-pyridine-*N*-oxide-DL, was studied by LC–ESI–MS/MS and LC–APCI–MS/MS under a variety of conditions. Experiments were performed by varying (1) the temperature of the heated capillary (150–350°C for LCQ Classic) and ion transfer tube (200–590°C for TSQ Quantum) in ESI mode and (2) the temperature of the

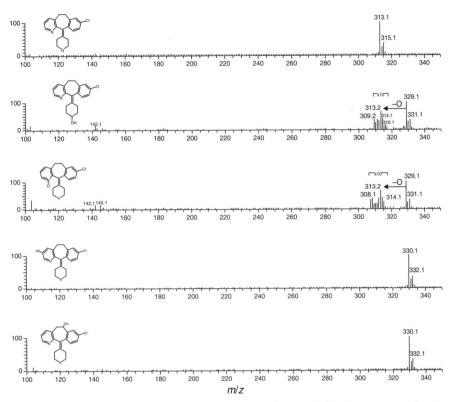

Figure 9.4. *LC–APCI–MS spectra of DL, 6-OH-DL, 3-OH-DL, N-OH-DL, and 1-pyridine-N-oxide-DL generated using a TSQ Quantum triple-quadrupole mass spectrometer and deuterated mobile phase.*

vaporizer (250–550°C for LCQ Classic and 200–590°C for TSQ Quantum) in APCI mode.

The ESI and APCI ion sources did not lead to any significant differences in the spectra obtained. Diagnostic MS/MS fragment ions were observed for all compounds. 6-OH-DL is easily distinguished from 3-OH-DL due to the formation of $[M + H - H_2O]$ at m/z 309, which is only favored with aliphatic hydroxylation (Ramanathan et al., 2000) (Fig. 9.5a). N-Oxide and hydroxylamine formed $[M + H - NH_3]^+$ and $[M + H - OH]^+$ ions at m/z 310, respectively, along with low abundance $[M + H - H_2O]^+$ ions at m/z 309 (Fig. 9.5a). The observed loss of water was more pronounced when the LCQ Classic was used (data not shown).

In contrast to the LC–APCI–MS experiments, loss of atomic oxygen in MS/MS spectra for N-oxide and hydroxylamine was not observed, further indicating that deoxygenation is a result of thermal degradation (Tong et al., 2001). Therefore, based on the extent of the water loss in MS/MS mode, aliphatic hydroxylation (6-OH-DL) can be differentiated from the aromatic one (3-OH-DL) as well as from

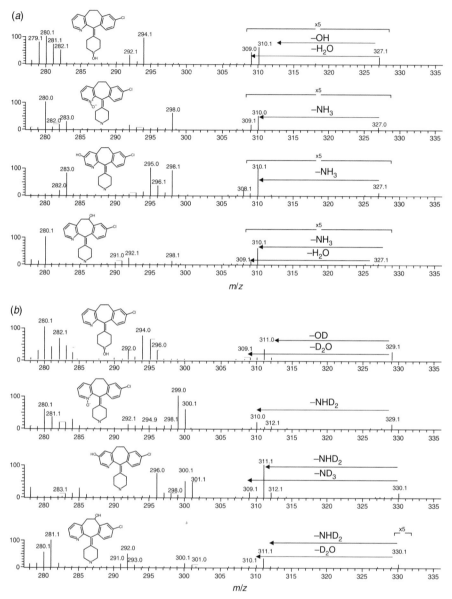

Figure 9.5. LC–APCI–MS/MS spectra of 6-OH-DL, 3-OH-DL, N-OH-DL, and 1-pyridine-N-oxide-DL generated using a TSQ Quantum triple-quadruple mass spectrometer with (a) conventional or (b) deuterated mobile phase.

the N-oxide (1-pyridine-N-oxide-DL) and hydroxylamine (N-OH-DL) structural isomers. A summary of in-source fragmentation and diagnostic MS/MS fragments generated using a TSQ Quantum mass spectrometer for four M+16 isomers of DL is presented in Table 9.1.

TABLE 9.1. A Summary of in-Source Fragmentation and Diagnostic MS/MS Fragments Generated Using a TSQ Quantum Mass Spectrometer

Compound	ESI–MS	APCI–MS	ESI–MS/MS or APCI–MS/MS
6-OH-DL	$[M + H]^+$ $\mathbf{[M + H - H_2O]^+}$	$[M + H]^+$ $\mathbf{[M + H - H_2O]^+}$	$[M + H]^+$ $[M + H - NH_3]^+$ $\mathbf{[M + H - H_2O]^+}$
3-OH-DL	$[M + H]^+$	$[M + H]^+$	$[M + H]^+ [m + H - NH_3]^+$
1-Pyridine-N-oxide-DL	$[M + H]^+$	$[M + H]^+$ $\mathbf{[M + H - O]^+}$ $[M + H - H_2O]^+$	$[M + H]^+$ $[M + H - NH_3]^+$ $[M + H - H_2O]^+$
N-OH-DL	$[M_H + H]^+$	$[M + H]^+$ $\mathbf{[M + H - O]^+}$ $[M + H - H_2O]^+$	$[M_H + H]^+$ $\mathbf{[M_H + H - OH]^+}$ $[M_H + H - H_2O]^+$

Even though a diagnostic fragment ion $[M + H - OH]^+$ was observed for hydroxylamine, N-OH-DL was indistinguishable from the N-oxide because $[M + H - NH_3]^+$ and $[M + H - OH]^+$ have the same m/z at unit mass resolution. Differentiation can be easily achieved with either accurate mass measurements or by conducting MS/MS experiments with deuterated mobile phase (Fig. 9.5b). After HDX, a $[M + D - OD]^+$ ion at m/z 311 was observed in the MS/MS spectrum for hydroxylamine, while for N-oxide, the loss of NHD_2 resulted in the formation of a corresponding ion at m/z 310. Since the loss of a hydroxyl moiety was not observed for any other compound, unequivocal assignment of the hydroxylation site was possible for hydroxylamine. Similar to nondeuterated conditions, $[M + D - D_2O]^+$ was observed for 6-OH-DL but not for 3-OH-DL, thereby providing means to differentiate the presence of an aliphatic versus aromatic hydroxyl group. Therefore, for a given group of compounds, a single LC–MS/MS experiment in ESI or APCI mode after HDX is necessary to differentiate between four isomeric oxidative metabolites and derivatives of DL.

9.3.3   Plasma Experiments

To demonstrate that the proposed methods are suitable for structural elucidation of isomeric metabolites and derivatives in biological matrices, human plasma was spiked with the mixture of DL, 6-OH-DL, 3-OH-DL, *N*-OH-DL, and 1-pyridine-*N*-oxide-DL. The resulting sample was extracted and analyzed by LC–MS and LC–MS/MS in ESI and APCI modes as described above. HDX was successfully performed online when the extract was injected directly onto the HPLC column without drying and reconstituting the sample in a deuterated solvent. In general, there were no differences between the results obtained for the spiked plasma extract and for the mixture of the standard compounds, which indicates that the LC–MS methods with HDX described here are applicable for the analysis drug-derived material in plasma or other biological matrices.

9.3.4   Other Examples

With a large variety of compounds in development it is expected that the method described here would not work for all of them. Chemical bonds which break under elevated temperatures or low collision energy can lead to compound-specific in-source fragmentation and interfere with characteristic fragments necessary for structural elucidation of isomers.

Another important fact to consider before conducting such experiments is type of mass spectrometer and source design to be used. Manufacturers introduce new instruments or improved ion source designs on a regular basis. For example, a clear difference in fragmentation could be observed with LCQ Classic and TSQ Quantum (Fig. 9.6). With the LCQ, 6-OH-DL and 3-OH-DL can be distinguished in LC–ESI–MS mode since the aliphatic OH group shows a pronounced loss of water (Fig. 9.6). By comparison, this fragmentation is not detected with the TSQ Quantum mass spectrometer (Fig. 9.1). 6-OH-DL and 3-OH-DL can also be distinguished in MS/MS. Spectra obtained on the LCQ in LC–ESI–MS mode share some features observed with the APCI ion source on TSQ Quantum due to a heated capillary in the front end of the LCQ. Since the LCQ is a very popular instrument for metabolite profiling, this feature can be considered advantageous for structural elucidation of isomeric monooxymetabolites.

For selected compounds, the site of oxidation can still be assigned properly even when the characteristic fragment ions that they produce are somewhat different from those provided in Table 9.1. A good example is *N*-oxide of DL with m/z 323, which shows a prominent loss of OH under LC–ESI–MS and MS/MS conditions on LCQ Classic mass spectrometer (Fig. 9.7).

In our laboratory, it is not unusual to encounter situations where the *N*-oxide of a parent drug is readily detected while the formation of *N*-hydroxylamines is absent. It is not uncommon for *N*-oxide to be observed as a circulating pharmacologically active metabolite, thereby requiring positive identification early in a drug development program. Carefully designed LC–MS experiments are necessary for identification to avoid time-consuming isolation and purification of large quantities of a

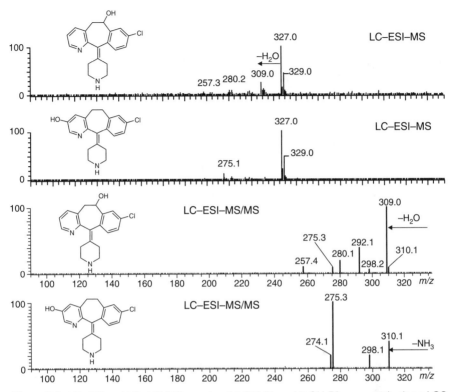

Figure 9.6. *LC–MS and LC–MS/MS spectra of 6-OH-DL and 3-OH-DL generated using a LCQ Classic mass spectrometer.*

compound for NMR analysis. Two compound substructures often subjected to N-oxidation are pyridine and pyperidine moieties; the LC–MS method described here was successfully applied for unequivocal identification of metabolites formed with both of them without having to resort to an orthogonal spectroscopic procedure:

For a compound which contains dimethylpyperidine substructure (shown on the left), isomeric metabolites oxygenated at positions 1, 2, and 3 can be easily distinguished without NMR by conducting a single LC–APCI–MS experiment on the TSQ Quantum. Loss of 16 is observed for oxidation at position 1 (*N*-oxide), loss of 18 (H₂O) for aliphatic hydroxylation at position 3, while for aromatic hydroxylation at position 2 no significant in-source fragmentation is observed.

In general, for approaches as the one described here, when experimental conditions and source design are critical for obtaining conclusive data, it is advisable

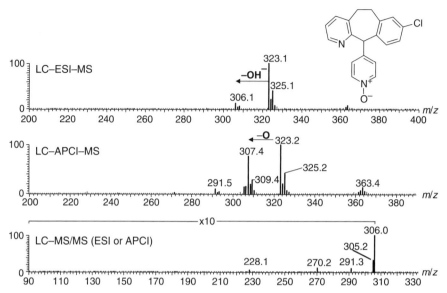

Figure 9.7. *LC–MS and LC–MS/MS spectra of DL-N-oxide generated using a LCQ Classic mass spectrometer.*

to test the system with a well-known mixture containing compounds hydroxylated at various atoms (aromatic, aliphatic hydroxylation and N-oxidation) before applying this method to unknown samples. Due to diversity in compound structures, this method is expected to be structure dependent and in some cases may provide results different from described here.

9.4 SUMMARY

LC–MS and LC–MS/MS methods combined with HDX experiments were able to unambiguously differentiate structures of four monooxygenated isomeric analogs of DL. Importantly, HDX is required to differentiate between hydroxylamine and N-oxide derivatives. This method can provide critical information for identification of novel metabolites.

REFERENCES

Bickel, M. H. (1969). The pharmacology and biochemistry of *N*-oxides. *Pharmacol. Rev.* **21:**325–355.

Cook, B. A., Sinnhuber, J. R., Thomas, P. J., Olson, T. A., Silverman, T. A., Jones, R., Whitehead, V. M., and Ruymann, F. B. (1983). Hepatic failure secondary to indicine *N*-oxide toxicity. A Pediatric Oncology Group Study. *Cancer* **52:**61–63.

Eichhorn, P., Ferguson, P. L., Perez, S., and Aga, D. S. (2005). Application of ion trap-MS with H/D exchange and QqTOF-MS in the identification of microbial degradates of trimethoprim in nitrifying activated sludge. *Anal. Chem.* **77:**4176–4184.

Gibson, G. G., and Skett, P. (2001). Pathways of drug metabolism, In *Introduction to Drug Metabolism* (Gibbson, G. G., and Skett, P., Eds.). Stanley Thornes, Cheltenham, UK, pp. 1–34.

Kiese, M. (1966). The biochemical production of ferrihemoglobin-forming derivatives from aromatic amines, and mechanisms of ferrihemoglobin formation. *Pharmacol. Rev.* **18:**1091–1161.

Kostiainen, R., Kotiaho, T., Kuuranne, T., and Auriola, S. (2003). Liquid chromatography/atmospheric pressure ionization-mass spectrometry in drug metabolism studies. *J. Mass Spectrom.* **38:**357–372.

Lam, W., and Ramanathan, R. (2002). In electrospray ionization source hydrogen/deuterium exchange, LC-MS and, LC-MS/MS for characterization of metabolites. *J. Am. Soc. Mass Spectrom.* **13:**345–353

Lin, J., Sahakian, D. C., de Morais, S. M., Xu, J. J., Polzer, R. J., and Winter, S. M. (2003). The role of absorption, distribution, metabolism, excretion and toxicity in drug discovery. *Curre. Topics Med. Chem.* **3:**1125–1154.

Liu, D. Q., and Hop, C. E. (2005). Strategies for characterization of drug metabolites using liquid chromatography-tandem mass spectrometry in conjunction with chemical derivatization and on-line H/D exchange approaches. *J. Pharm. Biomed. Anal.* **37:**1–18.

Liu, D. Q., Hop, C. E. C. A., Beconi, M. G., Mao, A., and Chiu, S. H. L. (2001). Use of on-line hydrogen/deuterium exchange to facilitate metabolite identification. *Rapid Commun. Mass Spectrom.* **15:**1832–1839.

Liu, D. Q., Hop, C. E. C. A., Beconi, M. G., Mao, A., and Chiu, S. H. L. (2003). Use of on-line hydrogen/deuterium exchange to facilitate metabolite identification. [Erratum to document cited in, CA136:79200]. *Rapid Commun. Mass Spectrom.* **17:**264.

Ma, S., Chowdhury, S. K., and Alton, K. B. (2006). Application of mass spectrometry for metabolite identification. *Curr. Drug Metab.* **7:**503–523.

Mattocks, A. R. (1971). Hepatotoxic effects due to pyrrolizidine alkaloid *N*-oxides. *Xenobiotica* **1:**563–565.

Mattocks, A. R., and Bird, I. (1983). Pyrrolic and *N*-oxide metabolites formed from pyrrolizidine alkaloids by hepatic microsomes in vitro: Relevance to in vivo hepatotoxicity. *Chem. Biol. Interact.* **43:**209–222.

Miao, X. S., March, R. E., and Metcalfe, C. D. (2003). A tandem mass spectrometric study of the *N*-oxides, quinoline *N*-oxide, carbadox, and olaquindox, carried out at high mass accuracy using electrospray ionization. *Int. J. Mass Spetrom.* **230:**123–133.

Nassar, A. E. (2003). Online hydrogen-deuterium exchange and a tandem-quadrupole time-of-flight mass spectrometer coupled with liquid chromatography for metabolite identification in drug metabolism. *J. Chromatogr. Sci.* **41:**398–404.

Ohashi, N., Furuuchi, S., and Yoshikawa, M. (1998). Usefulness of the hydrogen–deuterium exchange method in the study of drug metabolism using liquid chromatography-tandem mass spectrometry. *J. Pharm. Biomed. Anal.* **18:**325–334.

Olsen, M. A., Cummings, P. G., Kennedy-Gabb, S., Wagner, B. M., Nicol, G. R., and Munson, B. (2000). The use of deuterium oxide as a mobile phase for structural elucidation by HPLC/UV/ESI/MS. *Anal. Chem.* **72:**5070–5078.

Penner, N., Ramanathan, R., Alvarez, N., Chowdhury, S., Alton, K., and Patrick, J. (2003). LC-MS methods for distinguishing hydroxylamine from *N*-oxide and hydroxy metabolites, In *Proceedings of the 51st, ASMS Conference on Mass Spectrometry and Allied Topics*, Montreal, Canada.

Prakash, C., Shaffer, C. L., and Nedderman, A. (2007). Analytical strategies for identifying drug metabolites. *Mass Spectrom Rev.* **26**:340–369.

Ramanathan, R., Su, A. D., Alvarez, N., Blumenkrantz, N., Chowdhury, S. K., Alton, K., and Patrick, J. (2000). Liquid chromatography/mass spectrometry methods for distinguishing *N*-oxides from hydroxylated compounds. *Anal. Chem.* **72**:1352–1359.

Reilly, T. P., Bellevue, F. H., 3rd, Woster, P. M., and Svensson, C. K. (1998). Comparison of the in vitro cytotoxicity of hydroxylamine metabolites of sulfamethoxazole and dapsone. *Biochem. Pharmacol.* **55**:803–810.

Sugimura, T., Okabe, K., and Nagao, M. (1966). The metabolism of 4-nitroquinoline-1-oxide, a carcinogen. 3. An enzyme catalyzing the conversion of 4-nitroquinoline-1-oxide to 4-hydroxyaminoquinoline-1-oxide in rat liver and hepatomas. *Cancer Res.* **26**:1717–1721.

Tong, W., Chowdhury, S. K., Chen, J. C., Zhong, R., Alton, K. B., and Patrick, J. E. (2001). Fragmentation of *N*-oxides (deoxygenation). in atmospheric pressure ionization: Investigation of the activation process. *Rapid Commun. Mass Spectrom.* **15**:2085–2090.

Zgoda-Pols, J., Ramanathan, R., Chowdhury, S., and Alton, K. (2004). Evaluation of H/D exchange and, L. C.,-MS for identification of hydroxylamine, *N*-oxide, and hydroxyl analogs of desloratadine. In *Proceedings of the 52nd, ASMS Conference on Mass Spectrometry and Allied Topics*, Nashville, TN.

10

Turbulent-Flow LC–MS: Applications for Accelerating Pharmacokinetic Profiling and Metabolite Identification

Joseph L. Herman

Children's Hospital of Philadelphia, Philadelphia, Pennsylvania

Joseph M. Di Bussolo

ThermoFisher Scientific, West Chester, Pennsylvania

10.1 INTRODUCTION

The development of high-throughput screening (HTS) assays has completely changed the way that the pharmaceutical industry investigates new chemical entities (NCE) for possible druglike characteristics. The bottleneck to decision making has shifted from sample generation to sample analysis. The introduction of atmospheric pressure ionization (API) techniques to interface liquid chromatography with mass spectrometry (LC–MS) has drastically reduced the time needed to analyze druglike compounds in a variety of samples. In particular, API techniques in conjunction with multiple-reaction monitoring (MRM) have allowed for extremely fast LC–MS/MS methods (1–2 min per sample). Such methods rely on the mass spectrometer's ability to isolate ionized analytes by mass-to-charge ratio, fragment each precursor ion, and then isolate and detect a product ion characteristic of the precursor ions' structure. Consequently, LC–MS/MS methods do not require lengthy separations to quantify analytes selectively as the more traditional chromatographic methods do.

LC–MS/MS techniques moved the bottleneck for analyzing NCEs from sample analysis to sample preparation. Several methodologies have been developed to reduce sample preparation time. Online extraction methods are becoming more popular and have many advantages over offline techniques. For example, online sample preparation is easier to automate and minimizes recovery issues.

In this chapter, one such online extraction technique, turbulent-flow liquid chromatography (TFLC), is explored, removes protein and similar material from native biological samples. Very little sample preparation is needed, other than centrifuging to remove particulates or spiking with an internal standard for quantitative work. Avoiding sample preparation steps, from the analysis of biological matrices, greatly reduces the total sample analysis time. Removing proteinlike material also reduces ion suppression effects and lowers maintenance time by keeping the mass spectrometers cleaner.

Isocratic focusing techniques, in conjunction with turbulent-flow chromatography, are also explored in this chapter. Isocratic focusing is used to improve chromatographic peak shape, enhance sensitivity, and optimize reproducibility. Isocratic focusing has also led to the development of generic methodologies that eliminate the need for lengthy method development in HTS, thereby accelerating sample throughput. These methods are also used to concentrate samples online or to improve the retention of extremely hydrophilic compounds, such as metabolites.

In this chapter, the theory of TFLC and its application toward accelerating pharmacokinetic profiling and metabolite identification studies are discussed (Ayrton et al., 1997a; Brignol et al., 2000a; Wu et al., 2000; Herman, 2002b; Hopfgartner et al., 2002; Takino et al., 2003; Ceglarek et al., 2004; DeMaio et al., 2005; Xu et al., 2005b; Zhou et al., 2005; Black et al., 2006; Smalley et al., 2006, 2007; Xu et al., 2007).

10.2 BACKGROUND

High-performance liquid chromatography typically involves pumping a mobile phase at a laminar-flow velocity. This velocity averages around 0.1 cm/s for a column with an inside diameter of 0.2–0.46 cm. The column is tightly packed with small porous particles that range in size from 3 to 5 μm. The surfaces inside the pores of these particles support the stationary phase. Under laminar flow, the mobile phase does not travel in a uniform velocity through the column. Because of friction, the velocity decreases as the mobile phase approaches any surface, including the outer portion of the packing particles and the walls of the column and connecting tubing. A population of solute molecules traveling through the column tends to disperse as individual molecules diffuse between fast and slow streams of mobile phase within the column and through connected tubing. Solute molecules that diffuse to slower streams lag behind others that diffuse to faster ones. These processes contribute to the overall broadening and tailing of a chromatographic peak, which represents the population of those solute molecules emerging from the column.

The dispersion of solute populations reduces as the flow of mobile phase becomes turbulent. The kinetic energy of turbulent-flow causes a tremendous amount of mixing; thereby eliminating differences in flow velocities between and around the particles packed in the column (Giddings, 1965). Paradoxically, such random motion results in a uniform or "plug" velocity profile across the diameter of the column, which maintains a concentrated band of solute molecules as it travels through the column. The resulting concentration gradient promotes diffusion of solute molecules into the pores of the packing particles. However, if the mobile-phase velocity is high enough and the diameter of the pores of the packing material is small enough, then the opportunity for solutes to enter the pores of the packing material is limited by the molecular size. Large solutes, such as proteins, diffuse slowly. Consequentially, larger solutes are pushed past the entrances of the pores by the turbulent-flow of the mobile phase. Smaller solutes, which diffuse quickly, have sufficient opportunity to enter the pores of the packing material and effectively diffuse to the stationary phase at the internal surfaces. This rapid segregation of small and large molecules by TFLC facilitates extraction of a wide range of analytes, such as drugs and metabolites, from various biological samples, such as plasma, bile, and urine (Oberhauser et al., 2000).

10.3 THEORY AND HISTORY

Over 100 years ago, British physicist Osborne Reynolds discovered that the flow of a fluid through a conduit becomes turbulent when the momentum of the fluid exceeds its resistance to flow by a factor of 2000–3000 (Reynolds, 1883). The ratio of these opposing forces, known as the Reynolds number (Re), is expressed as

$$\text{Re} = \frac{v \rho D_f}{\eta} \tag{10.1}$$

The momentum, which is a certain amount of mass moving at a certain velocity (v), takes into account the fluid's density (ρ) and the diameter of the tube (D_f). The momentum of the fluid can be increased by increasing the velocity or the diameter of the tube or both. The resistance to flow is expressed as the absolute or dynamic viscosity of the fluid (η), which is in units of grams per centimeter-second, or centipoise (cP). The transition from laminar to turbulent-flow occurs as Re increases past a critical value between 2000 and 3000 in a straight tube having a smooth internal surface.

In the case of a tube packed with particles, such as a chromatographic column, the diameters of all of the flow paths cannot be measured, but the average diameter of the packing particles (d_p) can be measured. The spaces between particles are proportional to the average size of the particles. Therefore, columns packed with large particles hold a greater mass of mobile phase than those packed with smaller particles. Equation (10.2) describes a special case (Re′) of the Reynolds number for columns packed with particles:

$$\text{Re}' = \frac{v \rho d_p}{\eta} \tag{10.2}$$

according to which turbulence occurs more readily in columns packed with particles ranging in size from 35 to 75 μm than those packed with particles ranging in size from 3 to 10 μm. Experimental observations of solute band broadening relative to column retention have indicated that the transition from laminar to turbulent-flow occurs as Re′ increases beyond a value greater than 1 and that virtually all of the flow paths within a column become turbulent as Re′ exceeds a value of 10 (Knox, 1966). Solute band broadening relative to column retention, is measured as the height equivalent to a theoretical plate (H). Typically, H increases from a minimum as the mobile-phase velocity increases as long as the flow remains laminar. As the flow through more and more channels in the packed column becomes turbulent, H begins to decrease and reaches a minimum when all of the flow paths in the column become turbulent. Compared to flow through a straight hollow tube, transition from laminar to turbulent-flow is much more gradual in a packed column as the flow rate increases.

Turbulent-flow chromatography (TFC) was first described in the 1960s. Sternberg and Poulson (1964) showed how turbulent-flow increases the effective diffusivity of solutes and improves mass transfer between mobile and stationary phases. Giddings, Pretorius, and Knox elucidated the theory of turbulent-flow gas and liquid chromatography (Giddings, 1965; Knox, 1966; Pretorius and Smuts, 1966). However, several investigators pointed out its practical limitations in LC as it relates to solvent consumption and pressure drop. The technique was dormant for almost 20 years, until Guiochon experimented with capillary TFC in the 1980s (Martin and Guiochon, 1982). In the mid-1990s, Quinn rediscovered the technique and set out to optimize and commercialize columns and systems for TFLC (Ayrton et al., 1997b; Quinn et al., 1998, 1999; Quinn and Takarewski, 1998, 1999). Early publications of TFLC methods for extracting biological fluids began to appear in the late 1990s (Jemal et al., 1998a, 1999a, b; Zimmmer et al., 1999). From then on, the number of presentations and publications involving TFLC in various bioanalytical applications has grown exponentially.

10.4 PRACTICAL CONSIDERATIONS

TFLC involves pumping a mobile phase at an average velocity of at least $10 \, cm/s$ through a column having an inside diameter (i.d.) of $0.05-0.1$ cm. TFLC columns, which are typically 5 cm long, are packed with large particles ($35-75 \, \mu m$) that have average pore sizes of $60-120$ Å. The flow of the mobile phase becomes turbulent at flow rates above $4 \, mL/min$ in the 0.1-cm-i.d. columns and above $1 \, mL/min$ in the 0.05-cm-i.d. columns. Such columns have void volumes of 20 and 5 μL, respectively. Therefore, only a few seconds are required to pump several column volumes of mobile phase through these columns at previously stated flow rates. In order to achieve efficient extraction of analytes using TFLC, the stationary and mobile phases selected must provide a retention factor of at least 50 column volumes, preferably $100-200$ column volumes. That way, unwanted sample components, which are less retentive, have ample time to elute to waste via a switching valve during the loading step of the extraction.

As stated earlier, TFLC selectively limits access to the pores of the column packing material for large solutes. The diffusivity of a molecule is inversely proportional to its molecular weight. The diffusion coefficient for small molecules in water is from 1 to $1.5 \times 10.5 \, cm^2/s$, while that for a midsized protein is 10-fold less ($1 \times 10^{-6} \, cm^2/s$). Small sample components have high diffusivity and can quickly diffuse into the pores of the column packing. Large components, such as proteins, have low diffusivity and do not have time to diffuse into the pores. Therefore, proteins are quickly swept away by the turbulent-flow. The kinetic energy of the turbulent-flow, along with the chemistry of the loading mobile phase, quenches much of the protein–analyte interactions that may occur, thereby enhancing the separation efficiency.

The mechanism of separation between large and small molecules is what distinguishes TFLC from other methods of sample cleanup. Protein precipitation is often used to remove much of the sample matrix. While much of the proteins present in a given matrix can be "crashed" out of solution by the addition of an

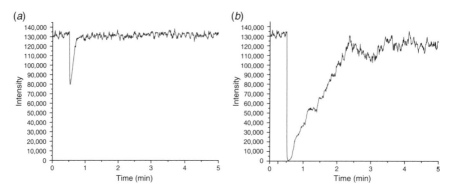

Figure 10.1. *Ion suppression effects from protein precipitated plasma: (a) blank injection (50/50 methanol–water); (b) Injection of protein precipitated rat plasma.*

organic solvent, there are still many components remaining in solution that can cause interference during analysis. Figure 10.1 illustrates an example of the ion suppression observed from a sample that has undergone protein precipitation. By comparing the LC–MS or LC–MS/MS signal following infusion of a standard postcolumn during a chromatographic run of a blank to that of protein precipitated plasma, one can detect ion suppression as a loss in signal strength in relation to the blank. Typically, a chromatographic method has to be developed that elutes the analytes of interest in a region where ion suppression does not occur. This involves a lot of time and effort. Extraction methods are much better at cleaning samples than protein precipitation, as is shown in Fig. 10.2. However, even after optimizing for a given sample type and for the analytes of interest, extraction methods are not 100% efficient at removing all matrix interferences. This is because extraction methods are based on equilibrium partitioning between two phases. While this can approach 100% for some compounds, it usually is somewhat lower and certainly will be different for every compound depending on the nature of the two phases being used.

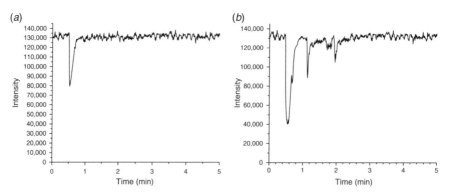

Figure 10.2. *Ion suppression effects from solid-phase extraction: (a) blank injection (50/50 methanol–water); (b) injection of extracted rat plasma in 50/50 methanol–water.*

As described earlier, TFLC performs an initial separation based on molecular weight. In fact, the amount of separation and the mass range excluded are dependent on the linear velocity of the mobile phase (Quinn and Takarewski, 1998). During the loading step, proteins, lipoproteins, and other large sample components are excluded from the pores of the packing material and rinsed into the waste. The turbulent flow also rinses away sample components less retentive than the analytes of interest, such as salts and sugars. The samples of interest are eluted after the initial cleanup has occurred. More retentive sample components, such as lipids, remain in the extraction column until they are washed away with strong organic solvents. Thus, a single TFLC column performs two-dimensional chromatography. The first separation is by size, and the second separation is by partitioning between the stationary and mobile phases; only the molecules that diffuse into the pores of the packing material interact with the stationary phase. This two-dimensional process makes TFLC the superior method for cleaning up biological matrices. Figure 10.3 shows that there is no distinguishable difference between a blank injection and a plasma injection after using TFLC. Therefore, there are no ion suppression effects from the sample matrix when using TFLC. Keep in mind that there are still other sources of ion suppression to worry about. For instance, excipients in the drug formulation may have a pharmacokinetic profile that superimposes on top of the analyte of interest and cause further suppression.

10.4.1 Single-Column Methods

The simplest application of TFLC is to use a single column to perform both the extraction and elution of the analyte of interest. Samples are injected into the turbulent-flow column under turbulent conditions. The analytes of interest are sequestered by a suitable stationary phase in the column, while less retentive sample components and excluded proteins are rinsed away to a waste line via a switching valve. Once the undesirable components have been eluted to waste, the switching valve reverses the flow through the column and directs the flow toward a highly selective detector such

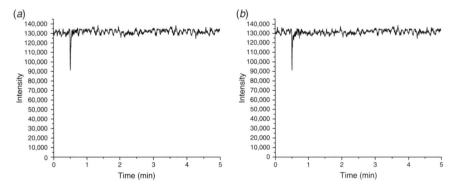

Figure 10.3. Ion suppression effects from turbulent-flow LC: (a) blank injection of buffered water; (b) neat rat plasma.

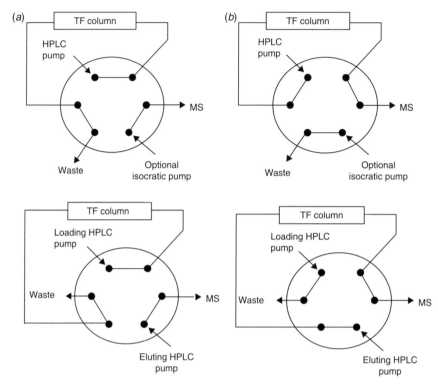

Figure 10.4. Single-column method configurations: (a) sample load and cleanup; (b) sample elution.

as a tandem mass spectrometer. Selective elution of the desired analytes proceeds with a mobile phase of suitable strength under laminar-or turbulent-flow conditions. Many bioanalytical assays utilize this so-called single-column methodology. A typical example of a single-column approach is shown in Fig. 10.4. Table 10.1 lists some of the applications of single-column methods found in the literature.

10.4.2   Dual-Column Methods

While single-column methods work quite well, these methods have two main drawbacks. First, when coupling turbulent-flow directly to a mass spectrometer, the mobile-phase effluent (4 or 5 mL/min) has to be split to make the effluent flow compatible for mass spectrometers (<1 mL/min). Although narrow-bore TFLC columns (0.5 mm i.d.) can be operated at 1–1.5 mL/min, only some mass spectrometers are capable of operating at these flow-rates. The splitting will result in lower detection limits. The lower sensitivity is not due to the detection limit of the mass spectrometer, which acts as a concentration detector, but to the fact that the analyte is more diluted when contained in the higher TFLC flows. There is also more mobile-phase waste to dispose. Of course, this may not be a major drawback, if one were to collect fractions

TABLE 10.1. Selected Application Using Single-Column TFLC and LC–MS/MS Detection

Application	Matrix	Cycle Time (min)	Reference
Pharmacokinetics	Serum	2.5	Ayrton et al., 1997a
Pharmacokinetics	Rat plasma and serum	4	Jemal et al., 1998b
Pharmacokinetics	Plasma	2.5	Zimmmer et al., 1999
Pharmacokinetics	Plasma	1.2	Ayrton et al., 1998a
Pharmacokinetics	Plasma	2	Ayrton et al., 1999
Metabolism	Human plasma	2	Hopfgartner et al., 2002
Pharmacokinetics and Metabolism	Bovine kidney	6	Van Eeckhout et al., 2000
Metabolism	P450 microsomes	2	Ayrton et al., 1998c
Pharmacokinetics	Plasma	2	Ayrton et al., 1998b
Metabolism	Plasma	2	Hopfgartner and Bourgogne, 2003

identified by the mass spectrometer for further study or if a different detector, such as UV, is desired. Second, using laminar flow to elute into the mass spectrometer from a turbulent-flow column typically results in broad and tailed chromatographic peaks due to the lower number of theoretical plates exhibited by columns packed with large particles.

In most common turbulent-flow chromatography experiments, a second conventional high-performance liquid chromatography (HPLC) column is employed. The conventional HPLC column, usually referred to as the analytical column, is used for peak sharpening and additional separation prior to MS detection. This approach requires an additional valve and pump to control the final HPLC separation. A typical dual-column methodology is shown in Fig. 10.5.

Samples are injected onto the turbulent-flow column similar to single-column methods. The analytes of interest are retained in the turbulent-flow column while the large macromolecules are eluted to waste. Once the analytes are separated from the matrix, the samples are then eluted into the analytical column. The characteristics of the analytical column determine the peak shape and separation seen at the MS detector. Flow rates which are compatible with the mass spectrometer can then be used and the chromatograms are based on conventional HPLC parameters. The key to dual-column methods is that the retentive properties of the analytical column must be sufficiently stronger than that of the turbulent-flow column; the dual-column approach is performed in such a manner so that the mobile-phase composition needed to elute the analyte from the turbulent-flow column does not elute the analyte from the analytical column. The sample is then focused at the head of the analytical column until the mobile-phase conditions are changed to elute the analyte. The choice of columns is critical to the success of dual-column methods. Table 10.2 lists some of the applications of dual-column methods found in the literature.

After elution of the desired analytes, the turbulent-flow extraction column is back-flushed and washed with a suitable solvent in an effort to remove more retentive

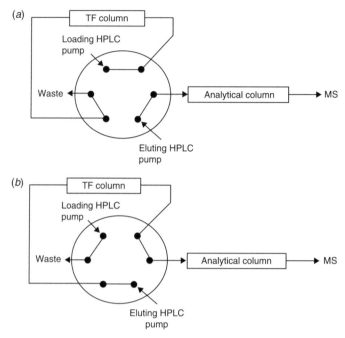

Figure 10.5. Dual-column method configuration: (a) sample load and cleanup; (b) sample elution.

TABLE 10.2. Selected Applications Using Dual-Column TFLC and LC–MS/MS Detection (Zeng et al., 2002)

Application	Matrix	Cycle Time (min)	Reference
Pharmacokinetics	Plasma	5	Hsieh et al., 2000
Pharmacokinetics	Human serum	5	Jemal et al., 1998b
Pharmacokinetics	Human plasma	2	Jemal et al., 1999b
Pharmacokinetics	Human plasma	1.6	Jemal et al., 1999a
Pharmacokinetics	Human plasma	8	Jemal and Xia, 2000
Pharmacokinetics	Plasma	8	Zeng et al., 2002
Pharmacokinetics	Human plasma	6	Kollroser and Schober, 2002
Metabolite screening	Microsomes	8	Lim et al., 2001
Pharmacokinetics	Human serum	0.4 and 0.8	Hsieh et al., 2004
Pharmacokinetics	Plasma	5	Brignol et al., 2000b
Pharmacokinetics	Human plasma	2	Ramos et al., 2000
Multicomponent screening	Dog plasma	10	Wu et al., 2000
Pharmacokinetics	Human serum	5	Du et al., 2005
Pharmacokinetics	Plasma	1.5	Zhou et al., 2005
Metabolism	Dog plasma	8	Wu et al., 2004
Pharmacokinetics	Human plasma		Ceglarek et al., 2004
Metabolism	Plasma	18.5	Takino et al., 2003
Pharmacokinetics	Plasma	1.2	Zeng et al., 2003
Pharmacokinetics	Plasma	2.5	Jemal et al., 2000
Metabolism	Plasma and brain	2 and 4	Ong et al., 2004

sample components and reduce carryover. Appropriate solvents are then pumped through the extraction column (turbulent-flow column) as well as the optional HPLC column (analytical column) to prepare the system for the next injection. Care must be taken to allow complete rinsing and equilibration with an aqueous mobile phase before injecting plasma or other protein-containing samples in order to prevent protein precipitation within the system.

10.4.3 Isocratic Focusing

While single- and dual-column methodologies work extremely well, in almost every case reported in the literature, there is a fair amount of peak tailing observed. The peak tailing is due to the rate-limiting diffusion of analytes deep within the pores of the large particles packed in the turbulent-flow columns during the elution step. Although acceptable quantitative results can be achieved, there is some loss in sensitivity and reproducibility associated with peak tailing. One can minimize peak tailing by coupling the extraction column to a much more retentive HPLC column. The mobile-phase strength sufficient for elution from the extraction column will not cause significant elution through the HPLC column, causing the analytes to "focus" into a sharp band at the head of the HPLC column. Although this approach effectively sharpens analyte peaks eluted from the turbulent-flow extraction column, the approach requires time and effort to test a suitable HPLC column and optimize mobile-phase conditions for the transfer step and final separation.

Isocratic focusing is a much more effective technique for improving the peak shape observed in most online sample cleanup methodologies. Isocratic focusing introduces a second mobile-phase stream that merges and mixes with the effluent stream from the turbulent-flow column prior to reaching the analytical column. Typically, the second mobile-phase stream is 100% aqueous. The result is that the strength of the mobile phase eluting from the turbulent-flow column, which carries the extracted analytes, is diluted before reaching the analytical column. The analytes focus into a sharp band at the head of the HPLC column even when the columns have similar retentive properties. The advantage of this approach is that the analytical column can be chosen based on desired chromatography rather than having to consider the column chemistry between the turbulent-flow and analytical columns. In addition, isocratic focusing greatly reduces the amount of time required for method development because retention characteristics between the turbulent-flow and analytical columns are not required to be addressed.

The general methodology of isocratic focusing dual-column cleanup methods is illustrated in Fig. 10.6 (Herman, 2004). The sample is injected onto a turbulent-flow column under turbulent-flow conditions (high flow rates) with 100% aqueous mobile phase (Fig. 10.6a). Small molecules are retained, while proteinlike material is washed to waste. Once the analytes are separated from the biological matrix, the analytes are eluted from the turbulent-flow column with organic mobile phase (Fig. 10.6b). The flow from the turbulent-flow column is combined with a 100% aqueous flow from a second HPLC pump prior to reaching an analytical column. During this transfer step, the analyte molecules accumulate at the head of the

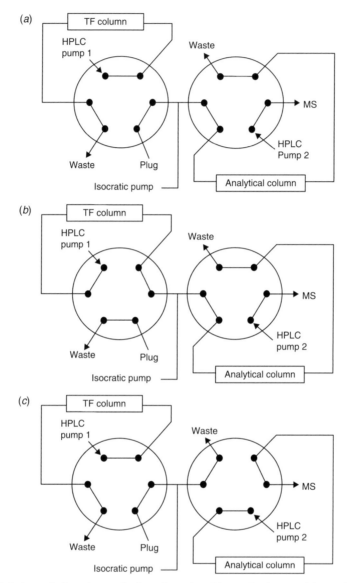

Figure 10.6. *Isocratic focusing method configuration: (a) sample cleanup; (b) sample transfer; (c) sample elution.*

HPLC column as they are carried by the diluted mobile phase. After the transfer and focusing are complete, the analytes are eluted from the HPLC column with a fast gradient (Fig. 10.6c). A comparison of the dual-column methods with and without isocratic focusing is shown in Fig. 10.7 and Table 10.3. Figure 10.7 demonstrates a greater than fourfold increase in signal height when using isocratic focusing and

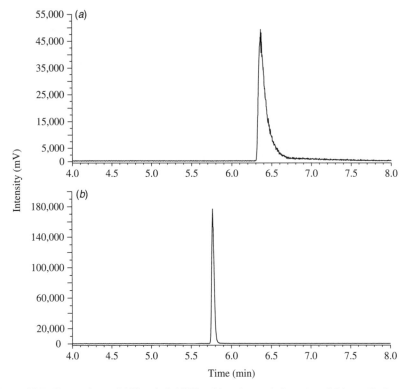

Figure 10.7. *Comparison of 100 ng/mL K252a: (a) no isocratic focusing; (b) isocratic focusing (Herman, 2002).*

Table 10.3 shows a reduction in the percent relative standard deviation (%RSD) by a factor of two for 10 replicate injections.

Sample run times, of isocratic focusing methods, are reduced by two improvements to the valve configuration. First, incorporating a transfer loop on the valve connected to the turbulent flow column, which is prefilled with eluant prior to injection of a sample, greatly reduces the sample run time; this is accomplished by eliminating the time needed to clear the void volume between the mixing chamber of the loading pump and the turbulent flow column. The same result can be achieved by adding an additional pump that is placed into the valve in the same way as the loop, but is more expensive and uses more of the mobile phase. Second, using an internal tee

TABLE 10.3. Comparison of 10 Replicate Injections of 100 ng/mL K252a over Time with and without Isocratic Focusing (Herman, 2002).

Injection no.	1–10	100–109	500–509	1000–1009	No Isocratic Focusing, 11–19
Area counts	2,1450	21,860	21,140	22,110	17,330
Std. dev.	393	415	412	487	745
% RSD	1.83	1.90	1.95	2.20	4.30

in Valve 2 reduces the void volume between the two valves, thereby reducing the transfer time needed to elute samples from the turbulent-flow column into the HPLC column. Figure 10.8 illustrates the configuration for this approach. Table 10.4 lists some of the applications of isocratic focusing methods found in the literature.

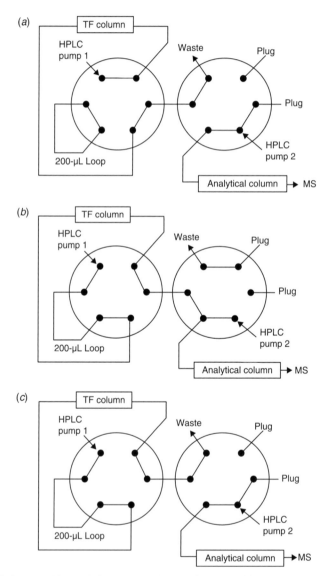

Figure 10.8. *Improved isocratic focusing method configuration: (a) sample cleanup; (b) sample transfer; (c) sample elution (Herman, 2002).*

TABLE 10.4. Selected Applications Using Isocratic Focusing TFLC and LC–MS/MS Detection

Application	Matrix	Cycle Time (min)	Reference
ADME screening	Plasma, urine, CSF, liver, brain, S9 fractions, microsomes, PAMPA	4.8	Herman, 2002a
Pharmacokinetics	Rat plasma	4	Mallet et al., 2001
Pharmacokinetics	Human plasma	4 and 1.5	Grant et al., 2002
Pharmacokinetic screening	Plasma, urine, CSF, liver, brain	6	Herman, 1999
Pharmacokinetic screening	Plasma, urine, CSF, liver, brain	8	Herman, 2004
Metabolism	Human plasma	5	Zeng et al., 2004
Pharmacokinetics	Plasma	4.5	Xu et al., 2005a
Metabolite identification	Rat plasma and bile	18.5	Herman, 2005
Pharmacokinetics	Plasma	4.5	Yndall and Hansen, 2003
Metabolism	Rat plasma	4	Hendricks et al., 2005

Abbreviations: ADME, Adsorption distribution metabolism excretion; CSF, Cerebrospinal fluid; PAMPA, Parallel artificial membrane permeability assay.

10.4.4 Generic Methods

The incorporation of online sample cleanup shifted the bottleneck in analysis time from sample preparation to method development. When screening large libraries for new chemical series, many new compounds must be evaluated every week with the various ADME screens already in place. If new methods were needed for each compound, then there would not be enough time left to actually assay the samples. Therefore, to truly take full advantage of high-throughput absorption, distribution, metabolism, and elimination (ADME) screening, it becomes necessary to develop a generic method that eliminates the need to develop a new method for each new chemical entity being screened.

The importance of removing method development time cannot be overstated. The largest time savings in early development comes not from the length of the assay but from the use of a generic method. For example, it typically takes 24–48 h to develop an isocratic focusing method to improve the injection-to-injection run time to 2–3 min. At that point, 192 samples can be assayed in 8 h. A generic method, once developed, takes no additional development time when changing to NCEs. If the assay takes 5 min from injection to injection, then 192 samples can be run in 16 h. Therefore, the faster method took half the time to run the same number of samples as the generic method. However, when the 24 h of development time is added to the faster method, the faster method actually takes 32 h to run 192 samples. Therefore, even though the injection-to-injection run time of the nongeneric method is twice as fast as the generic method, it actually took twice as long to analyze the samples. Keep in mind that this approach is ideal for early screening when several samples are assayed each week and may never be encountered again. Later in development, when the same compound is investigated for several months or even years, the extra time spent on developing a faster method becomes worth the time invested.

Once isocratic focusing was used to improve peak shape, it quickly became apparent that isocratic focusing had several other advantageous properties that resulted in the development of a generic method (Herman, 2002a). The isocratic focusing resulted in increasing the lifetimes of the turbulent-flow extraction columns. The longer column lifetimes are due to the fact that the shape of the peak eluting from the extraction column becomes irrelevant to what is observed at the MS detector. Therefore, if the shapes of peaks eluting from the turbulent-flow column start to deteriorate, then it is still possible to use this column. Figure 10.9 illustrates this point.

The refocusing of the peak eluting from the extraction column by adding the isocratic flow is responsible for the development of a generic method. This is due to the fact that the peak shape eluting from the cleanup column (turbulent-flow column) becomes irrelevant to the peak seen by the MS detector eluting from the analytical column. One only has to guarantee that the entire eluting peak is collected during the transfer from the extraction column to the analytical column. Therefore, by choosing the appropriate percent organic in the mobile phase for elution from the extraction

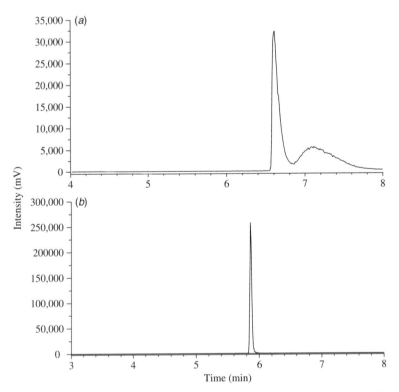

Figure 10.9. *100 ng/mL K252a after 1000 injections of plasma: (a) without isocratic focusing; (b) with isocratic focusing (Herman, 2002).*

column, the hydrophilic compounds will elute quickly while hydrophobic ones will elute as broader peaks over time. Both peak shapes will be less than ideal as they elute from the turbulent-flow column, but both types of compounds will produce excellent responses at the detector because of the isocratic focusing done at the head of the analytical column. A wide range of hydrophobicities can be analyzed by the same method as long as the entire peak is collected during the transfer step. To accomplish complete transfer of the eluting peak, it was necessary to increase the size of the transfer loop from 50 to 200 µL.

The utility of a generic approach has been demonstrated previously (Herman, 2002a) and was revised to incorporate the narrower bore (0.5-mm-i.d.) turbulent-flow columns that reduced the solvent consumption and reduced the method cycle time (Herman, 2004). Further refinements have been made to improve carryover and reduce the cycle run times even further. The valve configuration for the generic focusing method is the same as that illustrated in Fig. 10.8 for isocratic focusing. Table 10.5 describes the current generic method used by the authors. Figure 10.10 shows chromatograms of six compounds (log P values from 2 to 6) spiked into rat plasma and obtained using the generic method in Table 10.5. All the compounds showed excellent peak shapes and $>90\%$ recoveries. The nature of the biological matrix had no effect on the utility of the method. LC–MS responses from 100 ng/mL of K252a in plasma, urine, cerebrol spinal fluid, brain homogenate,

TABLE 10.5. Generic Method

Time (min)	A%	B%	C%	Valve Position	Flow (mL/min)
			Pump 1[a]		
0.00	100	0	0	Load	1.5
0.50	100	0	0		1.5
0.58	60	40	0	Inject	0.3
1.58	60	40	0		0.3
1.58	60	40	0	Load	1.5
2.50	0	0	100		1.5
3.00	100	0	0		1.5
4.50	100	0	0		1.5
			Pump 2[b]		
0.00	100	0	Out	1.2	
0.58	100	0	In	1.2	
1.58	100	0	Out	1.0	
2.58	5	95		1.0	
3.00	5	95		1.0	
3.01	100	0		1.2	
4.50	100	0		1.2	

[a]Turbulent flow column, Cyclone P HTLC 0.5×50 mm; injection volume, 25 µL; solvent A, 0.05% formic acid in H_2O; solvent B, 0.05% formic acid in ACN; solvent C, 40/40/15/5 ACN/IPA/acetone/formic acid.
[b]Analytical column, Eclipse XDB C18, 4.6×15 mm, 3 µm, 120 A; solvent A, 0.05% formic acid in H_2O; solvent B, 0.05% formic acid in ACN.

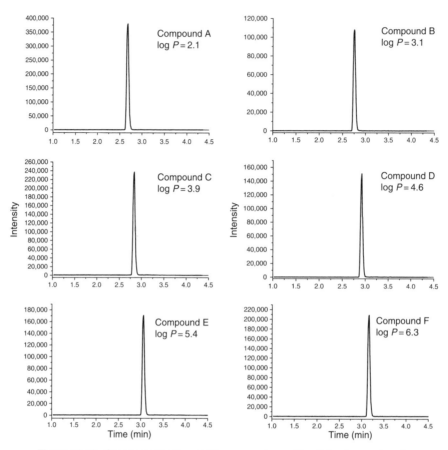

Figure 10.10. *Generic method versatility, comparison of log P to observed peak.*

intestinal perfusate, and liver homogenate are shown in Fig. 10.11. There was no significant difference observed in the response due to the biological matrix.

Compounds that cannot be analyzed using the generic methodology generally fall into two categories. First, very hydrophilic compounds may not be retained on either the Cyclone P turbulent-flow column or the analytical column at 8% organic. Approximately 2–3% of the compounds tested are not completely retained on either the turbulent-flow or analytical HPLC column. Compounds that are not completely retained on the head of the analytical column will have peak fronting due to partial migration of the compound through the column isocratically with the 8% organic seen during the transfer step. In extreme cases this can result in peak splitting or compound elution during the transfer step. The peak fronting introduces more error (less reproducibility) and lowers sensitivity due to peak broadening. However, if sensitivity and reproducibility requirements [limit of quantitation (LOQ) of 1 ng/mL and all standards and quality controls (QCs) within 20% of expected] are met, then the method is used with no revisions. The peak fronting can be eliminated by transferring the compound from the turbulent-flow-column to the analytical column with less

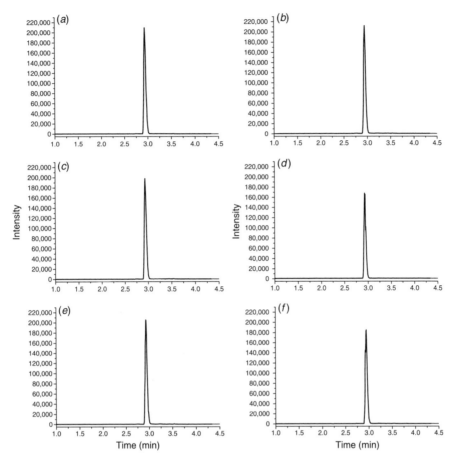

Figure 10.11. MRM chromatogram of 100 ng/mL K252a in (a) plasma, (b) urine, (c) cerebral spinal fluid (CNF), (d) brain homogenate, (e) liver homogenate, and (f) intestinal perfusate (Herman, 2002).

organic solvent in the mobile phase, changing to a weaker organic solvent such as methanol, or selecting an analytical column that will retain hydrophilic compounds more strongly. If the compound is not retained by the turbulent-flow column, a different turbulent-flow stationary phase is needed. However, there will be some extremely hydrophilic compounds that will not be retained by any of the available packing materials.

Second, extremely hydrophobic compounds may not be completely transferred to the analytical column at 40% organic, thereby reducing recovery and thus sensitivity. Only 0.3% of the compounds tested did not meet our sensitivity requirements due to this problem. Increasing the percent organic in the loop used to transfer the sample onto the analytical column or using a less retentive turbulent-flow column will usually solve this problem. It should be duly noted that compounds in this category would not conform as good drug candidates from a drug delivery perspective. These compounds would have extremely low aqueous solubilities.

10.5   ONLINE SAMPLE CONCENTRATION

Isocratic focusing can also be used to concentrate analytes online (Herman, 2005). Analytes from several injections can be extracted, cleaned, and transferred to the HPLC column before performing the gradient elution step. Analytes will be retained on the analytical column because the column is only exposed to a high aqueous mobile phase during the clean and transfer steps. The gradient elution step is performed once the appropriate number of injections has been concentrated. This methodology is quite useful for low-level metabolite identification (Herman, 2005).

The general methodology for online cleanup and concentration is the same as that used for isocratic focusing when assaying for known compounds, but sample cleanup and transfer steps are repeated as many times as necessary to concentrate the sample. Once the concentration is complete, elution from the analytical column occurs. LC–MS responses of a single standard injection to one that has been concentrated 10 times (10 injections) are compared in Fig. 10.12. Note that the response remains linear as long as the trapping analytical column is not over loaded.

To unambiguously identify unknowns, full-scan mass spectral data for each of the separated components are required. Complete baseline separation is not needed because these data are qualitative. However, the mass spectrum of each peak in the chromatogram must be from a single component. The 15-mm columns normally used for HTS are still needed during the concentration steps because the back pressure would be too high on a 2.1×150- or 4.6×150-mm column at the 1.5 mL/min flow used during the transfer step. Therefore, in order to introduce the longer analytical column needed for good HPLC separation, a third valve is added to the typical isocratic focusing method. Figure 10.13 illustrates the general methodology used for metabolite ID (Herman, 2005). The third valve is used to divert the mobile phase to waste during the cleanup and transfer steps. Once the concentration step is completed, the third valve is switched and the analytes are eluted into a 150-mm HPLC column prior to gradient elution into the mass spectrometer for analysis. In this case, the 4.6×15-mm column is essentially a guard column that is used to concentrate the sample before elution onto the analytical column. Figure 10.14 compares the mass spectrum from the constant neutral loss scan (NLS) of 78 from a single

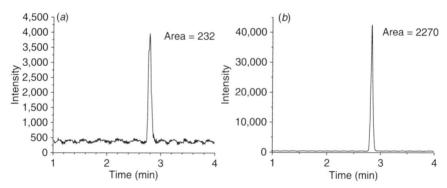

Figure 10.12. *Comparison of 10 ng/mL standard in rat plasma: (a) single injection vs. (b) 10 injections concentrated online with isocratic focusing (Herman, 2005).*

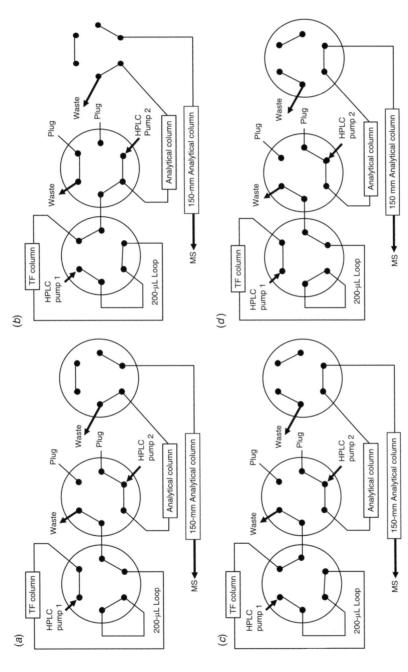

Figure 10.13. Online cleanup and concentration method: (a) sample cleanup; (b) sample transfer; (c) sample elution, clean turbulent-flow column, and fill loop; (d) sample elution, equilibrate turbulent-flow column (Herman, 2005).

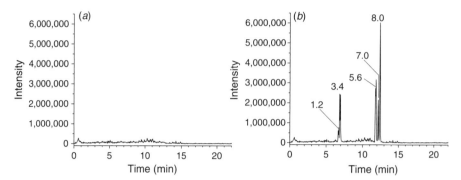

Figure 10.14. *Chromatograms from constant-neutral loss scans for 78 obtained using 2 h post-dose rat plasma following: (a) single injection; (b) 20 injections concentrated online with isocratic focusing (Herman, 2005).*

injection of unprocessed rat plasma to that of 25 injections concentrated online (Herman, 2005). No metabolites were observed from a single injection but eight were identified in the concentrated sample.

10.6   ONLINE SAMPLE CONDITIONING

Analytes are often isolated from various sample preparations using reversed-phase extraction columns. However, extraction efficiency decreases as the organic solvent

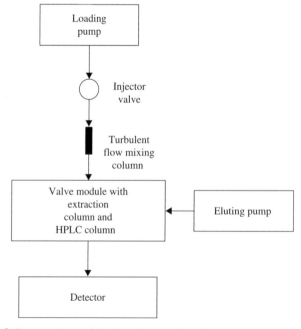

Figure 10.15. *Online sample conditioning using turbulent-flow mixing column (Hendricks, 2005).*

content and injection volume of the samples increase, thereby decreasing assay sensitivity. Microsomal and "crashed" plasma preparations typically contain 50–70% acetonitrile. To assure adequate retention of analytes from large injection volumes, it becomes necessary to either evaporate the excess organic solvent or minimize its effect by diluting each sample prior to reaching the extraction column. The process is more convenient to dilute each injection online by adding extra volume between the injector and the extraction column. The effectiveness of two online dilution approaches—laminar flow through large-volume tubing and turbulent-flow through mixing columns packed with inert beads—was recently investigated (Hendricks et al., 2005). Both approaches dramatically improved extraction efficiency of large injections of acetonitrile-rich samples resulting in greater assay sensitivity. Optimum extraction efficiencies were achieved for injections up to 100 μL into a Cyclone 0.5 × 50-mm turbulent-flow column by allowing the injections to pass through two 250-μL mixing columns installed between the injector and valve module while pumping an aqueous mobile phase under turbulent-flow conditions. A schematic of online sample conditioning is illustrated in Fig. 10.15.

10.7 CARRYOVER REDUCTION

One major source of carryover, excluding the obvious issues such as worn rotor seals, is protein precipitation in the autosampler. This situation is more problematic to those who inject raw biological matrices. Before injecting any biological fluid and immediately after the injection, it is imperative that the autosampler needle, syringe, and injection valve be rinsed with an aqueous solvent to avoid protein precipitation in those components. To quench interactions between sample components, especially basic compounds, and the surfaces within these components, the aqueous solvent should contain some acid (0.1% formic or acetic are often used) or some other reagent, such as the disodium salt of ethylenediaminetetraacetic acid (EDTA), which competes for active sites. Postinjection aqueous rinses should be followed by rinses with strong organic solvent to dissolve and wash hydrophobic sample components. Methanol or acetonitrile works well for most compounds. However, for some applications, solvent mixtures are more effective in reducing carryover. These include 1 : 1 mixtures of isopropanol and acetonitrile, to which is often added 10% acetone or DMSO.

The second major source of carryover is from the turbulent-flow column itself. Thorough cleaning of the column is necessary since the mobile phase may not be sufficiently strong to remove residual sample components between runs. Changing the organic mobile phase can alleviate this problem but may adversely effect the chromatography. Quaternary pumps are preferred over binary pumps for operating the turbulent-flow columns, permitting washes with strong solvents or solvent mixtures other than those used for the mobile phase (Herman, 2004). Precipitated protein trapped in the frits or pores of the turbulent-flow column is a common source of carryover. Back flushing to waste for a few seconds before transfer to the analytical column is one way to remove large particles trapped at the head of the extraction column.

Unfortunately, this is not permitted in isocratic focusing setups unless the transfer loop is removed. Doing so would result in excessive method time needed to permit the eluting solvent to clear the dwell volume of the loading pump. Back flushing the turbulent-flow column with turbulent flow of aggressive solvents immediately after the transfer step effectively reduces carryover. Zeng et al. (2004) have demonstrated that 15% acetic acid works very well for solubilizing proteins and similar material. However, only polymer columns are stable at such a low pH. The same group also showed that an additional wash with tetrahydrofuran (THF) not only eliminated carryover but also greatly extended the lifetimes of the turbulent-flow columns. The drawback to using THF is that it degrades polyetheretherketone (PEEK) tubing and requires that all HPLC connections be stainless steel. An alternative to THF is a mixture of $40/40/15/5$ acetonitrile–isopropanol–acetone–formic acid. This has been used with PEEK tubing with no deleterious effects as long as the wash times are performed in short durations.

The last source of carryover that is present in any LC system that has solvent changes and valves is found in the dead spaces between all the connections but especially in the valves. Cycling the valves back and forth a couple of times during the wash cycle will greatly reduce this source of carryover.

10.8   OTHER CONSIDERATIONS

Since turbulent-flow chromatography can be used on neat biological samples, there are several issues that need to be addressed beyond the LC–MS/MS methodology. First, the stability of the analytes of interest needs to be determined. This is not an issue in the good laboratory practices (GLP) environment where thorough validation is necessary for every method. However, in the early screening, there is no time to perform stability studies on every compound being investigated in every matrix being tested. Second, analytes that have high protein binding may be carried to waste along with the proteins. A low-pH mobile phase is usually sufficient to unfold most proteins and release the protein-bound analytes. However, quenching protein interactions may not be 100% complete. Third, there could be problems making standards and QCs for some compounds whose aqueous solubility is extremely low. Therefore, if an analyte of interest is stable in the biological matrix being used, is not protein bound in the mobile phase, and is sufficiently water soluble to make QCs and standards, then there is no issue using neat biological matrices. If any of these criteria are not met, then some sample preparation is necessary.

An alternative to investigating each of the criteria listed above is to take a generic approach to sample preparation that will work for all compounds. For quantitative analysis, an internal standard must be added to the biological matrix. Most plasma samples come frozen and, when thawed, there is some solid proteinlike material present. Tissue homogenates also have solid proteinlike material present. Therefore, all biological samples need to be filtered or centrifuged to remove the solid material before analysis. By adding internal standards and QCs in an organic solvent, such as methanol or acetonitrile, protein precipitation will occur. The protein precipitation will kill all of the enzymes present in the biological matrix, release all the protein-bound analyte, and ensure that low-solubility compounds stay in solution. The samples can then be centrifuged and the supernatant harvested for analysis. This procedure does not add any additional time or steps to the analysis since the solid material needs to be removed even when running neat biological samples. As shown in Fig. 10.1, protein precipitation alone does not remove all of the ion suppressors or clean the sample very well. In addition, when protein precipitation is used for sample cleanup, a $4:1$ or $5:1$ ratio of organic to biological matrix is normally used. When using turbulent-flow chromatography, the samples are cleaned by the turbulent-flow column so well that thorough protein precipitation is not necessary. A $1:1$ ratio of organic to biological matrix is all that is necessary to kill the enzymes, release the protein-bound analytes, and improve solubility of compounds with low aqueous solubility. Keep in mind that when these issues are not a problem, unprocessed biological matrices can be used without reservation.

10.9 SUMMARY

TFLC in conjunction with API mass spectrometry is an ideal technique for analyzing small druglike molecules in the presence of biological matrices. TFLC takes advantage of the size exclusion properties created by turbulent flow and the conventional HPLC properties of the TFLC columns to perform two-dimensional separations. Large proteinlike material is separated from the small molecules and washed to waste during the initial loading step due to their exclusion from the pores of the stationary phase under turbulent-flow conditions. Small molecules are less affected by the turbulent-flow and are still able to migrate into the pores of the stationary phase. The loading step is carried out with a 100% aqueous mobile phase such that the TFLC column retains the small analytes of interest. The retention of the sample molecules present in the sample is based on conventional HPLC partitioning. Therefore, sample components that are more or less retentive than the compound of interest are separated by conventional HPLC. It is the two-dimensional aspect of the TFLC column that makes it extremely efficient at achieving online sample cleanup.

TFLC–MS methodologies have many applications in both pharmacokinetics and metabolism. Quantitative and qualitative analysis have been performed with TFLC–MS in GLP and non–GLP environments. In comparison to other sample extraction methods, TFLC is more efficient at separating the analyte of interest from interfering matrix components. No sample preparation is necessary to use TFLC, although centrifugation or filtration may be desirable to remove solid components if present. TFLC is also more generic than other online extraction methods, which reduce methods development time. Methods employing TFLC–MS are easy to automate and do not suffer from recovery issues during the methods optimization stage because the separation is occurring online.

Isocratic focusing effects are used to (1) enhance peak shape; (2) allow generic approaches to rapid pharmacokinetic screening, (3) perform online concentration for low level metabolite identification, and (4) provide for reducing the amount of organic solvent present in the mobile phase of microsomal incubations to allow retention of hydrophilic components. The increasing use of TFLC–MS is evident in the literature, presentations at international conferences, and the growing availability within contract organizations. The growth in the use of TFLC–MS in the pharmaceutical industry clearly indicates that TFLC–MS is becoming one of the methods of choice for investigating analytes in the presence of biological matrices.

REFERENCES

Ayrton, J., Clare, R. A., Dear, G. J., Mallett, D. N., and Plumb, R. S. (1999). Ultra-high flow rate capillary liquid chromatography with mass spectrometric detection for the direct analysis of pharmaceuticals in plasma at sub-nanogram per millilitre concentrations. *Rapid Commun. Mass Spectrom.* **13**:1657–1662.

Ayrton, J., Dear, G. J., Leavens, W. J., Mallett, D. N., and Plumb, R. S. (1997a). The use of turbulent-flow chromatography/mass spectrometry for the rapid, direct analysis of a novel pharmaceutical compound in plasma. *Rapid Commun. Mass Spectrom.* **11**:1953–1958.

Ayrton, J., Dear, G. J., Leavens, W. J., Mallett, D. N., and Plumb, R. S. (1998a). Optimization and routine use of generic ultra-high flow-rate liquid chromatography with mass spectrometric detection for the direct online analysis of pharmaceuticals in plasma. *J. Chromatogr. A* **828**:199–207.

Ayrton, J., Dear, G. J., Leavens, W. J., Mallett, D. N., and Plumb, R. S. (1998b). Use of generic fast gradient liquid chromatography-tandem mass spectroscopy in quantitative bioanalysis. *J. Chromatogr. B Biomed. Sci. Appl.* **709**:243–254.

Ayrton, J., Dear, G. J., Leavens, W. J., Mallett, D. N., and Plumb, R. S. (1997b). The use of turbulent-flow chromatography/mass spectrometry for the rapid, direct analysis of a novel pharmaceutical compound in plasma. *Rapid Commun. Mass Spectrom.* **11**:1953–1958.

Ayrton, J., Plumb, R., Leavens, W. J., Mallett, D., Dickins, M., and Dear, G. J. (1998c). Application of a generic fast gradient liquid chromatography tandem mass spectrometry method for the analysis of cytochrome P450 probe substrates. *Rapid Commun. Mass Spectrom.* **12**:217–224.

Black, K., Erol, H., and Di Bussolo, J. M. (2006). Dealing with detergents and dosing vehicles in samples submitted for LC–MS analysis. In *Proceedings of the 54th ASMS Conference on Mass Spectrometry and Allied Topics*, Seattle, WA.

Brignol, N., Bakhtiar, R., Dou, L., Majumdar, T., and Tse, F. L. (2000a). Quantitative analysis of terbinafine (Lamisil) in human and minipig plasma by liquid chromatography tandem mass spectrometry. *Rapid Commun. Mass Spectrom.* **14**:141–149.

Brignol, N., Bakhtiar, R., Dou, L., Majumdar, T., and Tse, F. L. S. (2000b). Quantitative analysis of terbinafine (Lamisil®) in human and minipig plasma by liquid chromatography tandem mass spectrometry. *Rapid Commun. Mass Spectrom.* **14**:141–149.

Ceglarek, U., Lembcke, J., Fiedler, G. M., Werner, M., Witzigmann, H., Hauss, J. P., and Thiery, J. (2004). Rapid simultaneous quantification of immunosuppressants in transplant patients by turbulent-flow chromatography combined with tandem mass spectrometry. *Clin. Chim. Acta.* **346**:181–190.

DeMaio, W., Di Bussolo, J. M., Lloyd, T., Kandoussi, H., and Talaat, R. E. (2005). On-line extraction and concentration techniques using turbulent-flow hplc to analyze human plasma for metabolite identification by LC/MS. In *Proceedings of the 53rd ASMS Conference on Mass Spectrometry and Allied Topics*, San Antonio, TX.

Du, L., Musson, D. G., and Wang, A. Q. (2005). High turbulence liquid chromatography online extraction and tandem mass spectrometry for the simultaneous determination of suberoylanilide hydroxamic acid and its two metabolites in human serum. *Rapid Commun. Mass Spectrom.* **19:**1779–1787.

Giddings, J. C. (1965). Diffusion and kinetics and chromatography. In *Dynamics of Chromatography: Principles and Theory* (Giddings, J. C. Ed.). Marcel Dekker, New York, pp. 217–225.

Grant, R. P., Cameron, C., and Mackenzie-McMurter, S. (2002). Generic serial and parallel on-line direct-injection using turbulent-flow liquid chromatography/tandem mass spectrometry. *Rapid Commun. Mass Spectrom.* **16:**1785–1792.

Hendricks, J., Navaline, K., and Di Bussolo, J. (2005). On-line dilution techniques to maximize analyte extraction from acetonitrile-rich samples in turbulent-flow LC–MS/MS systems. In *Proceedings of the 53rd ASMS Conference on Mass Spectrometry and Allied Topics*, San Antonio, TX.

Herman, J. L. (1999). Generic LC/MS method for analysis of drug substances in neat biological matrices. *Technical Session on Chromatography, Eastern Analytical Symposium*, Sommerset, NJ.

Herman, J. L. (2002a). Generic method for on-line extraction of drug substances in the presence of biological matrices using turbulent-flow chromatography. *Rapid Commun. Mass Spectrom.* **16:**421–426.

Herman, J. L. (2002b). Generic method for on-line extraction of drug substances in the presence of biological matrices using turbulent-flow chromatography. *Rapid Commun. Mass Spectrom.* **16:**421–426.

Herman, J. L. (2004). Generic approach to high throughput ADME screening for lead candidate optimization. *Int. J. Mass Spectrom.* **238:**107–117.

Herman, J. L. (2005). The use of turbulent-flow chromatography and the isocratic focusing effect to achieve on-line cleanup and concentration of neat biological samples for low-level metabolite analysis. *Rapid Commun. Mass Spectrom.* **19:**696–700.

Hopfgartner, G., and Bourgogne, E. (2003). Quantitative high-throughput analysis of drugs in biological matrices by mass spectrometry. *Mass Spectrom. Rev.* **22:**195–214.

Hopfgartner, G., Husser, C., and Zell, M. (2002). High-throughput quantification of drugs and their metabolites in biosamples by LC–MS/MS and CE-MS/MS: Possibilities and limitations. *Ther. Drug Monit.* **24:**134–143.

Hsieh, S., Tobien, T., Koch, K., and Dunn, J. (2004). Increasing throughput of parallel on-line extraction liquid chromatography/electrospray ionization tandem mass spectrometry system for GLP quantitative bioanalysis in drug development. *Rapid Commun. Mass Spectrom.* **18:**285–292.

Hsieh, Y., Bryant, M. S., Gruela, G., Brisson, J. M., and Korfmacher, W. A. (2000). Direct analysis of plasma samples for drug discovery compounds using mixed-function column liquid chromatography tandem mass spectrometry. *Rapid Commun. Mass Spectrom.* **14:**1384–1390.

Jemal, M., Huang, M., Jiang, X., Mao, Y., and Powell, M. L. (1999a). Direct injection versus liquid-liquid extraction for plasma sample analysis by high performance liquid chromatography with tandem mass spectrometry. *Rapid Commun. Mass Spectrom.* **13:**2125–2132.

Jemal, M., Ouyang, Z., and Powell, M. L. (2000). Direct-injection LC-MS-MS method for high-throughput simultaneous quantitation of simvastatin and simvastatin acid in human plasma. *J. Pharm. Biomed. Anal.* **23:**323–340.

Jemal, M., Ouyang, Z., Xia, Y. Q., and Powell, M. L. (1999b). A versatile system of high-flow high performance liquid chromatography with tandem mass spectrometry for rapid direct-injection analysis of plasma samples for quantitation of a beta-lactam drug candidate and its open-ring biotransformation product. *Rapid Commun. Mass Spectrom.* **13:**1462–1471.

Jemal, M., and Xia, Y. Q. (2000). Bioanalytical method validation design for the simultaneous quantitation of analytes that may undergo interconversion during analysis. *J. Pharm. Biomed. Anal.* **22:**813–827.

Jemal, M., Xia, Y. Q., and Whigan, D. B. (1998a). The use of high-flow high performance liquid chromatography coupled with positive and negative ion electrospray tandem mass spectrometry for quantitative bioanalysis via direct injection of the plasma/serum samples. *Rapid Commun. Mass Spectrom.* **12:**1389–1399.

Jemal, M., Yuan, Q., and Whigan, D. B. (1998b). The use of high-flow high performance liquid chromatography coupled with positive and negative ion electrospray tandem mass spectrometry for quantitative bioanalysis via direct injection of the plasma/serum samples. *Rapid Commun. Mass Spectrom.* **12:**1389–1399.

Knox, J. H. (1966). Evidence for turbulence and coupling in chromatographic columns. *Anal. Chem.* **38:**253–261.

Kollroser, M., and Schober, C. (2002). Direct-injection high performance liquid chromatography ion trap mass spectrometry for the quantitative determination of olanzapine, clozapine and *N*-desmethylclozapine in human plasma. *Rapid Commun. Mass Spectrom.* **16:**1266–1272.

Lim, H. K., Chan, K. W., Sisenwine, S., and Scatina, J. A. (2001). Simultaneous screen for microsomal stability and metabolite profile by direct injection turbulent-laminar flow LC-LC and automated tandem mass spectrometry. *Anal. Chem.* **73:**2140–2146.

Mallet, C. R., Mazzeo, J. R., and Neue, U. (2001). Evaluation of several solid phase extraction liquid chromatography/tandem mass spectrometry on-line configurations for high-throughput analysis of acidic and basic drugs in rat plasma. *Rapid Commun. Mass Spectrom.* **15:**1075–1083.

Martin, M., and Guiochon, G. (1982). Influence of retention on band broadening in turbulent-flow liquid and gas chromatography. *Anal. Chem.* **54:**1533–1540.

Oberhauser, C. J., Niggebrugge, A. E., Lachance, J. D., Takarewski, J., and Quinn, H. M. (2000). Directions in discovery: Turbulent-flow LC for LC-MS and LC-MS-MS bioanalysis. *LC-GC* **12:**716–724.

Ong, V. S., Cook, K. L., Kosara, C. M., and Brubaker, W. F. (2004). Quantitative bioanalysis: An integrated approach for drug discovery and development. *Int J. Mass Spectrom.* **238:**139–152.

Pretorius, V., and Smuts, T. W. (1966). Turbulent flow chromatography. A new approach to faster analysis. *Anal. Chem.* **38:**274–281.

Quinn, H. M., Menapace, R. A., and Oberhauser, C. J. (1998). High performance liquid chromatography method and apparatus. U. S. Patent 5,795,469, Cohesive Technologies, Acton, MA.

Quinn, H. M., Menapace, R. A., and Oberhauser, C. J. (1999). High performance liquid chromatography method and apparatus. U. S. Patent 5,968,367, Cohesive Technologies, Acton, MA.

Quinn, H. M., and Takarewski, J. J. (1998). High performance liquid chromatography method and apparatus. U. S. Patent 5,772,874, Cohesive Technologies, Acton, MA.

Quinn, H. M., and Takarewski, J. J. (1999). High performance liquid chromatography method and apparatus. U. S. Patent 5,919,368, Cohesive Technologies, Acton, MA.

Ramos, L., Brignol, N., Bakhtiar, R., Ray, T., McMahon, L. M., and Tse, F. L. S. (2000). High-throughput approaches to the quantitative analysis of ketoconazole, a potent inhibitor of cytochrome P450 3A4, in human plasma. *Rapid Commun. Mass Spectrom.* **14**:2282–2293.

Reynolds, O. (1883). An experimental investigation of the circumstances which determine whether the motion of water in parallel channels shall be direct or sinuous and of the law of resistance in parallel channels. *Philos. Trans. Roy. Soc.* **174**:935–982.

Smalley, J., Kadiyala, P., Xin, B., Balimane, P., and Olah, T. (2006). Development of an on-line extraction turbulent-flow chromatography tandem mass spectrometry method for cassette analysis of Caco-2 cell based bi-directional assay samples. *J. Chromatogr. B Ana. Technol. Biomed. Life Sci.* **830**:270–277.

Smalley, J., Marino, A. M., Xin, B., Olah, T., and Balimane, P. V. (2007). Development of a quantitative LC-MS/MS analytical method coupled with turbulent-flow chromatography for digoxin for the in vitro P-gp inhibition assay. *J. Chromatogr. B Ana. Technol. Biomed. Life Sci.* **854**:260–267.

Sternberg, J. C., and Poulson, R. E. (1964). Particle-to-column diameter ratio effect on band spreading. *Anal. Chem.* **36**:1492–1502.

Takino, M., Daishima, S., Yamaguchi, K., and Nakahara, T. (2003). Quantitative liquid chromatography-mass spectrometry determination of catechins in human plasma by automated on-line extraction using turbulent-flow chromatography. *Analyst* **128**:46–50.

Van Eeckhout, N., Perez, J. C., and Van Peteghem, C. (2000). Determination of eight sulfonamides in bovine kidney by liquid chromatography/tandem mass spectrometry with on-line extraction and sample clean-up. *Rapid Commun. Mass Spectrom.* **14**:2331–2338.

Wu, J. T., Zeng, H., Qian, M., Brogdon, B. L., and Unger, S. E. (2000). Direct plasma sample injection in multiple-component LC-MS-MS assays for high-throughput pharmacokinetic screening. *Anal. Chem.* **72**:61–67.

Wu, S. T., Xing, J., Apedo, A., Wang-Iverson, D. B., Olah, T. V., Tymiak, A. A., and Zhao, N. (2004). High-throughput chiral analysis of albuterol enantiomers in dog plasma using on-line sample extraction/polar organic mode chiral liquid chromatography with tandem mass spectrometric detection. *Rapid Commun. Mass Spectrom.* **18**:2531–2536.

Xu, R. N., Fan, L., Rieser, M. J., and El-Shourbagy, T. A. (2007). Recent advances in high-throughput quantitative bioanalysis by LC-MS/MS. *J. Pharm. Biomed. Anal.* **44**:342–355.

Xu, X., Yana, K. X., Songa, H., and Loa, M. W. (2005a). Quantitative determination of a novel dual PPAR α/γ agonist using on-line turbulent-flow extraction with liquid chromatography–tandem mass spectrometry. *J. Chromatogr. B Anal. Technol. Biomed. Life Sci.* **814**:29–36.

Xu, X. S., Yan, K. X., Song, H., and Lo, M. W. (2005b). Quantitative determination of a novel dual PPAR alpha/gamma agonist using on-line turbulent-flow extraction with liquid chromatography-tandem mass spectrometry. *J. Chromatogr. B Anal. Technol. Biomed. Life Sci.* **814**:29–36.

Yndall, L., and Hansen, S. H. (2003). On-line turbulent-flow chromatography–high-performance liquid chromatography–mass spectrometry for fast sample preparation and quantitation. *J. Chromatogr. A* **1020**:59–67.

Zeng, H., Deng, Y., and Wu, J. T. (2003). Fast analysis using monolithic columns coupled with high-flow on-line extraction and electrospray mass spectrometric detection for the direct and simultaneous quantitation of multiple components in plasma. *J. Chromatogr. B Anal. Technol. Biomed. Life Sci.* **788**:331–337.

Zeng, H., Wu, J. T., and Unger, S. E. (2002). The investigation and the use of high flow column-switching LC/MS/MS as a high-throughput approach for direct plasma sample analysis of single and multiple components in pharmacokinetic studies. *J. Pharm. Biomed. Anal.* **27**:967–982.

Zeng, W., Fisher, A. L., Musson, D. G., and Wang, A. Q. (2004). High-throughput liquid chromatography for drug analysis in biological fluids: Investigation of extraction column life. *J. Chromatogr. B Anal. Technol. Biomed. Life Sci.* **806**:177–183.

Zhou, S., Zhou, H., Larson, M., Miller, D. L., Mao, D., Jiang, X., and Naidong, W. (2005). High-throughput biological sample analysis using on-line turbulent-flow extraction combined with monolithic column liquid chromatography/tandem mass spectrometry. *Rapid Commun. Mass Spectrom.* **19**:2144–2150.

Zimmmer, D., Pickard, V., Czembor, W., and Muller, C. (1999). Comparison of turbulent-flow chromatography with automated solid-phase extraction in 96-well plates and liquid–liquid extraction used as plasma sample preparation techniques for liquid chromatography–tandem mass spectrometry. *J. Chromatogr. A* **854**:23–35.

11

Desorption Ionization Techniques for Quantitative Analysis of Drug Molecules

Jason S. Gobey, John Janiszewski, and Mark J. Cole

Pfizer Global Research and Development, Department of Pharmacokinetics, Dynamics and Metabolism, Groton, Connecticut

Mass Spectrometry in Drug Metabolism and Pharmacokinetics. Edited by Ragu Ramanathan
Copyright © 2009 John Wiley & Sons, Inc.

11.1 INTRODUCTION

Currently, high-performance liquid chromatography (HPLC) combined with atmospheric pressure ionization (API) triple-quadrupole mass spectrometry (MS) is the predominate quantitative technique used in modern pharmaceutical bioanalysis. The key technological achievement in API–MS was the efficient ionization in a liquid stream and transference of ions from atmosphere to vacuum. Of the API approaches developed, electrospray ionization (ESI) is the most commonly used. ESI provides an efficient means of soft ionization amenable to most molecules encountered in a drug discovery setting. An alternative soft ionization approach is the use of desorption ionization (DI) techniques. The major distinguishing feature of DI techniques is that ions are typically produced from dried samples.

Electrospray ionization and matrix-assisted desorption ionization were both introduced around the same time, in the late 1980s. In fact matrix-assisted laser desorption ionization (MALDI) was first mentioned in the literature in 1987 (Karas et al., 1987). In the years prior to that, there were limited reports of the application of laser desorption MS. Early developments in MALDI focused primarily on macromolecules, particularly peptides and proteins. Historically, MALDI ion sources have predominantly been coupled to time-of-flight (TOF) instruments. TOF requires precise timed ionization events, and since ions are generated in MALDI by a pulsed desorption, this combination is complementary. Mass spectra generated by MALDI can be relatively simple, containing predominantly singly charged ions. The importance of both ESI and MALDI are well proven in the analysis of biomolecules, and both techniques were awarded the Nobel Prize for chemistry in 2002 (Chapter 1).

Earlier DI techniques include fast atom bombardment (FAB), secondary ionization mass spectrometry (SIMS), plasma desorption, and field desorption. Since their applications are primarily qualitative, they will not be discussed here.

Combinatorial and rapid synthesis techniques are fundamentally changing the analytical workflow in modern drug discovery. In early-stage drug discovery, scientists assess the metabolic, safety, toxicity, and potency profiles of the relevant chemical space in order to guide synthesis. This assessment is done with hundreds of molecules at a time, placing ever-increasing demands on liquid chromatography (LC)–MS/MS bioanalysis. As the demand for small-molecule quantification has increased, high-capacity systems have been conceived and implemented to attain the necessary throughput (Janiszewski et al., 2001). Certainly most, if not all, examples of high-throughput quantitative bioanalysis of small molecules are based on ESI–triple-quadrupole MS.

A number of factors have contributed to renewed interest in DI techniques for quantitative analysis. Since the reasonable throughput in routine use of the LC–ESI–triple-quadrupole mass spectrometer seems to have peaked at about 3-4 samples per minute, there is renewed interest in finding alternate ways to improve quantitative throughput. Recent developments in DI mass spectrometry (DIMS) have realistically made routine small-molecule quantification feasible. This chapter will briefly outline the issues encountered when quantifying drug molecules by DIMS. The latest developments will be described in relation to historical problems that limited the application of DI in small-molecule bioanalysis. The potential for a dramatic increase in throughput is of greatest interest in the application of these technologies. In particular, the throughput of absorption, distribution, metabolism, and elimination (ADME) screening is reaching limits imposed by current LC–MS/MS approaches. Thus, increased bioanalytical throughput in the drug discovery arena will be the main theme of this chapter.

Considering all the reported applications using MALDI–MS, the mechanisms of ion generation and desorption have, until recently, been poorly understood. Lately, there has been promising progress in modeling the various complicated interactions occurring at the molecular level, and readers are referred to these references for further coverage of this topic (Zenobi and Knochenmuss, 1999; Knochenmuss, 2003; Knochenmuss and Zenobi, 2003).

11.2 QUANTITATIVE ANALYSIS BY TRADITIONAL MALDI–MS

The literature reveals a limited number of reports of quantitative analysis of typical small-molecule pharmaceuticals. Quantification of some low-molecular-weight antibiotics (Ling et al., 1998) was demonstrated with a MALDI–TOF mass spectrometer. This first MALDI–TOF quantification study entailed determining suitable matrices for the 30 compounds of interest. In this study, α-cyano-4-hydroxycinnamic acid (CHCA) and 2,5-dihydroxybenzoic acid (DHB) were the initially tested matrices. Three compounds did not ionize with either matrix. Most of the compounds had molecular weights below 300 Da, and had been identified as having possible low-mass ion interferences with the preferred matrices. Alternate matrices were evaluated with limited success. Only 5 of these 20 compounds generated protonated molecular ions, although in most cases sodium or potassium adducts were observed. Three compounds were selected as model compounds, and quantification data were reported. A structurally close analog was used as an internal standard, and linearity was shown over two orders of magnitude.

The quantitative determination of amperozide in plasma using MALDI–MS was demonstrated (Jespersen et al., 1995). This study required addition of a stable-isotope-labeled amperozide as an internal standard and a typical liquid–liquid extraction procedure with dry down and reconstitution. The sample was then mixed with the matrix and analyzed. Linearity was achieved over a range of 2.5–40 μM. Other examples have been demonstrated (Duncan et al., 1993) which also required the use of stable-isotope analogs as internal standards. These examples also utilized TOF instruments with nitrogen lasers.

Recently, a charge derivatization method was applied (Lee et al., 2004) to successfully analyze small amine molecules in buffer. The compounds to be analyzed were primary and secondary amines including several antibiotics. The amines were mixed with a 10-fold excess of N-hydroxysuccinimide ester. Although the reaction mixture contained millimolar concentrations of buffer, the mixture was directly analyzed without a desalting step. While noteworthy, this technique adds another tedious step to the analysis, and linearity was limited to a single order of magnitude.

The common thread in the previous examples is the possibility to obtain good quantitative results by MALDI–MS. However, the additional steps involved that were necessary for success negate the potential for high throughput.

11.3 ISSUES HINDERING QUANTIFICATION BY MALDI–MS

While it has been demonstrated that small molecule quantitative analysis using traditional MALDI is possible, there remain several barriers to its routine use. Perhaps, the biggest problem for quantifying drug molecules by MALDI is the matrix itself. The most common matrices are photoactive organic acids that produce a rich spectra of peaks below molecular weights of about 500 Da. These peaks interfere with analyte measurements in this mass range. Figure 11.1 depicts a full-scan spectrum of one of the more commonly used matrices, CHCA, which shows the extent of the problem. Small-molecule quantitative bioanalysis methods using MALDI should incorporate some means of addressing the interferences in the lower molecular weight range caused by the MALDI matrix. Additional cleanup may be warranted to account for endogenous interferences present in biological samples.

A modern TOF instrument may well provide adequate resolution to allow quantification of small-molecule analytes, but examples in the literature report limited dynamic range. The latest TOF instruments provide for improved dynamic range in quantitative applications, but there are still other critical obstacles. These include sample preparation, selection of an internal standard, instrumental protocol, and

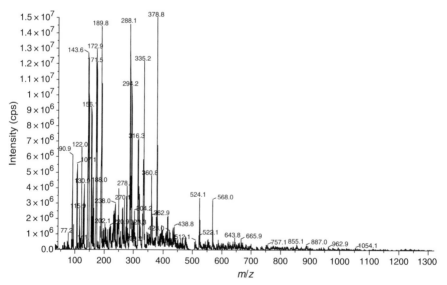

Figure 11.1. *Full scan of CHCA below* m/z *1300 obtained by directly analyzing matrix solution with the prototype MALDI triple-quadrupole mass spectrometer.*

ion suppression (Duncan et al., 1993; Muddiman et al., 1996; Bucknall et al., 2002; Cohen and Gusev, 2002). In general, it is preferable to have an internal standard that is as structurally similar to the analyte of interest as possible. In the drug discovery setting, where sample throughput is the goal, individual selection of internal standards is not practical. Traditional MALDI-related issues, such as spotting technique and matrix purity/composition, also limit small molecule quantification. Crystal quality and homogeneity tend to impact spot-to-spot reproducibility and become a major source of imprecision, limiting quantitative applications. If an internal standard is to be of any benefit, then it must assist in addressing all issues relating to matrix interferences.

Selection of an appropriate internal standard can also assist in correcting for ion suppression issues caused by matrix components. However, if the ion suppression is too severe, then inevitably sensitivity will suffer. Ion suppression must be limited to a degree sufficient to avoid sensitivity problems. To reduce ion suppression, a sample cleanup method is necessary. Moreover, proper co-crystallization is directly related to sample composition and therefore, sample cleanup is necessary for successful MALDI ionization.

Although CHCA works well with most drug molecules, there is no universal MALDI matrix. However, as with the internal standard, it is not practical to optimize matrix selection for each compound in broad drug discovery applications.

11.4 AN ALTERNATIVE APPROACH: MATRIXLESS MALDI

A recent development, within the field of DIMS for small-molecule quantification, is the introduction of desorption ionization on silicon (DIOS) (Wei et al., 1999). In this approach, the analyte of interest is deposited on a prepared porous silicon surface, and there is no matrix per se. The porous silicon exhibits a high absorbance of the ultraviolet light provided by the laser. The absorbed energy is subsequently imparted to the analyte, thereby generating ions (Shen et al., 2001). Like MALDI, DIOS provides an efficient means of soft ionization that is amenable to molecules through a broad mass range. Reportedly, DIOS performs optimally for molecules below 3000 amu (Lewis et al., 2003). An example of successful quantitative analysis of a relevant small molecule has been reported for steroids in human urine (Shen et al., 2001). Close analogs to the analyte of interest were utilized as internal standards, and a two-step liquid–liquid extraction was necessary, due to the degree of ion suppression likely caused by the biological matrix. DIOS–MS has been applied successfully to bioanalysis of enzyme inhibition reactions (Wall et al., 2004). The feasibility of atmospheric pressure (AP) DIOS for analysis of small molecules with high proton affinity was recently demonstrated (Huikko et al., 2003). The detection limit for midazolam by AP-DIOS was in the low femtomole range, but the method did not work well for certain neutral or acidic compounds. Acids, in particular, did not yield protonated molecules. For compounds with good proton affinity, linearity over three orders of magnitude was obtained, allowing for quantitative analysis.

11.5 ISSUES HINDERING QUANTIFICATION BY DIOS–MS

DIOS has some current limitations. Until recently, there was a lack of commercially available quantities of uniform quality prepared chips. Recently, DIOS chips suitable for small-molecule quantification applications have become commercially available (MassPREP DIOS target plate, Waters). Development of these targets, including the etching processes, has been described in detail (Credo et al., 2004). The etching process was based on the original process described by Shen et al. (2001). The performance of DIOS is highly sensitive to environmental contamination of the silicon chips. To minimize contamination, DIOS chips are either stored in a sealed container or packaged with an inert gas or stored in a Teflon-type container with isopropanol. The chips should be handled with special care. Using a pair of clean tweezers reserved for this purpose is best, and care should be exercised to minimize touching the chips with or without gloves. As with MALDI, DIOS quantification also requires use of an appropriate internal standard and an effective sample cleanup method. The universality of DIOS across pharmaceutical chemistries and its wider performance in relevant assays are areas that remain to be explored.

11.6 MALDI ON TRIPLE-QUADRUPOLE MASS SPECTROMETER

An approach to solving the problem of matrix interference is coupling a MALDI ion source to a tandem mass spectrometer (Hatsis et al., 2003; Gobey et al., 2005). The inherent selectivity of the triple-quadrupole mass spectrometer is used to bias against the matrix background observed in a full-scan spectrum. In the initial work, a prototype MALDI source was coupled to a modified API 3000 triple-quadrupole mass spectrometer (MDS Sciex, Canada). The standard interface was removed and modified to accept the orthogonal MALDI ion source. The samples of interest were deposited on a stainless steel target plate which was placed in the vacuum region of the source and exposed to within a few millimeters of the entrance quadrupoles. Figure 11.2a illustrates the difficulty in selecting a parent ion from a full-scan spectrum obtained by MALDI–MS. The protonated carbamazepine ion at m/z 237 is not easily resolved from the background of matrix and other interfering ions. In this example, the analyte standard was at a relatively high concentration of 4 ng/μL. However, a product ion scan, as depicted in Fig. 11.2b, shows a clean product spectrum containing the expected fragment ions and is similar to the product ion spectrum obtained with the ESI–MS/MS. Using the selected reaction-monitoring (SRM) mode, 25-ng/mL replicates of carbamazepine were acquired. These data are shown in Fig. 11.2c, with the replicates separated by blanks. In practical terms, product spectra are similar to those obtained by ESI–MS/MS, with increased potential for additional ion peaks due to fragmentation of isobaric matrix components. The use of SRM experiments provides the ability to discriminate analyte from background in most cases.

An additional component of this application is the use of a high-repetition-rate laser that delivers pulses at frequencies up to 1400 Hz. Historically, MALDI ion

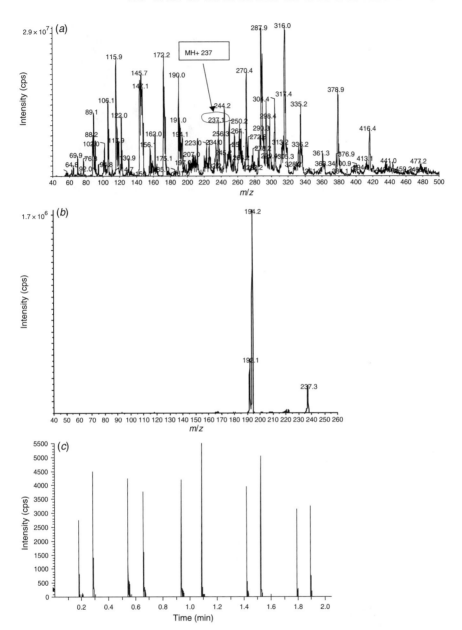

Figure 11.2. (a) Full-scan spectrum of 4 ng/μL of carbamazepine neat standard by MALDI–MS; (b) MALDI–MS/MS scan of carbamazepine (products of m/z 237); (c) 25 pg/μL replicates of carbamazepine separated by blanks (SRM 237→194).

sources have been coupled with TOF analyzers, and the mass measurement has been triggered off of the laser pulse. Since the triple-quadrupole mass spectrometer is a continuous-beam analyzer, there is no reliance on the timing event of the laser

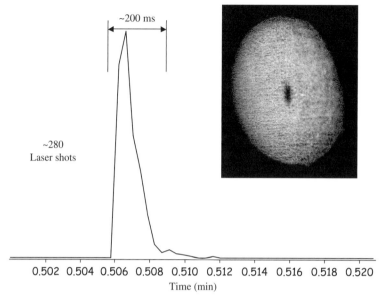

Figure 11.3. Hole drilled into sample spot by laser generating a "peak."

pulse, and as such, the quadrupole mass spectrometer is well suited to take advantage of higher repetition rate lasers. Figure 11.3 shows the profile collected when the laser, while stationary, completely ablates through the sample spot. At a repetition rate of 1400 Hz, this depth profile was generated in ~200 ms as ~280 laser shots were

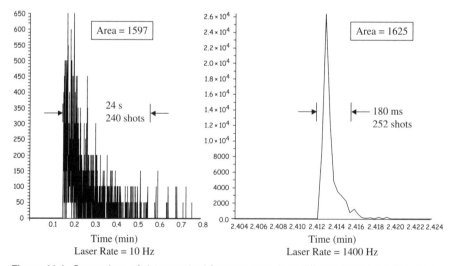

Figure 11.4. Comparison of data acquired for same sample at different laser firing frequencies (10 and 1400 Hz). The data acquisition time is much more rapid and measurable sensitivity is improved.

fired. The laser generates a power output of $\sim 16\,\mu J$ independent of the firing rate, up to a maximum frequency of 1500 Hz. In practice, the laser power that is actually incident on the sample spot after various transmission losses is $\sim 3-8\,\mu J$. Figure 11.4 shows two profiles derived from the same sample at the two extremes of frequency. Evidence shows that the high-repetition-rate laser will have an impact on throughput as well as improved sensitivity. At 10 Hz, 24 s was used to collect the profile, while at 1400 Hz, only 180 ms was necessary to collect the same profile. There is no measurable background in either of these profiles, so the signal-to-noise ratio is not a meaningful metric. The profile acquired at a laser frequency of 10 Hz appears rather ragged, but in fact the signal in this trace is fully attributed to the analyte. The sensitivity is a function of the detected signal rate, which is directly proportional to the laser repetition rate. This is analogous to the band compression achievable in chromatography. The end result of incorporating the high-repetition-rate laser in to the MALDI ion source of the triple quadrupole mass spectrometer is better sensitivity, better precision due to the large number of laser firings, and improved throughput.

11.7 DETERMINATION OF MS CONDITIONS FOR MALDI/MS/MS

Prior to analyses of assay samples by LC–ESI–MS/MS, appropriate precursor ions with the corresponding product ions must be determined. In drug discovery programs, the number of compounds that need MS conditions for support of high-throughput screening ranges in the magnitude of several hundred compounds per week. An automated workstation that facilitates the determination of such MS/MS conditions was developed (Whalen et al., 2000). The system utilizes a dual LC column/dual-injection system coupled to a triple-quadrupole mass spectrometer. Each injection port utilizes a 6-port, two-position switching valve. An additional 6-port valve directs the aqueous mobile-phase flow, while a 10-port valve directs the organic mobile phase, in addition to controlling column switching. A sequence of two injections was made onto the LC–MS/MS system. The first injection determined the appropriate polarity as well as the precursor ion, while a second injection determined the optimal collision energy and the product ion. This system allows for the determination of semioptimized MS/MS conditions for 90 compounds in less than 2 h.

Determination of MS/MS conditions using the MALDI ion source coupled to the triple-quadrupole mass spectrometer takes a different approach (Kovarik et al., 2003) which at present is not as easily automated. As mentioned earlier, some product ion spectra will contain additional fragment ions not related to the analyte due to isobaric matrix peaks. However, MS/MS conditions previously obtained by using an ESI–MS/MS system can be directly ported over to the MALDI triple-quadrupole mass spectrometer. These conditions provide the advantage of product ion selection for SRM of the analyte. In the typical high-throughput environment, individual methods for each chemical entity are not normally utilized. Rather, the semioptimized template-style methods referred to above are often used wherein a few values of collision energy are combined with the appropriate SRM and polarity. These methods are ported to the MALDI triple quadrupole mass spectrometer, and no further consideration is made for optimized MALDI matrix selection or laser power. In fact, all analyses were performed using a single MALDI matrix, CHCA, as this matrix has been found to be the most universal for small-molecule analysis. Furthermore, a single compound was selected to use as an internal standard for all small-molecule quantification. The molecule selected was prazosin, due to its selective ionization by MALDI and its representative nature associated with small druglike compounds.

11.8   MALDI TRIPLE-QUADRUPOLE MASS SPECTROMETER PERFORMANCE

The MALDI triple-quadrupole mass spectrometer generates calibration curves with linearity and dynamic ranges similar to those typically expected from a triple-quadrupole mass spectrometer. In most cases, linearity is established over three orders of magnitude with suitable accuracy and precision. Figure 11.5 depicts calibration curves obtained for some common drugs using neat standards.

The literature suggests that MALDI is less susceptible to ion suppression than the ESI process (Hatsis et al., 2003). Current research indicates significant ion suppression in MALDI–MS/MS. All efforts to directly analyze in vitro samples by MALDI without prior cleanup failed. Experimenting with various ratios of sample to matrix did not improve the results. However, strong spectra were obtained from in vitro samples following solid-phase extraction (SPE) and further cocrystallization of the clean samples on the MALDI target. Good crystals formed following a desalting step (SPE) and poor crystals formed upon direct analysis and are compared in Fig. 11.6.

If MALDI–MS/MS is to find a niche in a high-throughput discovery-screening laboratory, the technique must be capable of delivering data with comparable quality to established methodology. Inevitably, the data will be compared to that obtained by established LC–ESI–MS/MS methods. A selection of 53 diverse drug-like compounds was assayed through a standard human microsome incubation study (Gobey et al., 2003). The resultant samples were split, allowing for the analysis of one

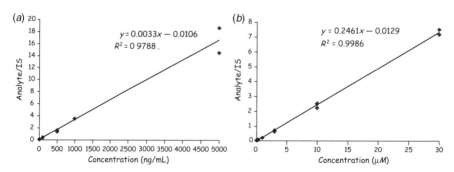

Figure 11.5. *Example calibration curves for (a) verapamil and (b) buspirone.*

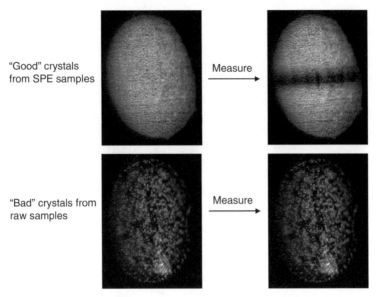

Figure 11.6. *Crystal formation is critical for good MALDI–MS/MS performance. Samples were desalted using SPE cleanup (top). Raw samples did not yield good crystals and, as a result, data (below).*

set by the standard LC–ESI–MS/MS system. Prior to analysis using the MALDI–MS/MS, the other set was processed using an SPE protocol scheme (Fig. 11.7). The eluant from the SPE cleanup step contained analyte, internal standard, MALDI matrix, and a mixture of methanol and isopropanol. The eluant was directly spotted on to the

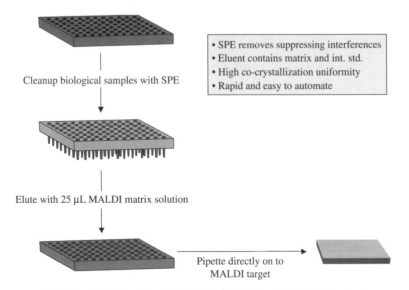

Figure 11.7. *Sample cleanup by simple SPE with MALDI matrix in eluant.*

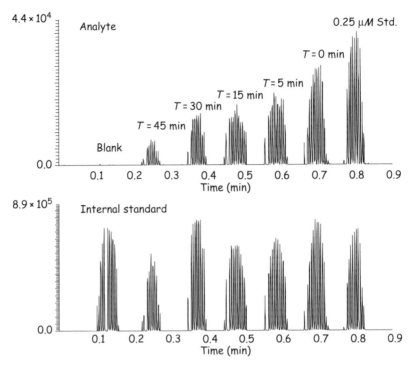

Figure 11.8. *Example of microsome incubate time course. All time points were acquired to a single data file for clarity and speed.*

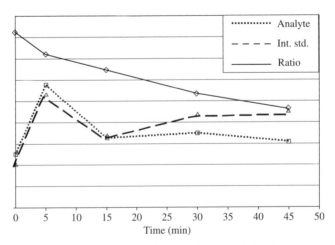

Figure 11.9. *Microsomal time course for typical analyte, internal standard, and analyte–internal standard ratio.*

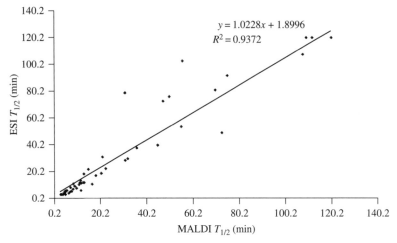

Figure 11.10. *Human microsome half-lives for 53 compounds: ESI vs. MALDI.*

MALDI target for analysis. All data points were collected into one data file for clarity and speed. An example of a MALDI target prepared by eluant spotting is shown in Fig. 11.8. Figure 11.9 shows a microsome time course for a typical analyte, internal standard, and analyte–internal standard ratio. This example illustrates the necessity of the internal standard for quantification. Half-lives were calculated for all compounds from both data sets. The half-lives obtained by the two ionization methods are plotted in Fig. 11.10. A good correlation was obtained between the two techniques.

11.9 QUANTITATIVE ANALYSIS OF DRUGS BY DIOS

In principle, DIOS should work with any commercially available MALDI mass spectrometer; however, difficulties arose with deriving any meaningful DIOS data from the experimental source used for MALDI on the triple-quadrupole mass spectrometer. Therefore, microsomal incubation samples from 15 compounds were also analyzed by DIOS using a Waters Micromass MALDI–linear reflection (LR)–TOF mass spectrometer (Gobey et al., 2004). The SPE cleanup procedure was essentially identical to the standard protocol described in the previous section. Figure 11.11 shows the comparison of data obtained for nine of the compounds using DIOS–MS and LC–ESI–MS/MS. Although the data set is relatively small, the correlation is good. Six of the compounds tested failed to give usable data by DIOS–MS.

A comparison of sensitivity of some drugs using DIOS–MS and MALDI–MS was performed. Table 11.1 contains the data, which show that on the Waters MALDI–LR–TOF mass spectrometer DIOS may be a preferable technique. These four drugs were 2–28 times more sensitive using DIOS than MALDI. Recent

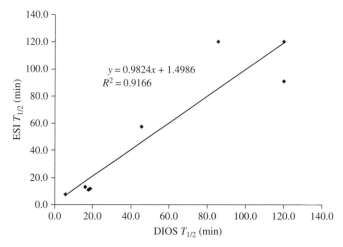

Figure 11.11. *Human microsome half-lives for nine compounds: ESI vs. DIOS.*

TABLE 11.1. Average Response for Selected Drugs at 10 μM Using MALDI versus DIOS

	MALDI	DIOS	DIOS/MALDI Ratio
Reserpine	10,217	41,582	4.1
Lidocaine	6,639	189,582	28.6
Procainamide	3,646	20,942	5.7
Amitryptyline	8,265	18,898	2.3

work using an AP-MALDI ion source also suggests that AP-DIOS may be 5–10 times more sensitive than AP-MALDI (Pihlainen et al., 2005).

11.10 SENSITIVITY AND SPEED CONSIDERATIONS

The data obtained from a set of samples by both MALDI–MS/MS and LC–ESI–MS/MS are shown in Fig. 11.12. ESI provides the opportunity to consume a comparatively large volume of sample, whereas with MALDI the sample consumption is small and relatively fixed. Typically, sample spots represent a pipetted volume of ~0.25 μL, and measuring across the sample only consumes about a 25-nL equivalent. In this example, the ESI injection volume was 25 μL. Even though the ESI analysis consumed 1000 times more sample, the analyte area produced was only 10 times greater, suggesting that the MALDI–MS/MS was 100 times more efficient in terms of ionization efficiency and/or ion transfer efficiency. Discrimination was not possible between the two, but this example suggests that assay performance equality between the two techniques could be possible if the MALDI samples were further concentrated during the sample cleanup step.

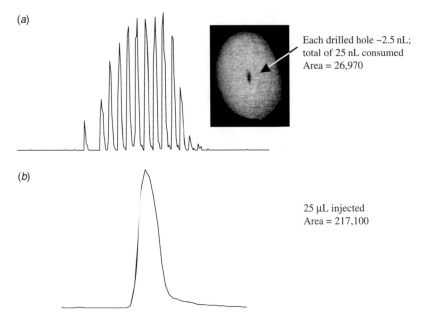

(a)

Each drilled hole ~2.5 nL;
total of 25 nL consumed
Area = 26,970

(b)

25 μL injected
Area = 217,100

Figure 11.12. *Measurements for same original sample by (a) MALDI–MS and (b) ESI–MS: MALDI is 100 times more efficient than ESI.*

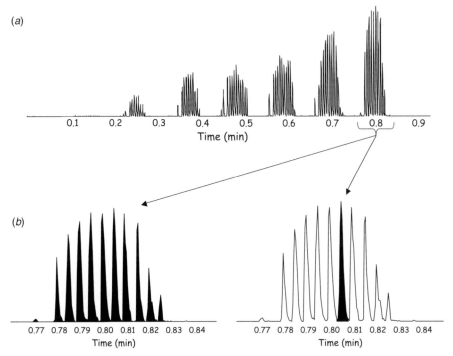

Figure 11.13. *(a) Acquisition of seven samples within 0.9 min. (b) Expansion of single sample measurement.*

The data depicted in Fig. 11.8 show data acquisition at a rate of eight samples per minute. Figure 11.13 also depicts the expansion of a single sample measurement, which actually consists of 8–10 individual measurements. The eight interior peaks were individually integrated and compared with the average. All the individual measurements had a precision of 10% or better, suggesting that, by incorporating faster motors and hopping between samples, after only drilling single holes within each sample spot, the data could be acquired much more rapidly, perhaps at greater than one sample per second. In this manner, the minimum data necessary for an accurate measurement are acquired, and the system moved as rapidly as possible to the next measurement, virtually eliminating analytical dead time. Using the setup described, a plate of 96 compounds could be analyzed in less than 2 min.

11.11   DIOS OR MALDI?

This question is premature and possibly irrelevant. As of this writing, the two techniques have application niches and are only now converging on small-molecule quantification. Both techniques require further characterization in terms of application space and may become complementary. Table 11.2 summarizes some of the key traits of MALDI as compared to ESI for quantitative analysis of drug molecules. The table specifically compares MALDI, for which the larger existing data set allows better comparison, but many of the observations apply directly to DIOS as well. The biggest drawback to using MALDI–MS/MS for small-molecule quantification is the necessary sample cleanup, which adds significant cost. There is not enough data to properly determine chemical library coverage by DIOS. The technique reportedly works well for many molecules, but in limited studies, some drugs did not work at all. Conversely, DIOS has been shown to be more sensitive than MALDI for some compounds, and the background signal is usually minimal. DIOS offers the same potential as MALDI for sample throughput and may have

TABLE 11.2. Summary of the Comparison of MALDI to ESI for Quantification of Drugs

	MALDI–MS/MS	ESI–MS/MS
Ion suppression	Yes	Yes
Sample preparation	SPE	ACN precipitation
Sufficient sensitivity for HT ADME	Yes	Yes
Coverage	Good	Excellent
Simple plumbing	Yes	No
Space constraints	No	Yes
Solvent use	<50 µL per sample	>1 mL per sample
Current speed	~8 samples/min	3–4 samples/min
Potential speed	1–8 samples/s or 3600–28,800 samples/h	6–8 samples/min or 360 samples/h

Abbreviations: ACN, Acetonitrile; ADME, Adsorption, distribution, metabolism, and excretion; HT, High throughput.

more flexibility in terms of sample preparation methods. However, the technique is dependent upon higher cost consumables that require precise and unforgiving sample-handling procedures.

11.12 SUMMARY

As of this writing, it remains to be seen whether the great speed potential of MALDI and DIOS for the quantitative analysis of drug molecules will be realized in practice. There are hurdles typical of any emerging application that need to be addressed. The need for providing rapid quantitative analysis of drugs is only becoming more acute, and as such, it is likely that DI techniques will become more prevalent in the area of small molecule bioanalytical MS.

REFERENCES

Bucknall, M., Fung, K. Y. C., and Duncan, M. W. (2002). Practical quantitative biomedical applications of MALDI–TOF mass spectrometry. *J. Amer. Soc. Mass Spectrom.* **13:**1015–1027.

Cohen, L. H., and Gusev, A. I. (2002). Small molecule analysis by MALDI mass spectrometry. *Anal. Bioanal. Chem.* **373:**571–586.

Credo, G., Hewitson, H., Benevides, C., and Bouvier, E. S. P. (2004). Development of a porous silicon product for small molecule mass spectrometry. *Mater. Res. Soc. Symp. Proc.* **808:**471–476.

Duncan, M. W., Matanovic, G., and Cerpa-Poljak, A. (1993). Quantitative analysis of low molecular weight compounds of biological interest by matrix-assisted laser desorption ionization. *Rapid Commun. Mass Spectrom.* **7:**1090–1094.

Gobey, J., Cole, M., Janiszewski, J., Covey, T., Chau, T., Kovarik, P., and Corr, J. (2005). Characterization and performance of MALDI on a triple quadrupole mass spectrometer for analysis and quantification of small molecules. *Anal. Chem.* **77:**5643–5654.

Gobey, J. S., Janiszewski, J., Cole, M. J., Corr, J. J., Kovarik, P., Chau, T. K., and Covey, T. R. (2003). Performance of MALDI/MS/MS for Small Molecule Quantitation. In *Proceedings of The 51st ASMS Conference on Mass Spectrometry and Allied Topics,* Montreal, Canada.

Gobey, J. S., Janiszewski, J., Credo, G. M., and Bouvier, E. S. P. (2004). A comparison of desorption ionization techniques for small molecule quantitation. In *Proceedings of The 52nd ASMS Conference on Mass Spectrometry and Allied Topics,* Nashville, TN.

Hatsis, P., Brombacher, S., Corr, J., Kovarik, P., and Volmer, D. A. (2003). Quantitative analysis of small pharmaceutical drugs using a high repetition rate laser matrix-assisted laser/desorption ionization source. *Rapid Commun. Mass Spectrom.* **17:**2303–2309.

Huikko, K., Oestman, P., Sauber, C., Mandel, F., Grigoras, K., Franssila, S., Kotiaho, T., and Kostiainen, R. (2003). Feasibility of atmospheric pressure desorption/ionization on silicon mass spectrometry in analysis of drugs. *Rapid Commun. Mass Spectrom.* **17:**1339–1343.

Janiszewski, J. S., Rogers, K. J., Whalen, K. M., Cole, M. J., Liston, T. E., Duchoslav, E., and Fouda, H. G. (2001). A high-capacity LC/MS system for the bioanalysis of samples generated from plate-based metabolic screening. *Anal. Chem.* **73:**1495–1501.

Jespersen, S., Niessen, W. M. A., Tjaden, U. R., and van der Greef, J. (1995). Quantitative bioanalysis using matrix-assisted laser desorption/ionization mass spectrometry. *J. Mass Spectrom.* **30:**357–364.

Karas, M., Bachmann, D., Bahr, U., and Hillenkamp, F. (1987). Matrix-assisted ultraviolet laser desorption of non-volatile compounds. *Int. J. Mass Spectrom. Ion Process.* **78:**53–68.

Knochenmuss, R. (2003). A quantitative model of ultraviolet matrix-assisted laser desorption/ionization including analyte ion generation. *Anal. Chem.* **75:**2199–2207.

Knochenmuss, R., and Zenobi, R. (2003). MALDI ionization: The role of in-plume processes. *Chem. Rev.* **103:**441–452.

Kovarik, P., Corr, J. J., and Covey, T. R. (2003). Application of orthogonal MALDI for quantitation of small molecules using a triple quadrupole mass spectrometer. In *Proceedings of The 51st ASMS Conference on Mass Spectrometry and Allied Topics*, Montreal, Canada.

Lee, P. J., Chen, W., and Gebler, J. C. (2004). Qualitative and quantitative analysis of small amine molecules by MALDI–TOF mass spectrometry through charge derivatization. *Anal. Chem.* **76:**4888–4893.

Lewis, W. G., Shen, Z., Finn, M. G., and Siuzdak, G. (2003). Desorption/ionization on silicon (DIOS) mass spectrometry: Background and applications. *Int. J. Mass Spectrom.* **226:**107–116.

Ling, Y.-C., Lin, L., and Chen, Y.-T. (1998). Quantitative analysis of antibiotics by matrix-assisted laser desorption/ionization time-of-flight mass spectrometry. *Rapid Commun. Mass Spectrom.* **12:**317–327.

Muddiman, D. C., Gusev, A. I., and Hercules, D. M. (1996). Application of secondary ion and matrix-assisted laser desorption-ionization time-of-flight mass spectrometry for the quantitative analysis of biological molecules. *Mass Spectrom. Rev.* **14:**383–429.

Pihlainen, K., Grigoras, K., Franssila, S., Ketola, R., Kotiaho, T., and Kostiainen, R. (2005). Analysis of amphetamines and fentanyls by atmospheric pressure desorption/ionization on silicon mass spectrometry and matrix-assisted laser desorption/ionization mass spectrometry and its application to forensic analysis of drug seizures. *J. Mass Spectrom.* **40:**539–545.

Shen, Z., Thomas, J. J., Averbuj, C., Broo, K. M., Engelhard, M., Crowell, J. E., Finn, M. G., and Siuzdak, G. (2001). Porous silicon as a versatile platform for laser desorption/ionization mass spectrometry. *Anal. Chem.* **73:**612–619.

Wall, D. B., Finch, J. W., and Cohen, S. A. (2004). Comparison of desorption/ionization on silicon (DIOS) time-of-flight and liquid chromatography/tandem mass spectrometry for assaying enzyme-inhibition reactions. *Rapid Commun. Mass Spectrom.* **18:**1482–1486.

Wei, J., Buriak, J. M., and Siuzdak, G. (1999). Desorption-ionization mass spectrometry on porous silicon. *Nature (Lond.)* **399:**243–246.

Whalen, K. M., Rogers, K. J., Cole, M. J., and Janiszewski, J. S. (2000). AutoScan: An automated workstation for rapid determination of mass and tandem mass spectrometry conditions for quantitative bioanalytical mass spectrometry. *Rapid Commun. Mass Spectrom.* **14:**2074–2079.

Zenobi, R., and Knochenmuss, R. (1999). Ion formation in MALDI mass spectrometry. *Mass Spectrom. Rev.* **17:**337–366.

12

MALDI Imaging Mass Spectrometry for Direct Tissue Analysis of Pharmaceuticals

Yunsheng Hsieh and Walter A. Korfmacher

Schering-Plough Research Institute, Department of Drug Metabolism and Pharmacokinetics, Kenilworth, New Jersey

Mass Spectrometry in Drug Metabolism and Pharmacokinetics. Edited by Ragu Ramanathan
Copyright © 2009 John Wiley & Sons, Inc.

12.1 INTRODUCTION

Mass spectrometry (MS)–based bioanalytical support is an essential element in the discovery process when looking for new medicines to improve the quality of human life. Rapid assessment of the druglike properties of new chemical entities (NCEs) is a critical step for enhancing the success rate of drug candidates entering into drug development. High-performance liquid chromatography (HPLC) coupled with tandem mass spectrometry (HPLC–MS/MS) has totally replaced HPLC methods that use UV or fluorescent detectors for most drug analysis applications. HPLC–MS/MS methodologies have become the standard tool for both qualitative and quantitative determination of pharmaceuticals in various in vitro and in vivo samples for most drug discovery and drug development studies (Korfmacher et al., 1997; Ackermann et al., 2002; Hopfgartner et al., 2002; O'Connor, 2002; Hopfgartner and Bourgogne, 2003; Hsieh et al., 2003a; Hsieh et al., 2003b; Korfmacher, 2003; Hsieh and Wang, 2004; Korfmacher, 2005a; Xu et al., 2005; Sugiura et al., 2007). However, HPLC–MS/MS approaches are not able to provide answers to certain questions regarding the distribution of a drug within individual organs or tissues obtained from laboratory animal experiments.

Conventional methods such as autoradiography or fluorescence spectroscopy are still the most common methods for determining the distribution of pharmaceuticals in various organs and tissues from laboratory animal dosing studies. While these techniques are very powerful, they do have some limitations. First, these techniques require the synthesis of a radioactive isotope or a fluorescent tag for the test compound—a time-consuming and costly step. Typically, radiolabeled versions of new test compounds are not available at the lead optimization stage. Second, distinguishing the dosed drug from its metabolite signals is difficult with these techniques. Thus, autoradiographic techniques allow for the visualization of total drug-related materials but cannot differentiate between the distribution of the administrated drug and its metabolites.

Alternatively, imaging mass spectrometry (IMS) is capable of detecting known compounds with no additional isotopic labels or affinity tags required. IMS encompasses a variety of ionization techniques such as ion-beam-induced desorption for secondary ion mass spectrometry (SIMS) and matrix-assisted laser desorption/ionization (MALDI) that can enable the chemical imaging of analytes directly from biological tissues (Rubakhin et al., 2005). In SIMS, analyte ions are desorbed from the surface of samples through the bombardment of a primary energetic ion beam with high spatial resolution at the cellular level. However, the use of SIMS for imaging of drug molecules in tissue samples has been hampered due to low secondary yield and extensive fragmentation for large (>100 Da) organic molecules (Sjovall et al., 2003; Altelaar et al., 2005). Several efforts like the modifications of the variety of primary ion beams and the use of matrix-enhanced SIMS have been investigated to increase the survival yield and ionization efficiency of intact biomolecular species. It still remains to be seen if these improvements can advance the SIMS field to be competitive with MALDI imaging for IMS application.

Historically, MALDI–IMS has been extensively employed to measure macromolecules such as peptides and proteins directly from the tissue sections (Stoeckli et al., 2001; Todd et al., 2001; Chaurand et al., 2004). Although MALDI–MS has been applied extensively as an analytical tool for biopolymers, the potential of using this technique for both qualitative and quantitative determination of small molecules (<1500 Da) such as enzyme substrates (Kang et al., 2000), acetylcholine, 3,4-dihydroxyphenylalanine (Duncan et al., 1993), benzodiazepines (Hatsis et al., 2003), vitamins (Chen and Ling, 2002), amino acids (Dally et al., 2003), caffeine (McCombie and Knochenmuss, 2004), phosphonium cations (Cheng et al., 2005), and others (Ayorinde et al., 1999; Cohen and Gusev, 2002; Donegan et al., 2003; Owen et al., 2003; Sleno and Volmer, 2005; Li et al., 2007; Sugiura et al., 2007) has been demonstrated. In this chapter, we focus on the IMS technique using MALDI to map the distribution of pharmaceuticals in thin tissue sections. The principles and fundamental aspects of the MALDI–IMS technique along with its application to tissue imaging of pharmaceuticals and in small animal studies are discussed in detail.

12.2 PRINCIPLES OF MALDI–IMS

The concept of MALDI–IMS was introduced in 1997 by Caprioli and co-workers for rapid and direct profiling of analytes within a tissue section (Caprioli et al., 1997). The MALDI–IMS technique employs a matrix solution that is placed on the sample and dries to produce a layer of crystals on top of the sample. The application of the matrix solution serves two purposes: (1) the matrix deposition step serves as an analyte extraction step and the extracted analytes are included in the crystals that are formed on the surface of the tissue and (2) the matrix crystals that are formed absorb energy at the wavelength of the laser beam resulting in desorption and ionization of the analytes. Using various techniques, the matrix is uniformly deposited on the tissue sections, as shown in Fig. 12.1. Next, the sample on a MALDI plate is assayed by moving the plate stepwise in a predetermined two-dimensional array so that a stationary laser beam can be directed at the sample to generate a series of MS observations that become the pixels in the MS image, as depicted in Fig. 12.2. Typically, the localization of the laser beam on the sample is accurate to within ~5 μm. The movement of the sample in the x and y directions is controlled by the software that is also used to specify the edges of the tissue sample and to define the exact region of interest for assay (see Fig. 12.1).

During an imaging mass spectrometric experiment, the resulting MS data are first obtained as a function of the acquisition time, which are associated with the location in an array of pixels, as given in Fig. 12.3. Using software tools, two-dimensional ion maps are reconstructed from a given m/z value that was monitored in each mass spectrum. In this type of image, the color of each pixel represents the intensity of the selected ion, as shown in Fig. 12.4. In order to detect small molecules in a complex biological tissue section using MALDI–IMS, MS/MS was required in order to generate signals from the compound of interest that could easily be distinguished from the background interference produced by the matrix (Reyzer et al.,

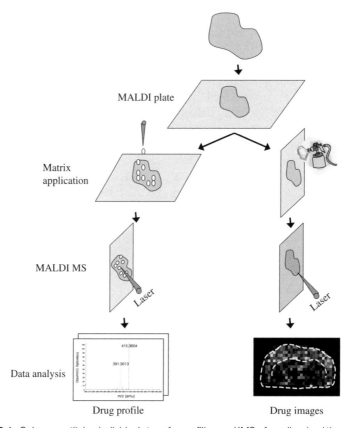

Figure 12.1. *Scheme outlining individual steps for profiling and IMS of small-animal tissue samples.*

2003; Reyzer and Caprioli, 2005a). In most MS/MS systems, two analyzers are connected together by a collision cell. The first analyzer is used to select the protonated molecule or "precursor ion" of interest. This precursor ion passes through a collision cell, where it is fragmented into product ions using a higher density of gas (typically Ar) and voltage (called collision energy). Analysis of the resulting product ions by the second analyzer generates a mass spectrum of the product ions that can be used to provide structural information. Therefore, by monitoring the transition of a selected precursor ion to its product ions, MS/MS systems give researchers a tool for distinguishing between isobaric compounds (analytes with the same nominal mass-to-charge ratio), such as matrix ions and the small-molecule drug candidate compounds of interest (Reyzer and Caprioli, 2005a).

12.2.1 Tissue Preparation

Inappropriately handled tissue samples may cause problems such as contamination or degradation of the analytes. Cryostats are normally used as a standard tool for slicing

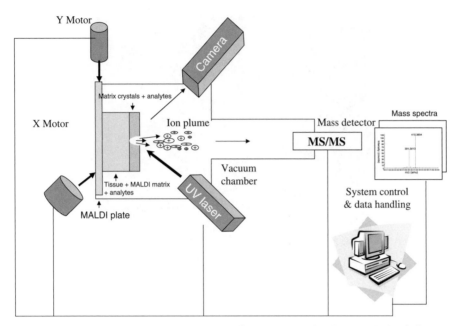

Figure 12.2. Diagram for automated MALDI–IMS system used for direct analysis of pharmaceuticals in tissues.

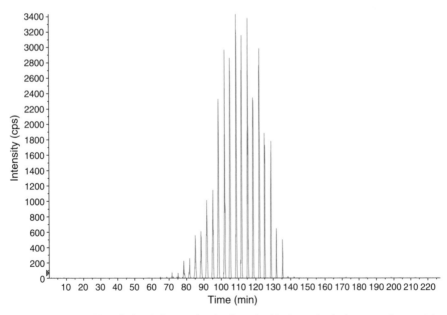

Figure 12.3. MALDI–MS signals from a droplet deposited in the rat brain tissue section containing a drug candidate are plotted against acquiring time which is directly related to the locations of each pixel.

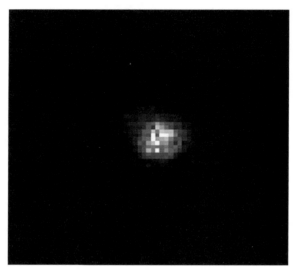

Figure 12.4. *An ion image of a droplet in the rat brain slice was obtained after reconstruction of MALDI signals from Figure 12.2.*

frozen tissue to reduce sample contamination. Contamination of the tissue surface due to the use of OCT (optimal cutting temperature) polymer as an embedding medium for stabilizing the organ specimens should be avoided because it would lead to ionization suppression in the MALDI–IMS analysis. If the analyte of interest is sensitive to photodegradation, then the tissue sample should be maintained in a dark container as much as possible during the processes. Typically, the tissue slice has a thickness of 10–20 μm.

There are several direct ways to transfer tissue slices to the sample plate. First, the frozen tissue slice is gently positioned on the cold target plate ($-15°C$) using an artist's brush and thawmounted onto the plate by quickly moving it out of the cryostat chamber; the frozen section then adheres to the MALDI plate when held at room temperature. As an alternative procedure, one can attach the section to a double-sided transparent tape, which is then glued on the sample plate (Schwartz et al., 2003).

Blotting the tissue onto cellulose or carbon-conducting membranes has been reported as an indirect way to transfer the section (Bunch et al., 2004). For these blotting experiments, a low transfer yield of analytes from the tissue onto the blotting membrane may result in poor-quality images (Prideaux et al., 2005). The extraction performance using a standard blotting procedure was found to be specific to the properties of the analyte. For example, the corticosteroid hydrocortisone, an anti-inflammatory measure in the treatment of eczema and dermatitis, with low water solubility presented poor blotting efficiency (Atkinson et al., 2005).

12.2.2 Sample Preparation

Once the tissue slice is mounted on the sample plate, the sample preparation becomes the critical step in the process. In general, two strategies are employed to apply matrix directly onto a tissue surface. For profiling to obtain low-resolution images, the tissue sections are spotted with small droplets of matrix on specific morphological regions of the tissue using a pipette. For imaging, the sections can be coated by electrospray deposition, airspray deposition, or using robotics to deposit matrix on the tissue surface before MALDI analysis (Fig. 12.1) (Schwartz et al., 2003; Kenny et al., 2005; Reyzer and Caprioli, 2005a; Yang et al., 2005). Tissue profiling allows for a rapid characterization of the drug content within a tissue section. High-resolution images are achieved with the matrix solution as a continuous coating or as a high density of low-volume droplets.

The basic requirement during the sample preparation stage of a MALDI–IMS experiment is to deposit matrix uniformly onto the sample using a technique to prevent redistribution of analytes and to obtain sufficient cocrystallization of both the matrix and analyte. For the airspray approach, in order to produce a homogeneous crystal layer over the tissue surface, spraying multiple coats of matrix across the surface of the tissue is recommended. One coating cycle consists of passing the sprayer two to three times across the surface of the tissue and allowing the sample to dry between cycles. The sample plate is held vertically about 20–30 cm from the sprayer nozzle. The sample is allowed to dry for at least 2–5 min before the next coating cycle to avoid deposition of a large quantity of solvent in any one region of the tissue. This step serves to redissolve and recrystallize the matrix, enhancing the incorporation efficiency of analytes into the crystals. The combination of the spray deposition rate and spray distance should be adjusted to avoid excessive wetting of the tissue section, which could result in dispersing the analytes within the section. This process typically requires more than 10 cycles.

Contaminants such as salts and phospholipids in tissues may significantly inhibit the cocrystallization step and result in low MALDI intensities of the analytes. Reports suggest that washing the samples with weak organic solvents such as 70% ethanol may help to eliminate endogenous materials from the native tissue prior to matrix application and can improve the crystallization process and the ionization efficiency for biopolymers on tissue (Caprioli et al., 1997). However, this step has the potential of altering the spatial integrity of drug-related components on tissues, so using this washing step during sample preparation for small-molecule assays may become problematic.

The selection of matrix and matrix solution conditions such as solvent composition, pH, and rate of cocrystal growth can affect the quality of mass spectra for both macromolecules and small molecules (Cohen and Chait, 1996). The success of the MALDI application to an analyte in the tissue strongly depends on the choice of appropriate matrix materials. The common matrices used for MALDI analysis are UV-absorbing molecules that are benzoic acid–based components with low molecular weights ($<$ 500 Da); three common compounds that are used as matrices are sinapinic acid (SA), α-cyano-4-hydroxycinnamic acid (CHCA), and dihydroxybenzoic acid (DHB). These matrix compounds dominate the low-mass range for a typical MALDI–MS spectrum, which is a challenge for the advancement of MALDI for the analysis of small molecules (Reyzer et al., 2003). Although the discovery of porphyrin matrices had been reported to be valuable for the detection of low-mass analytes with a minimum of mass interference from matrix signals (Chen and Ling, 2002), poor ion production yield of drug molecules in tissues was observed when these porphyrin matrices were employed for MALDI–IMS (Hsieh et al., 2006a). Other alternatives have been reported, such as using ionic matrices (Lemaire et al., 2005) and transparent matrices (Rubakhin and Sweedler, 2005) to replace regular crystalline matrices when using MALDI for the analysis of small molecules. Unfortunately, the proposed benefits were not observed when these matrices were tested on tissue samples (Hsieh et al., 2006a). Because of the nature of biological tissues, the growth of crystal is more complicated on tissue than on a metal plate where a small volume of matrix is mixed with a neat drug solution. Strong acids such as 0.1% trifluoroacetic acid (TFA) that are normally added to the matrix solution to assist the protonation of proteins were found to have a marginal effect on the ionization efficiency for small molecules when using the MALDI–IMS technique on tissue samples. As a general rule, higher matrix densities tend to produce better quality mass spectral data. Solvent composition also has a substantial effect on the ionization responses of the targeted compounds; presumably, this is dependent on the effectiveness of analyte extraction and crystal formation.

MALDI is considered a surface analysis technique and a relatively nondestructive ionization technique (Page and Sweedler, 2002). Typically, a sample spot on the target can be assayed multiple times because only a small fraction is vaporized for each laser pulse. However, no further spectra are detected after a completed ablation by the laser in MALDI. The depleted amounts were observed to be associated with the sample identity, sample spot size, and MALDI matrices (Page and Sweedler, 2002).

One of the purposes of using matrix solvent is to extract analytes from the interior to the surface of the tissue sections prior to MALDI analysis. Crossman et al. (2006) have investigated the depth profile of analyte extraction from different tissue types (Crossman et al., 2006). Here, prazosin, a small drug molecule, and cytochrome C, a 12-kDa protein, were used as model compounds to investigate the thickness of the tissue section on the extraction efficiencies using both MALDI–MS/MS and MALDI–MS. The spotting technique has been reported to provide better sensitivity than the coating method for the detection of small molecular components (Reyzer et al., 2003). For the tissue sections that are 20 μm or less in thickness, diffusion of analyte from the interior to the surface was fast (Crossman et al., 2006). The MALDI signal intensities were found to be a function of the thickness of the tissue section. A higher organic content of the matrix solution provided more complete analyte extraction; on the other hand the rate of solvent evaporation increases with increasing organic solvent content of the matrix solution. Therefore, care must be taken so that the solution does not dry too rapidly when using higher organic solvent concentrations.

12.2.3 Ion-Imaging Considerations

MALDI imaging resolution is governed by both crystal size and laser diameter. Smaller crystal sizes provide an opportunity for better imaging resolution. With crystal diameters smaller than the laser beam, typically 50–200 μm depending on the instrument, imaging resolution is generally limited to the laser diameter (Schwartz et al., 2003).

Laser wavelength, fluence, and spot size have a strong impact on the desorption/ionization process (Feldhaus et al., 2000; Fournier et al., 2003; Sleno and Volmer, 2005). In general, MALDI suffers some degree of signal irreproducibility due to crystallization behavior and laser properties such as energy profile and firing repetition rate (Sleno and Volmer, 2005). The quality of a MALDI mass spectrum greatly depends on the morphology of the irradiated matrix sample crystals (Garden and Sweedler, 2000). Variations in peak intensity and mass accuracy of analytes may be seen when the laser focuses on different regions of the tissue samples. The spectrum quality from a tissue sample can be optimized through tuning the instrument settings to maximize signal quality of analytes with homogeneous crystals produced from the diluted stock solution. The optimized instrument parameters for a given drug component are then employed for further MALDI–IMS experiments. As a general rule, higher laser energies are required to generate mass spectra from tissue samples than what is needed to generate mass spectra from a neat solution of the same compound spotted directly on the MALDI plate. The relation between laser radiation power and the MALDI–MS responses from the tissue surface is compound dependent. For clozapine, the MALDI–MS signal amplitude improved with an increase in laser beam exposure up to around 70 µJ and then declined steadily with higher laser power. For norclozapine, the ion production was roughly proportional to the increasing laser power under the same experimental condition (Fig. 12.5) (Hsieh et al., 2006a).

For imaging of low-molecular-weight compounds using MALDI–IMS, MS/MS instruments are the platforms of choice since regular organic matrices for MALDI are likely to result in background matrix ions that appear at every mass in the low-m/z

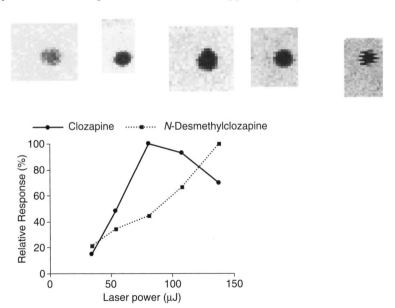

Figure 12.5. (Top) MALDI–MS images of clozapine in rat brain slice as a function of laser fluence; (bottom) relative MALDI responses of clozapine and norclozapine in rat brain slice as a function of laser fluence.

region and in much higher abundance than the analyte ions. Therefore, imaging of compounds with molecular weights <500 Da becomes challenging using a mass spectrometer that does not have MS/MS capabilities. Also, the matrix is generally present in great excess ($\sim$1000 : 1) over the analyte of interest, and matrices are effective at self-protonation. Therefore, one of the major concerns with MALDI–TOF technology is the spectral noise in the low-mass region generated from the matrix, matrix clusters, and matrix-related fragment ions. The problem becomes even more severe when drugs in tissues are the target for the assay. This is because many endogenous components can also be desorbed and produce interfering signals in the mass spectra. In addition, there is a relatively high concentration of salt (Na^+ and K^+) on native tissue sections. As a result, many matrix cluster ions can form with multiple matrix molecules clustered about one alkali metal ion. As an example of the problem, SCH 226374 with a calculated protonated mono-isotopic molecular weight of 695.35 was assayed using MALDI–MS (Fig. 12.6)

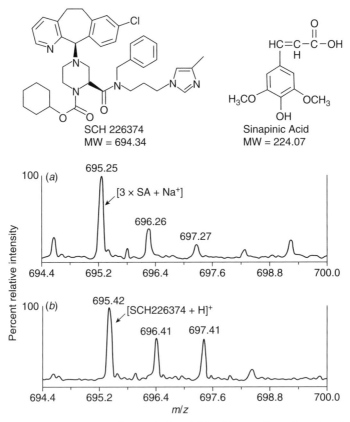

Figure 12.6. *Top: Structures of SCH 226374 and SA with monoisotopic molecular weights. Bottom: Standards run on Voyager STR in reflectron mode. (a) SA with added NaH$_2$PO$_4$. (b) SCH 226374, 5 pmol on plate, with applied SA. The presence of ^{13}C and ^{37}Cl isotope signals at m/z 696.41 and 697.41 confirms the presence of the drug (Reprinted with permission from Reyzer et al., 2003).*

and can be used as an example for the need to use MALDI–MS/MS for small-molecule analysis (Reyzer et al., 2003). As shown in Fig. 12.6, a SA matrix cluster of the type $3 \times SA + Na^+$ has a calculated monoisotopic molecular weight of 695.20, less than 0.2 Da from the mass of protonated SCH 226374. The signals from SCH 226374 and the matrix cluster could not readily be differentiated in a single MS mode even with the higher mass resolution provided by the TOF–MS system. SA forms a $3 \times SA + Na^+$ cluster that dominates the spectrum at m/z 694–700. In the same mass range, MALDI of 5 pmol of SCH 226374 with

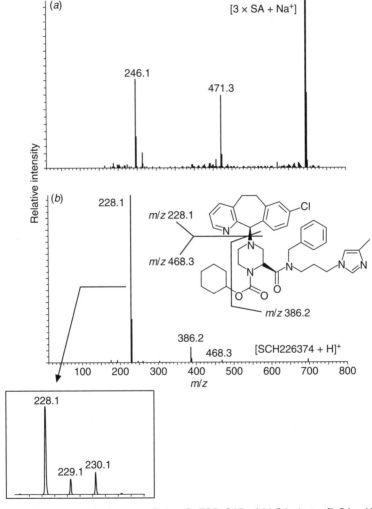

Figure 12.7. CAD of standards run on Pulsar QqTOF. CAD of (a) SA cluster $[3 \, SA + Na^+]$ and (b) $[M + H]^+$ of SCH 226374. Inset is a close-up of the fragment ion at m/z 228.1 showing presence of distinctive ^{13}C and ^{37}Cl isotope signals (Reprinted with permission from Reyzer et al., 2003).

SA as matrix reveals a dominant signal at m/z 695.42, corresponding to (SCH 226374 + H)$^+$.

Figure 12.7 shows the fragmentation pattern of the SA cluster ion ($3 \times$ SA + Na$^+$) and that of the (M + H)$^+$ ion from MALDI of 5 pmol SCH 226374 with SA. As shown, the SA cluster dissociates into two main fragments, one at m/z 471.3, corresponding to a loss of one SA molecule, and the other at m/z 246.1. The protonated drug ion dissociates into several fragments, with one dominant fragment at m/z 228.1, corresponding to a chlorinated tricyclic moiety. A closeup view of the fragment at m/z 228.1 shows two additional signals at m/z 229.1 and 230.1, consistent with ^{13}C and ^{37}Cl isotope signals. The key piece of information obtained from these dissociation patterns is that none of the fragment signals overlap, and thus SCH 226374 is readily differentiated from SA upon collision-activated dissociation (CAD) of the ion packet at m/z 695.

12.2.4   Data Acquisition Aspects

The image acquisition time for tissues is dependent on several instrumental parameters such as spatial resolution requirements, the laser repetition rate, spot-to-spot sample repositioning transfer time, and data processing. In order to get higher resolution, one needs more pixels (each pixel is one MS data point) on a given tissue slice with the resulting longer acquisition times. Lasers with faster repetition rates and improved electronics can reduce acquisition times from hours to minutes (Chaurand et al., 2004). A commercially available component of oMALDI Server developed using LabVIEW software can be used for the automated imaging data acquisition. The oMALDI Server allows the user to move the MALDI plate within the source. The full plate pane represents the complete accessible area of the sample plate. A course adjustment to define and resize the exact region of interest (x and y coordinates) is achieved by the crosshairs. Turning the laser on and off as well as adjusting the laser power can be performed within this software, which allows the user to decide the horizontal and vertical spacing between imaging spots with a defined two-dimensional array and to start an automated imaging acquisition run. The process image dialog within the same software allows the user to select an imaging data file, to build a peak list, and to treat the data to construct a visual two- or three-dimensional image for each selected mass peak. The process image dialog also allows the user to revisit the plate location where a displayed spectrum was collected and to review the previously processed results. The intensity of the drug signal is indicated by the color. The three-dimensional visualization capability allows graphical files to be imported and overlaid on the MALDI image, which allows the user to compare histological features and optical images with the MALDI images. The data-processing time for the sample plate containing 10,000 pixels is less than 3 min. As other vendors create software for MALDI–IMS, getting the MS data into a display format that provides for a useful MS image will become easier.

12.3 RELIABILITY OF MALDI–IMS IMAGES FOR DRUGS IN TISSUE SECTIONS

12.3.1 Semiquantitation

MALDI signals were found to be proportional to the concentrations of pharmaceuticals in tissues, although different regions within the same organ section and different types of tissues might demonstrate different surface properties (Prideaux et al., 2005; Hsieh et al., 2006a, 2007). Bunch and co-workers (2004) reported that calibration graphs for the determination of ketoconazole using the sodium matrix adduct ion as a standard showed a relatively good degree of linearity. This linearity relationship permitted calibration of the concentration of ketoconazole at differing skin depths which were examined by individual spectra acquired from a MALDI–IMS assay. The linearity of calibration curves of clozapine and its metabolite norclozapine obtained by depositing several droplets with increasing concentrations on a blank tissue section and assaying by MALDI–IMS also suggested that concerns about possible ionization suppression variations due to the heterogeneity within a tissue section were not substantial. Good linearity data also suggested that the impact of ionization suppression due to the difference of one location to another location within a given tissue section is not significant.

The feasibility of the MALDI–IMS technique could be evaluated by mapping the location of the spiked analytes within a blank tissue section. For example, the letters S and P were inscribed by connecting microdroplets containing clozapine and norclozapine using a microsyringe, respectively (Hsieh et al., 2006a, 2007). The precursor-to-product ion transitions from m/z 327 to m/z 192 and from m/z 313 to m/z 192 were used for clozapine and norclozapine, respectively. The MALDI–IMS images of clozapine and norclozapine clearly indicated that these two letters could be seen on the tissue sections, as demonstrated in Fig. 12.8. These experiments suggested that the redistribution of surface analytes after coating with the SA solution by air spraying was minimal. The resulting MS imaging distribution of either clozapine or norclozapine was not skewed by overlapping with other compound signals.

12.3.2 Correlation between MALDI–QqTOF and LC–MS Data

Accumulation of the drug in a particular tissue section is likely to be nonuniform, and so correlation between the quantitative amounts determined via LC–MS for a whole brain and relative intensities obtained for the drug in all areas of a single brain section is not expected. However, one could assume that as the overall amount of drug in the whole tissue increases, the average MALDI signal response for tissue sections would reflect that trend. In order to test that assumption, the MALDI–QqTOF–MS response for a series of samples was compared to the quantitation results obtained with HPLC–MS/MS analysis. As reported by Reyzer et al. (2003), the animals were administrated with SCH A at 1 mg/kg, 5 mg/kg and 2.5 mg/kg, 25 mg/kg through IV and oral routes, respectively. The brains were removed at either 1 or 4 h after dosing. Half of the brains (two samples per dose per time) were analyzed by HPLC–MS/MS,

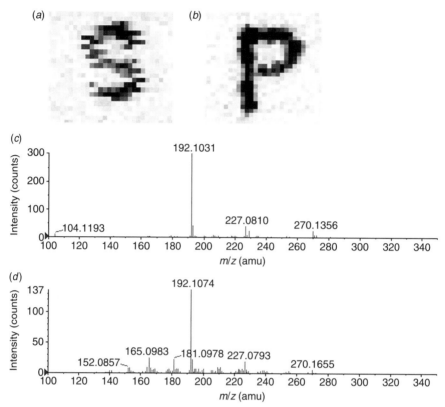

Figure 12.8. *MALDI–MS images of letters (a) S and (b) P containing norclozapine and clozapine in rat brain sections, respectively. MALDI–MS/MS spectra of (c) clozapine and (d) norclozapine in rat brain sections (Hsieh et al., 2006a).*

while the other half were sectioned and analyzed by MALDI–QqTOF–MS. The brain concentrations of SCH A were expressed as nanograms of test compound per gram of brain tissue, while the MALDI results were recorded as the absolute intensity of the dominant product ion averaged over all spots examined on one tissue section. In order to account for variability among animals, the average concentration of SCH A in the brains at one dose was plotted against the average amount of drug found in the plasma of the same animals. Further, to compare the HPLC–MS/MS and MALDI–QqTOF analyses of brain samples directly, the responses from each method were normalized to the maximum response in each case, and the resulting percentages were plotted on the *y* axis. As reported, the HPLC–MS/MS and MALDI–IMS responses were found to be linear and well correlated (Reyzer et al., 2003).

In a separate experiment, two regions of rat brain tissue sections that contained different levels of clozapine based on the MALDI–IMS results were isolated. The isolated tissue sections were transferred into plastic test tubes and mixed with 300 μL of 95:5 methanol–water solution for protein precipitation. After centrifugation, a 10-μL aliquot of the supernatant from each sample was injected for HPLC–MS/MS analysis. The HPLC was operated under a fast gradient condition using a cyano column. The reconstructed HPLC–APCI–MS/MS chromatograms of clozapine showed that the clozapine concentration in the more intense region (based on the MALDI–IMS response) was around 10 times greater than that in the less intense region within the same tissue section; therefore, these data agreed with the MALDI–IMS results (Hsieh et al., 2006a).

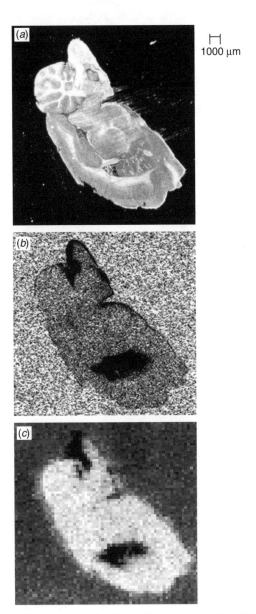

Figure 12.9. (a) Optical images, (b) radioautographic images, and (c) MALDI–MS/MS images from study rat brain tissue section (Hsieh et al., 2006a).

12.3.3 Correlation between MALDI–IMS and Autoradiography

The possibility of using MALDI–IMS methodologies for direct intact analysis of pharmaceuticals in tissues has been cross-validated by autoradiographic techniques (Reyzer et al., 2003). For a fair comparison between these molecular imaging technologies, a study of rat brains was designed in which the animals were administered ^{3}H-clozapine by direct infusion into the rat brain to avoid the possibility of biotrasformation. Rat brain was chosen as a tissue model due to its unique symmetry and well-defined anatomy. An optical image of a sagittal section of the brain defined the presence of the cortex, limbic system, cerebellum, brain stem, and ventricles (Fig. 12.9*a*).

For the autoradiographic measurement, the brain sections on MALDI plates were placed on a freshly erased BAS-TR2025 imaging plate for 14 days. After completion of imaging plate exposure in a lead-lined cabinet, the plate was scanned. Results from the autoradiography (Fig. 12.9*b*) suggested that clozapine was distributed throughout the brain, with the highest concentration found in the lateral ventricle. For the MALDI–IMS experiment, a hybrid quadrupole time-of-flight (QqTOF) mass spectrometer fitted with an orthogonal MALDI ion source and a nitrogen laser (337 nm) was used to acquire the MALDI–MS/MS spectra. The nitrogen laser (337 nm) was operated at 20 Hz with adjustable laser power. MS/MS methods were optimized on neat test compounds mixed with matrices and spotted on MALDI plates. Argon was used as the collision gas with a CAD gas pressure setting of 5 ($\sim$3–4 $\times$ 10^{-5} torr). The frozen tissues were cut into 12-μm-thick sections on a cryostat and the sections were transferred and thawed onto silver-coated stainless steel MALDI plates. The plates were desiccated for at least 1 h prior to matrix application. The matrix solutions on tissue were prepared as a 25-mg/mL solution in 80 : 20 acetonitrile–water or methanol–water unless otherwise noted and was coated over the entire surface of the tissue by spraying multiple coats of matrix across the surface of the tissue using the airspray technique. As the results in Fig. 12.9*c* indicate, the conclusion can be made that the MALDI–IMS result was in agreement with that obtained from autoradiography with regard to the distribution of clozapine in rat brain (Hsieh et al., 2006a).

12.4 APPLICATIONS OF MALDI–IMS TO DRUG DISCOVERY AND DEVELOPMENT

MALDI–IMS has been employed extensively to investigate the distribution of either commercially available or prospective drugs in tissues. Troendle et al. (1999) reported using MALDI in conjunction with a quadrupole ion trap equipped with a laser microprobe to detect the drug paclitaxel [molecular weight (MW) 853] from rat liver and human ovarian tumor tissue. The liver tissue was incubated with a solution of paclitaxel, while the ovarian tumor tissue was from an animal dosed with paclitaxel in vivo. The concentration of drug was approximately 50 mg/kg in each tissue of interest. In both cases, no localization of the drug was observed.

Reyzer and co-workers (2003) have demonstrated the imaging capability of using a MALDI–QqTOF instrument for direct analysis of drug discovery compounds in tissue samples. From a mouse dosed with the drug candidate, SCH 226374 at 80 mg/kg, a low-resolution mass spectral image of the drug in the tissue was obtained by moving the sample stage under the laser beam in a 10-spot × 10-spot grid, with each spot being 1 mm apart. Product ion spectra were acquired from each spot on the tissue, and the relative intensities of the fragment ion at m/z 228.1 (a product ion for SCH 226374) were plotted. The resulting image showed that the drug was present over most of the tumor section but in a higher concentration in the outer periphery (Fig. 12.10). For brain tissue, SCH A was a compound under development whose

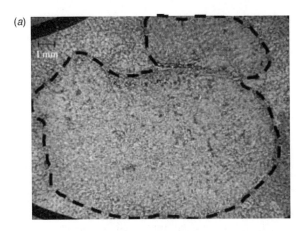

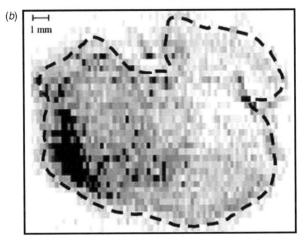

Figure 12.10. Image of SCH 226374 in mouse tumor tissue. The mouse was dosed at 80 mg/kg and the tumor was excised 7 h after the last dose. (a) Optical image of tissue section (after coating with SA). (b) Mass spectral image of SCH 226374 in tumor section utilizing selected reaction monitoring of transition m/z 695 > m/z 228. White indicates no signal, black indicates high signal. The drug appears to be localized in the periphery of the tumor section (Reprinted with permission from Reyzer et al., 2003).

target is in the striatum of the brain. A mass spectral image was obtained from a rat brain section intravenously dosed with SCH A at 5 mg/kg with the brain removed 1 h after dosing. For MALDI–MS/MS analysis, the transition from m/z 466 to m/z 225 was monitored from the laser raster over the brain section in a 30 × 15-spot grid with spots being 500 μm apart in both the x and y directions. The MALDI–IMS result showed that the cortex area appears darker than the striatum, reflecting a higher amount of SCH A detected in the cortex (Reyzer et al., 2003).

Bunch et al. (2004) mapped the distribution of a xenobiotic substance, ketoconazole, in skin using MALDI–IMS to demonstrate its capability of producing spatial data. Here, porcine epidermal tissue was treated with a medi-cated shampoo containing ketoconazole. A cross section of the drug-treated tissue was blotted onto a cellulose membrane precoated with CHCA by air-spray depo-sition. All samples were analyzed by MALDI–TOF–MS with an orthogonal MALDI ion source. The permeation of ketoconazole into the skin was investigated by imaging a cross-sectional imprint of treated tissue to construct a quantitative profile of drug in skin. By superimposing the MALDI–MS and histological images, the permeation depth profiling of the pharmaceutical in a "vertical" skin section was visualized with the greatest concentration in the microgram range present in the dermal skin layer. Imaging, of new chemical entities in animal models allows for the assessment of physiological parameters linked to the efficacy of a drug candidate, is likely to be a useful and important tool to guide the optimi-zation of new pharmaceuticals. Using MALDI–IMS, Signor et al. (2005) studied the presence of R0508231 and its metabolites in rat tissue sections from liver, spleen, and muscle. The experiments were performed on a MALDI–QqTOF mass spectrometer. The drug was administrated to rats at 5 mg/kg and tissues were sectioned with a cryomicrotome at a thickness of 30 μm and mounted onto MALDI plates. For MALDI analysis, tissues were spotted with 0.2-μL droplets of either SA or CHCA in a 50:50:0.1 mixture of acetonitrile–water–TFA (v/v/v). Here, SA was found to be more suitable than CHCA in terms of sensitivity for the analytes. The identity of all analytes was confirmed by the associated product ion spectra. The parent compound, R0508231, was detected in all tissues (liver, spleen, and muscle). The O-demethylated metabolites were only found in liver tissue.

The MALDI instrument utilizes a relatively large (5-cm×5-cm) target plate sufficient to support a mouse whole-body tissue slice. Recently, some researchers have extended the applications of MALDI–IMS from an isolated organ tissue to whole-body sections. As shown in Fig. 12.11, the presence of a drug candidate, SCH 206272, was directly monitored from the whole-body tissue sections of mouse liver, kidney, brain, and others at a very early time point postdose. The MALDI signal of SCH 206272 was detected primarily in the stomach at the early time point and moved into surrounding organs at late time points.

This same principle has been further adapted to determine the spatial distribu-tion of both drugs and metabolites in sagittal tissue sections of whole rat with MALDI–IMS (Khatib-Shahidi et al., 2005, 2006). Here the test model compounds (olanzapine, m/z 313 → 256); propranolol, m/z 260 → 155); and spiperone, m/z

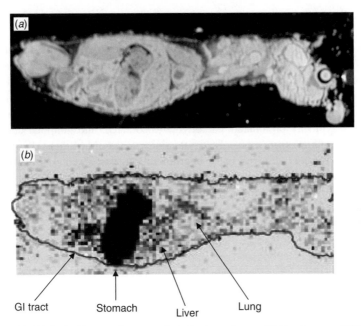

GI tract   Stomach   Liver   Lung

Figure 12.11. (a) Optical image of mouse whole-body section and (b) ion image of SCH 206272 in mouse whole-body section using MALDI–IMS.

396 → 165) were administered in Sprague-Dawley rats at a pharmacologically equiv-alent dose via oral administration. The rat whole-body section was divided into four parts, which were mounted onto four MALDI target plates. The MALDI images from individual sample plates were combined together as an integral whole-body image in rat. As an example, the olanzapine signal was detected primarily in the stomach at early time points and then was localized into surrounding organs at 6 and 12 h. The emergence of its metabolite was also detected in tissue and correlated to loss of parent drug signal. The MALDI–MS/MS signal intensities correlated well with published olanzapine quantitative whole-body autoradiography studies. One advan-tage of the MALDI–IMS images of the whole-body section is that, when monitoring for the dosed compound, the signal should only come from the dosed compound (and not metabolites in most cases).

Stoeckli et al. (2005) advanced the development of a protocol for the analysis of whole-body rat tissue sections using tape in a manner similar to the preparation for autoradiography. Due to the size of the sections, whole-body rat sections of an animal treated with a ^{14}C-labeled discovery compound were divided onto four plates and the CHCA matrix was sprayed onto the plates using a pneumatic sprayer. On top of the dry matrix, a thin gold film was sputter deposited. MS/MS images of selected analytes were acquired and analyzed using a custom software package (BioMap, www.maldi-msi.org). The resulting MALDI–IMS images revealed the presence of the protonated signal of the administrated compound in the heart and the liver.

12.5 SUMMARY

MALDI–IMS for the direct measurement of small molecules in tissue sections is still in a very early stage. There is room for improvements in sample preparation, instrumentation, and software within the coming years. Items such as matrix selection, analyte extraction, image resolution, and sensitivity need further development. In MALDI–MS, different classes of compounds are ionized differently. For example, proteins are typically protonated, carbohydrates sodiated, and nonpolar compounds ionized by adduct formation. The potential of the MALDI–IMS molecular imaging technique is believed to be enormous. The fundamental contributions of the methodology in rapidly providing compound-specific maps at relatively high resolution and sensitivity will provide a powerful tool for the investigation of biotransformation and disposition processes of pharmaceuticals for both drug discovery and drug development applications.

MALDI–IMS may also be beneficial as a diagnostic, screening, or discovery tool where tissue sections can be explored without knowing in advance what specific molecules have changed in a comparative study. Image comparison between control tissues and study tissues allows researchers to identify differences in resulting changes induced by drug candidates or associated metabolites to yield important in vivo biological information, which would also enhance the field of drug discovery. The knowledge of the spatial localization of the drug components and biomarkers [see the report by Reyzer and Caprioli (2005b) for a discussion of using MALDI–IMS for biomarker discovery] within all organs of the dosed animal by using MALDI–IMS could be valuable in the characterization of toxicity findings and may be used in the future to assist in reducing the compound attrition rate in drug development programs. Overall, it can be predicted that MALDI–IMS will become a commonly used technique in both the drug discovery and the drug development arenas.

REFERENCES

Ackermann, B. L., Berna, M. J., and Murphy, A. T. (2002). Recent advances in use of LC/MS/MS for quantitative high-throughput bioanalytical support of drug discovery. *Curr. Top. Med. Chem.* **2:**53–66.

Altelaar, A. F., van Minnen, J, Jimenez, C. R., Heeren, R. M., and Piersma, S. R. (2005). Direct molecular imaging of Lymnaea stagnalis nervous tissue at subcellular spatial resolution by mass spectrometry. *Anal. Chem.* **77:**735–741.

Atkinson, S. J., Clench, M. R., and Parkinson, D. (2005). Investigating hydrocortisone uptake in porcine tissue using a solvent extraction method for indirect analysis by imaging MALDI–MS. In *Proceedings of the 53rd ASMS Conference on Mass Spectrometry and Allied Topics*, San Antonio, TX.

Ayorinde, F. O., Hambright, P., Porter, T. N., and Keith, Q. L., Jr. (1999). Use of meso-tetrakis(pentafluorophenyl)porphyrin as a matrix for low molecular weight alkylphenol ethoxylates in laser desorption/ionization time-of-flight mass spectrometry. *Rapid Commun. Mass Spectrom.* **13:**2474–2479.

Bunch, J., Clench, M. R., and Richards, D. S. (2004). Determination of pharmaceutical compounds in skin by imaging matrix-assisted laser desorption/ionisation mass spectrometry. *Rapid Commun. Mass Spectrom.* **18:**3051–3060.

Caprioli, R. M., Farmer, T. B., and Gile, J. (1997). Molecular imaging of biological samples: Localization of peptides and proteins using MALDI-TOF MS. *Anal. Chem.* **69:**4751–4760.

Chaurand, P., Schwartz, J., and Caprioli, R. M. (2004). Profiling and imaging proteins in tissue sections by MS. *Anal. Chem.* **76:**87A–93A.

Chen, Y. T., and Ling, Y. C. (2002). Detection of water-soluble vitamins by matrix-assisted laser desorption/ionization time-of-flight mass spectrometry using porphyrin matrices. *J. Mass Spectrom.* **37:**716–730.

Cheng, Z., Winant, R. C., and Gambhir, S. S. (2005). A new strategy to screen molecular imaging probe uptake in cell culture without radiolabeling using matrix-assisted laser desorption/ionization time-of-flight mass spectrometry. *J. Nucl. Med.* **46:**878–886.

Cohen, L. H., and Gusev, A. I. (2002). Small molecule analysis by MALDI mass spectrometry. *Anal. Bioanal. Chem.* **373:**571–586.

Cohen, S. L., and Chait, B. T. (1996). Influence of matrix solution conditions on the MALDI–MS analysis of peptides and proteins. *Anal. Chem.* **68:**31–37.

Crossman, L., McHugh, N. A., Hsieh, Y., Korfmacher, W. A., and Chen, J. (2006). Investigation of the profiling depth in matrix-assisted laser desorption/ionization imaging mass spectrometry. *Rapid. Commun. Mass. Spectrom.* **20:**284–290.

Crossman, L. F., Cui, X., Knemeyer, I., Morrison, R., Hsieh, Y., and Korfmacher, W. (2007). Multilevel MALDI MS tissue imaging of pharmaceuticals, In *Proceedings of the 55th ASMS Conference on Mass Spectrometry and Allied Topics*, Indianapolis, IN.

Dally, J. E., Gorniak, J., Bowie, R., and Bentzley, C. M. (2003). Quantitation of underivatized free amino acids in mammalian cell culture media using matrix assisted laser desorption ionization time-of-flight mass spectrometry. *Anal. Chem.* **75:**5046–5053.

Donegan, M., Krishnan S., Hattan, S., Juhasz, P., and Martin, S. (2003). Sample preparation methods for MALDI analysis of small molecule metabolites. In *Proceedings of the 51st ASMS Conference on Mass Spectrometry and Allied Topics*, Montreal, Canada.

Duncan, M. W., Matanovic, G., and Cerpa-Poljak, A. (1993). Quantitative analysis of low molecular weight compounds of biological interest by matrix-assisted laser desorption ionization. *Rapid Commun. Mass Spectrom.* **7:**1090–1094.

Feldhaus, D., Menzel, C., Berkenkamp, S., Hillenkamp, F., and Dreisewerd, K. (2000). Influence of the laser fluence in infrared matrix-assisted laser desorption/ionization with a 2.94 microm Er: YAG laser and a flat-top beam profile. *J. Mass Spectrom.* **35:**1320–1328.

Fournier, I., Marinach, C., Tabet, J. C., and Bolbach, G. (2003). Irradiation effects in MALDI, ablation, ion production, and surface modifications. Part II. 2,5-Dihydrxybenzoic acid monocrystals. *J. Am. Soc. Mass Spectrom.* **14:**893–899.

Garden, R. W., and Sweedler, J. V. (2000). Heterogeneity within MALDI samples as revealed by mass spectrometric imaging. *Anal. Chem.* **72:**30–36.

Hatsis, P., Brombacher, S., Corr, J., Kovarik, P., and Volmer, D. A. (2003). Quantitative analysis of small pharmaceutical drugs using a high repetition rate laser matrix-assisted laser/desorption ionization source. *Rapid Commun. Mass Spectrom.* **17:**2303–2309.

Hopfgartner, G., and Bourgogne, E. (2003) Quantitative high-throughput analysis of drugs in biological matrices by mass spectrometry. *Mass Spectrom. Rev.* **22:**195–214.

Hopfgartner, G., Husser, C., and Zell, M. (2002). High-throughput quantification of drugs and their metabolites in biosamples by LC–MS/MS and CE–MS/MS: Possibilities and limitations. *Ther. Drug Monit.* **24:**134–143.

Hsieh, Y., Casale, R., Fukuda, E., Chen, J., Knemeyer, I., Wingate, J., Morrison, R., and Korfmacher, W. (2006a). Matrix-assisted laser desorption/ionization imaging mass spectrometry for direct measurement of clozapine in rat brain tissue. *Rapid Commun. Mass Spectrom.* **20:**965–972.

Hsieh, Y., Chen, J., and Korfmacher, W. A. (2007). Mapping pharmaceuticals in tissues using MALDI imaging mass spectrometry. *J. Pharmacol. Toxicol. Methods* **55:**193–200.

Hsieh, Y., Merkle, K., Wang, G., Brisson, J. M., and Korfmacher, W. A. (2003a). High-performance liquid chromatography-atmospheric pressure photoionization/tandem mass spectrometric analysis for small molecules in plasma. *Anal. Chem.* **75:**3122–3127.

Hsieh, Y., and Wang, G. (2004). Integration of atmospheric pressure photoionization interfaces to HPLC–MS/MS for pharmaceutical analysis. *Am. Pharm. Rev.* **7:**88–93.

Hsieh, Y., Wang, G., Wang, Y., Chackalamannil, S., and Korfmacher, W. A. (2003b). Direct plasma analysis of drug compounds using monolithic column liquid chromatography and tandem mass spectrometry. *Anal. Chem.* **75:**1812–1818.

Kang, M. J., Tholey, A., and Heinzle, E. (2000). Quantitation of low molecular mass substrates and products of enzyme catalyzed reactions using matrix-assisted laser desorption/ionization time-of-flight mass spectrometry. *Rapid Commun. Mass Spectrom.* **14:**1972–1978.

Kenny, D., Snel, M., Brown, J., Bateman, B., Coleman, J., Petrie, J. R., Laidlaw, H., and Ashford, M. (2005). Evaluation of a new aerosol matrix deposition method for atmospheric pressure and vacuum MALDI ion imaging. In *Proceedings of the 53rd ASMS Conference on Mass Spectrometry and Allied Topics*, San Antonio, TX.

Khatib-Shahidi, S., Andersson, M., Herman, J. L., Gillespie, T. A., and Caprioli, R. M. (2006). Direct molecular analysis of whole-body animal tissue sections by imaging MALDI mass spectrometry. *Anal. Chem.* **78:**6448–6456.

Khatib-Shahidi, S., Reyzer, M. L., Herman, J., Gillespie, T., and Caprioli, R. (2005). Detecting drug distribution in whole rat sagittal sections by imaging mass spectrometry. In *Proceedings of the 53rd ASMS Conference on Mass Spectrometry and Allied Topics*, San Antonio, TX.

Korfmacher, W. (2005a). Bioanalytical assays in a drug discovery environment. In *Using Mass Spectrometry for Drug Metabolism Studies* (Korfmacher, W., Ed.). CRC Press, Boca Raton, FL, pp. 1–34.

Korfmacher, W. (2005b). New strategies for the implementation and support of bioanalysis in a drug metabolism environment. In *Integrated Strategies for Drug Discovery using Mass Spectrometry* (Lee, M. S., Ed.). Wiley, Hoboken, NJ, pp. 359–378.

Korfmacher, W. A. (2003). Lead optimization strategies as part of a drug metabolism environment. *Curr. Opin. Drug Discov. Devel.* **6:**481–485.

Korfmacher, W. A. (2005c). Principles and applications of LC–MS in new drug discovery. *Drug Discov. Today* **10:**1357–1367.

Korfmacher, W. A., Cox, K. A., Bryant, M. S., Veals, J., Ng, K., and Lin, C. C. (1997). HPLC–API/MS/MS: A powerful tool for integrating drug metabolism into the drug discovery process. *Drug Discov. Today* **2:**532–537.

Lemaire, R., Tabet, J., Salzet, M., and Fournier, I. (2005). Paper presented at the 53rd ASMS Conference on Mass Spectrometry and Allied Topics, ASMS, San Antonio, TX.

McCombie, G., and Knochenmuss, R. (2004). Small-molecule MALDI using the matrix suppression effect to reduce or eliminate matrix background interferences. *Anal. Chem.* **76:**4990–4997.

O'Connor, D. (2002). Automated sample preparation and LC–MS for high-throughput ADME quantification. *Curr. Opin. Drug Discov. Devel.* **5:**52–58.

Owen, S. J., Meier, F. S., Brombacher, S., and Volmer, D. A. (2003). Increasing sensitivity and decreasing spot size using an inexpensive, removable hydrophobic coating for matrix-assisted laser desorption/ionisation plates. *Rapid Commun. Mass Spectrom.* **17:** 2439–2449.

Page, J. S., and Sweedler, J. V. (2002). Sample depletion of the matrix-assisted laser desorption process monitored using radionuclide detection. *Anal. Chem.* **74:**6200–6204.

Prideaux, B., Clench, M. R., Carolan, V. A., Morton, J., and Rajan-Sithamparanadarajah, B. (2005). Imaging matrix assisted laser desorption ionisation—mass spectrometry for the investigation of dermal absorption of chlorpyrifos. In *Proceedings of the 53rd ASMS Conference on Mass Spectrometry and Allied Topics*, San Antonio, TX.

Reyzer, M. L., and Caprioli, R. (2005a) MS imaging: New technology provides new opportunities. In *Using Mass Spectrometry for Drug Metabolism Studies* (Korfmacher, W., Ed.). CRC Press, Boca Raton, FL, pp. 305–324.

Reyzer, M. L., and Caprioli, R. M. (2005b). MALDI mass spectrometry for direct tissue analysis: A new tool for biomarker discovery. *J. Proteome Res.* **4:**1138–1142.

Reyzer, M. L., Hsieh, Y., Ng, K., Korfmacher, W. A., and Caprioli, R. M. (2003). Direct analysis of drug candidates in tissue by matrix-assisted laser desorption/ionization mass spectrometry. *J. Mass Spectrom.* **38:**1081–1092.

Rubakhin, S. S., Jurchen, J. C., Monroe, E. B., and Sweedler, J. V. (2005). Imaging mass spectrometry: fundamentals and applications to drug discovery. *Drug Discov. Today* **10:**823–837.

Rubakhin, S. S., and Sweedler, J. V. (2005). Paper presented at the 53rd ASMS Conference on Mass Spectrometry and Allied Topics, ASMS, San Antonio, TX.

Schwartz, S. A., Reyzer, M. L., and Caprioli, R. M. (2003). Direct tissue analysis using matrix-assisted laser desorption/ionization mass spectrometry: Practical aspects of sample preparation. *J. Mass Spectrom.* **38:**699–708.

Signor, L., Staack, R. F., Varesio, E., Hopfgartner, G., Starke, V., and Richter, W. (2005). Analysis of RO0508231 and its metabolites in rat tissue sections by MALDI-quadrupole-time of flight mass spectrometry. In *Proceedings of the 53rd ASMS Conference on Mass Spectrometry and Allied Topics*, San Antonio, TX.

Sjovall, P., Lausmaa, J., Nygren, H., Carlsson, L., and Malmberg, P. (2003). Imaging of membrane lipids in single cells by imprint-imaging time-of-flight secondary ion mass spectrometry. *Anal. Chem.* **75:**3429–3434.

Sleno, L., and Volmer, D. A. (2005). Some fundamental and technical aspects of the quantitative analysis of pharmaceutical drugs by matrix-assisted laser desorption/ionization mass spectrometry. *Rapid Commun. Mass Spectrom.* **19:**1928–1936.

Stoeckli, M., Chaurand, P., Hallahan, D. E., and Caprioli, R. M. (2001). Imaging mass spectrometry: A new technology for the analysis of protein expression in mammalian tissues. *Nat. Med.* **7**:493–496.

Stoeckli, M., Knochenmuss, R., McCombie, G., Staab, D., and Rohner, T. (2005). MALDI–MSI of compounds and metabolites in whole-body tissue sections. In *Proceedings of the 53rd ASMS Conference on Mass Spectrometry and Allied Topics*, San Antonio, TX.

Sugiura, Y., Shimma, S., Konishi, Y., Ageta, H., Nirasawa, T., and Setou, M. (2007). Imaging mass spectrometry revealed the distinct distribution and developmental change of ganglioside molecular species in the mouse hippocampus. In *Proceedings of the 55th ASMS Conference on Mass Spectrometry and Allied Topics*, Indianapolis, IN.

Todd, P. J., Schaaff, T. G., Chaurand, P., and Caprioli, R. M. (2001). Organic ion imaging of biological tissue with secondary ion mass spectrometry and matrix-assisted laser desorption/ionization. *J. Mass Spectrom.* **36**:355–369.

Troendle, F. J., Reddick, C. D., and Yost, R. A. (1999). Detection of pharmaceutical compounds in tissue by matrix-assisted laser desorption/ionization and laser desorption/chemical ionization tandem mass spectrometry with a quadruple ion trap. *J. Am. Soc. Mass Spectrom.* **10**:1315–1321.

Xu, X., Lan, J., and Korfmacher, W. (2005). Rapid LC/MS/MS Method Development for Drug Discovery. *Anal. Chem.* **77**:389 A–394 A.

Yang, M., James, A., Covey, T., and Kovarik, P. (2005). High-speed automated deposition of matrix onto tissue samples for small molecule imaging application using MALDI MS/MS. In *Proceedings of the 53rd ASMS Conference on Mass Spectrometry and Allied Topics*, San Antonio, TX.

Index

Mass Spectrometry in Drug Metabolism and Pharmacokinetics. Edited by Ragu Ramanathan
Copyright © 2009 John Wiley & Sons, Inc.